Chemical Reagents for
PROTEIN
MODIFICATION
Third Edition

Chemical Reagents for
PROTEIN
MODIFICATION
Third Edition

Roger L. Lundblad

CRC PRESS

Boca Raton London New York Washington, D.C.

Library of Congress Cataloging-in-Publication Data

Lundblad, Roger L.
    Chemical reagents for protein modification / by Roger L. Lundblad.—3rd ed.
       p. cm.
    Includes bibliographical references and index.
    ISBN 0-8493-1983-8 (alk. paper)
    1. Proteins—Chemical modification. I. Title.

QP551.L883 2004
572′.6—dc22
                                                           2004051973

Visit the CRC Press Web site at www.crcpress.com

© 2005 by CRC Press

No claim to original U.S. Government works
International Standard Book Number 0-8493-1983-8
Library of Congress Card Number 2004051973
States of America 1 2 3 4 5 6 7 8 9 0
rinted on acid-fr

Preface

This material first appeared in 1983 as *Chemical Reagents for Protein Modification*. The first version was prepared in collaboration with Dr. Claudia Noyes, who moved on to greener pastures, leaving me with sole responsibility for the later versions that appeared in 1991 and 1994. The current edition represents a major revision from previous editions. The chapters on amino acid analysis, peptide separation by HPLC, and amino acid sequence determination have been removed. The material on amino acid analysis has been set as Appendix II (determination of protein concentration), as amino acid analysis is increasingly regarded as a gold standard in the determination of protein concentration.

In the preface of the 1994 edition I stated, "While a consideration of major biochemical journals would suggest that the use of chemical modification to study the relationship between structure and function in proteins is no longer an active area..." I retract this statement for the present edition with the caveat that it is increasingly difficult to know whether chemical modification has been used in a given study without a thorough consideration of the manuscript. This is most notable in proteomics, wherein chemical modification is extensively used but never really discussed. The present material will be of significant value to investigators working in proteomics. Another area is the biochemistry of aging, wherein oxidation and the various reactions of nitric oxide are of great importance. The study of these phenomena would not be possible without considering previous work on the chemical modification of proteins.

I hope that this book will be of value to investigators with training in different disciplines. I strongly encourage the rigorous reporting of laboratory techniques. It is discouraging to read an otherwise useful paper in which temperature is referred to as room temperature, reaction time is not given, buffer details are not presented, and so on. If investigators wish to have their work seriously considered for therapeutic or diagnostic development, information such as precise temperature and reaction time is required.

Finally, I strongly encourage the careful consideration of Chapter 1 before embarking on the design of an experiment by using chemical reagents for the modification of proteins.

Acknowledgments

In the previous version of this material, *Techniques in Protein Modification*, I commented that major portions of the book had been written in various airports and at various altitudes. I am not sure whether this is a good way to write a book, and hence the present version has been written in Chapel Hill, NC. Thus, my first acknowledgment is to the libraries of the University of North Carolina at Chapel Hill. The staff and facilities of the Health Sciences Library, the Kenan Library in the Department of Chemistry, and the Brauer Library (Physics, Mathematics) were very helpful. I acknowledge Professor Bryce Plapp of the University of Iowa for his encouragement to the thermodynamically challenged and Professor Nicholas Price of the University of Glasgow for guidance regarding content. Other individuals making unique contributions include Professor William McClure of the University of Southern California and Professors Charles Craik and Robert Fletterick of the University of California at San Francisco. I am indebted to Professor Ralph Bradshaw of the University of California at Irvine for his guidance in both the murky world of proteomics and the continuing dominance of Duke basketball. Finally, I acknowledge the continuing force of the fifth floor of Flexner Hall at the Rockefeller University.

Roger L. Lundblad

Table of Contents

Author

Dr. Roger Lundblad is a native of San Francisco, CA, and resident of Chapel Hill, NC, where he is an independent consultant in biotechnology. He is the immediate past editor-in-chief of *Biotechnology and Applied Biochemistry* and at present a member of the editorial board. He is also an adjunct professor of pathology at the University of North Carolina at Chapel Hill. Dr. Lundblad received his B.S. in chemistry from Pacific Lutheran University in Tacoma, WA, and his Ph. D. degree from the University of Washington in Seattle. After several years of post-doctoral study at the Rockefeller Institute in New York, with Nobel laureates Stanford Moore and William Stein, he joined the faculty of the University of North Carolina at Chapel Hill in 1968. He rose quickly through the ranks and was promoted to professor of pathology, biochemistry and periodontics in 1977. In 1991, he was recruited by Baxter Healthcare as director of technology development at the Hyland Division Facility in Hayward, CA. Dr. Lundblad moved to southern California in 1992 to become director of science and technology development at Baxter Hyland in Duarte, CA. During his time at Baxter, he became a member of the Senior Leadership Group in the Baxter Technical Council and was chair of the committee on technical and organizational knowledge. Dr. Lundblad left Baxter in 2000 to become an independent consultant in biotechnology.

Dr. Lundblad is the author of more than 120 publications and is also the author of best-selling books on protein chemistry. He has edited several books in the area of biotechnology. Dr. Lundblad is recognized as an expert in the area of protein chemistry, thrombosis and hemostasis, biotechnology manufacturing, process validation, assay validation, GLP laboratory compliance, product development, and cGMP issues.

1 The Site-Specific Chemical Modification of Proteins

The advent of techniques such as site-specific mutagenesis (oligonucleotide-directed mutagenesis) and solution chemistry approaches such as nuclear magnetic resonance has been of great value to protein chemistry. Notwithstanding these remarkable advances in technology, the chemical modification of proteins continues to be quite useful in the study of proteins.[1–20]

Site-specific chemical modification has been very useful as a tool in proteomics. The chemistry for the application is well known and the remarkable advances in the use of site-specific chemical modification in proteomics are clearly a result of the combination of mass spectrometry and data processing.[21–29] The use of isotope-coded affinity tags (ICATs) was developed by Abersold and colleagues.[30–32] ICATs enabled the relatively specific introduction of a deuterium-labeled moiety on the sulfhydryl groups of a protein. The use of a chemically identical modifying reagent not containing deuterium allows the comparison of protein expression.[33] The presence of a biotin moiety permits the isolation of modified peptides. Subsequent work has refined this technique[34–36] and reagents that target residues other than cysteine have been developed.[37–39] The development of a reagent with an acid-labile link to a resin permits the facile purification of peptides.[40] Figure 1.1 shows a general description of the affinity purification of peptides by these techniques. A related approach uses the incorporation of stable isotope-labeled amino acids in cell culture (SILAC).[41]

Other examples of site-specific chemical modification in proteomics include the reaction of fluorogenic reagents for the modification of lysine residues, which provides high sensitivity in the analysis of proteins by two-dimensional capillary electrophoresis[42]; the selective modification of thiophosphorlated peptides at low pH with an derivative of iodoacetic acid[43]; and the oxidation of methionine residues with hydrogen peroxide.[44] The in situ oxidation of methionine[45] was used to identify methionine peptides in an earlier version of two-dimensional electrophoresis (diagonal electrophoresis).[46,47] Modification of active-site residues with fluorescent probes[48,49] or biotin-containing probes[49] has been reported. A related approach has allowed the identification of protein substrates for transglutaminases, using biotinylated amino- and acyl-donor probes.[50] It has been possible to modify 3-nitrotyrosine residues in complex mixtures by first reducing the protein mixture with dithiothreitol followed by alkylation with iodoacetic acid. The nitrotyrosine residues are then reduced to the aminotyrosine derivative with sodium dithionite and alkylated with a sulfosuccimidyl derivative.[51] The presence of a biotin tag allows the isolation of

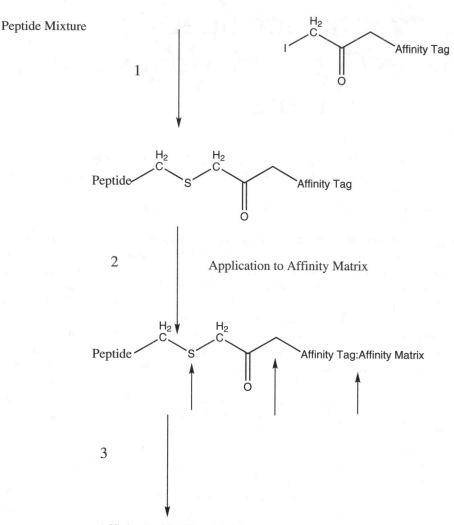

FIGURE 1.1 A general scheme for the purification of peptides by a combination of chemical modification and affinity purification. Step 1 is the chemical modification of a peptide mixture with a class-specific reagent; an iodoacetyl derivative alkylating a cysteine residue is shown as an example. Note that the reagent has an affinity tag (e.g., biotin), which can be used for purification of the modified peptides. Step 2 is the application to the affinity matrix; in the case of biotin, it would be an avidin matrix. Step 3 is the elution or recovery of the bound peptides from the affinity matrix. This could be accomplished by several approaches. First, the modified peptide including the affinity tag could be eluted from the matrix. Second, the bond between the affinity tag and the chemical modifier could, for example, be acid labile, in which case the modified peptide would be removed from the affinity tag during elution. Third, the bond formed between the chemical reagent and the modified peptide could be broken by chemical means. For example, in the case of the chemical modification of tryptophan peptides by sulfenyl chemistry, reduction would release the thiol-tryptophan-containing peptide.

the peptides containing the modified tyrosine residues. This is another example of the application of previous work[52,53] to proteomics. Clever approaches to isolate methionine or tryptophan peptides from complex mixtures have been developed.[54] Taking advantage of the selective reaction of methionine with α-keto alkyl halides at low pH,[55] methionine peptides react with a bromoacetyl function coupled to a bead; the methionine peptides are released with 2-mercaptoethanol at pH 8.5 to 8.8. This approach has been used to isolate methionine peptides from cell surface proteins from a mouse monocytes cell line.[56] Tryptophan peptides are isolated by reaction with a bead containing a link to a S-sulfenyl chloride function, which reacts with tryptophan in a manner similar to that described for the reaction with 2-nitrobenzyl-sulfenyl chloride.[57] The bound peptides are released by reduction of the disulfide linkage between the bead and the modified tryptophan residue with 2-mercapto-ethanol at pH 8.5 to 9.0.

The final subject in this brief discussion of the application of the site-specific chemical modification of proteins to proteomics concerns the development of protein microarrays. The same technologies can be used to attach proteins to biosensors for surface plasmon resonance. The reader is directed to several recent reviews that provide an excellent overview of protein-based microarrays.[58–63] Physical absorption to a surface or a surface modified with, for example, poly-L-lysine has been used for many years and is the basis for ELISA-based assay systems. The trend, however, is toward processes in which there is a chemical bond between the target protein and the microarray surface. Most technologies involve the reaction of lysine residues in the target protein with a suitable reactive group on the array surface. The chemistry is similar to that used to modify lysine residues in solution.[64] Attachment via other residues is also possible by using chemistries described previously. We specifically mention some novel approaches. Virtual mask lithography has been adapted to this purpose by first binding streptavidin protected by a photolabile group, nitroveratryl-oxycarbony.[65] This group is removed with UV light, permitting the binding of a biotin-labeled protein to the exposed streptavidin. One of the major issues in this type of analysis is the correct binding of the target protein to optimize interaction. Specific orientation of the target protein (capture agent) can be obtained by reaction with an oxidized carbohydrate function in a Fab' fragment.[66] Although not directly related to the chemical modification of the protein, the attachment of a hexahistidine tail to IgA permits the correct orientation of this protein on a surface.[67] The possibility to incorporate novel amino acids in proteins that can be converted to reactive sites[68] presents an opportunity to develop novel chemistries for coupling of target proteins to surfaces.

Site-specific chemical modification has been used to produce biotherapeutics.[69–74] Modification of proteins and peptides with poly(ethylene glycol) (PEG; Figure 1.2) is frequently used in the manufacture of biopharmaceuticals. There have been numerous reviews on the use of PEG in the past several years.[75–80] Successful modification of therapeutic proteins and peptides with PEG is associated with an extension of circulatory half-life and reduced or eliminated immunogenicity. It is thought that these properties arise from the physical blocking of the therapeutic from immunological surveillance and catabolic recognition. Reaction of activated PEG molecules with proteins usually occurs at primary amino groups, but the reaction

H-(CH$_2$CH$_2$)$_n$-OH

poly(ethylene)glycol

CH$_3$-(CH$_2$CH$_2$)$_n$-OH

monomethoxypoly(ethylene)glycol

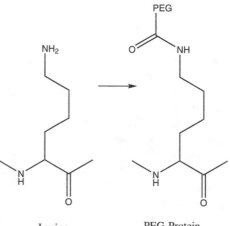

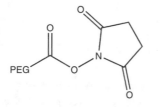

N-succinimidylcarbonylpoly(ethylene)glycol

Lysine   PEG-Protein

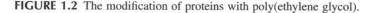

FIGURE 1.2 The modification of proteins with poly(ethylene glycol).

has been demonstrated to occur at other amino acid residues too.[81] Protein engineering has been used to introduce a free cysteine residue into an immunotoxin, which can be subsequently modified with PEG to yield a monosubstituted derivative.[82] Reaction at pH values below neutrality results in monosubstitution at amino-terminal amino groups[83,84] (Table 1.1). The use of cleavable linkers has permitted the design of PEG–protein conjugates with half-lives depending on *in vivo* rate of cleavage of the linkers between the PEG moiety and the protein.[85]

Other protein therapeutics that use site-specific chemical modification in the manufacturing process include various protein conjugates. Coupling of peptides and enzymes to albumin improves circulatory half-life in glucagon-like peptide 1.[86,87] Monoclonal antibody conjugates have also proved useful.[88,89] The Bowman–Birk inhibitor and an IgM antibody have been coupled to dextran via oxidation, permitting the targeted delivery of the Bowman–Birk inhibitor to tumor cells.[90] Urokinase was coupled to pulmonary surfactant protein by using a heterobifunctional cross-linker, S-sulfosuccinimidyl-4-(*p*-maleidophenyl-) butyrate.[91] Cross-linking of proteins with glutaraldehyde, polyethyleneglycol diacrylate, and formaldehyde is used to prepare hydrogels used extensively as biomaterials.[92–96] Carbamylation with potassium cyanate is used in the manufacturing process of some allergens.[97–100]

Site-specific chemical modification is extensively used for the selective fragmentation of proteins for determination of the primary structure of proteins, preparation of large fragments for characterization by mass spectrometry, and chemical synthesis of proteins. This includes reagents such as cyanogen bromide (CNBr) for the chemical cleavage of specific peptide bonds, citraconic anhydride for the reversible blocking

---

TABLE 1.1
Dissociation Constants for
Nucleophiles in Proteins

Potential Nucleophile	pK_a
Γ-Carboxyl (glutamic acid)	4.25
B-Carboxyl (aspartic acid)	3.65
A-Carboxyl (isoleucine)	2.36
Sulfhydryl (cysteine)	10.46
α-Amino (isoleucine)	9.68
Phenolic hydroxyl (tyrosine)	10.13
ε-Amino (lysine)	10.79
Imidazole (histidine)	6.00
Guanidine (arginine)	12.48

Source: Data from Mooz, E.D., In *Practical Handbook of Biochemistry and Molecular Biology* (Fasman, G.D., Ed.), CRC Press, Boca Raton, FL, 1989. Also see Dawson, R.M.C. et al., *Data for Biochemical Research*, Oxford University Press, Oxford, 1969, p. 2.

---

of lysine residues to restrict tryptic cleavage to arginine residues, and 1,2-cyclohexanedione for the reversible blocking of arginine residues to restrict tryptic cleavage to lysine residues. CNBr has been used for top–down characterization of proteins by mass spectrometry.[101,102] CNBr cleavage of proteins yields fragments larger than those obtained by tryptic digestion, facilitating structure assignments. In a similar approach, modification of lysine residues with citraconic anhydride restricts tryptic cleavage to arginine residues and yields fragments that, as with the CNBr fragments, provide greater sequence coverage of the protein.[103] CNBr fragments have been used in the engineering of semisynthetic proteins.[104] A 223-residue hybrid of *Streptomyces griseus* trypsin was synthesized by native chemical ligation[105] from a chemically synthesized amino-terminal fragment and a larger carboxy-terminal fragment obtained by CNBr cleavage. CNBr is also used for the cleavage of fusion proteins.[106–110]

The use of specific chemical modification to study changes in the environment has been studied over the past 30 years. The study by Kirtley and Koshland[111] provided the basis for the concept of using reporter groups to study changes in the microenvironment surrounding a site of modification. This study used 2-bromo-acetamido-4-nitrophenol to modify a limited number of sulfhydryl groups in glyceraldehyde-3-phosphate dehydrogenase. The modified protein has a λ_{max} at 390 nm ($\varepsilon = 7100$ M^{-1} cm^{-1}) between pH 7.0 and 7.6. The addition of the coenzyme NAD caused a marked change in the spectral properties (decrease in absorbance at ca. 375 nm and increase in absorbance at ca. 420 nm) of the modified enzyme, which is consistent with a change in the microenvironment around the modified residue (increase in polarity of medium, which results in increased formation of the nitro-phenolate ion). The reactions of 2-hydroxy-5-nitrobenzyl bromide with tryptophanyl residues to yield the 2-hydroxy-5-nitrobenzyl derivative[112] and that of tetranitro-methane with tyrosyl residues[113] to form the 3-nitrotyrosyl derivative were extensively used to study microenvironmental changes in the modified proteins.[114]

Fluorescent probes have been used extensively in the study of protein confor-mation. The chemistry used for the covalent modification of proteins is described in detail in other chapters. The majority of studies use either lysine or cysteine as a target residue for modification.[115] The insertion of cysteine into recombinant proteins[116] via oligonucleotide-directed mutagenesis[117,118] has provided opportunity for the attachment of fluorescent probes to specific protein domains.[119] Fluorescent energy transfer (FRET) with covalently attached fluorescent probes is proving increasingly useful in the study of protein conformation.[120–123] Fluorescent probes (dyes) have also been used to modify proteins prior to separation by two-dimensional electrophoresis.[124,125]

Spin-labeled reagents (Figure 1.3) have also been useful.[126] One early study used spin-labeled derivatives of diisopropylphosphorofluoridate to study the active-site environment of trypsin.[127] Subsequent studies used various spin-labeled derivatives (piperidinyl nitroxide, pyrrolidinyl nitroxide, and pyrrolinyl nitroxide substituent groups) of phenylmethylsulfonyl fluoride to compare microenvironments surround-ing the active sites in α-chymotrypsin and trypsin.[128,129] These reagents have been more recently used to study the active site of thrombin.[130,131] In subsequent studies, Nienaber and Berliner[132] obtained the crystal structures of thrombin modified at the active-site serine with two substituted pyrrolidone nitroxide derivatives, 4-(2,2,5,5-tetramethylpyrrolidone-1-oxyl)-*p*-(fluorosulfonyl) benzamidine and 3-(2,2,5,5-tetra-methylpyrrolidone-1-oxyl)-*m*-(fluorosulfonyl) benzamidine. The crystal structures confirmed the earlier observations on the topography of the extended active-site region of thrombin obtained by the previously cited electron spin resonance studies. The preparation of spin-labeled pepsinogen has been reported.[133] This study used a *N*-hydroxysuccinimide ester derivative, 3-[[(2,5-dioxo-1-pyrrolidiny)oxyl] carbonyl]-2,5,-dihydro-2,2,5,5-tetramethyl-1H-pyrrolyl-1-oxy, to modify lysyl residues in pep-sinogen. Coupling was accomplished at pH 7.0 (0.1 *M* sodium phosphate) for 7 h at 22°C, resulting in the derivatization of approximately three amino groups. Site-directed attachment of nitroxide spin labels to inserted cysteine residues has been used to study the conformation of S-adenosyl-methionine synthetase[134] and the mitochondrial oxoglutarate carrier.[135]

A spectrum of *in vivo* chemical modifications of proteins controls biological activity. One group of such *in vivo* chemical modifications is cotranslational and posttranslational reactions. This includes such reactions as sulfation, methylation, and phosphorylation, which are primarily the result of enzyme-catalyzed reactions. Note that the characterization of such reactions has benefited significantly from the excellent previous work on the organic chemistry of proteins. A meaningful discus-sion of cotranslational and posttranslational protein modification reactions is beyond of the scope of this treatise. The reader is directed to several recent discussions of this area of proteomic research.[136–138]

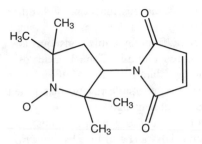

3-Maleimido-2,2,5,5-tetramethyl-1-pyrrolidinyloxyl

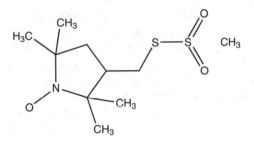

1-oxyl-2,2,5,5-tetramethylpyrolidin-3-yl methyl methanethiosulfonate

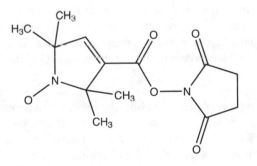

2,2,5,5-Tetramethyl-3-pyrrolin-1-oxyl-3-carboxylic acid *N*-hydroxysuccinimide

FIGURE 1.3 Some useful spin-labeled reagents that have been used for site-specific modification of proteins.

In addition to cotranslational and posttranslational modifications, a number of chemical, nonenzymatic *in vivo* modifications are important for the regulation of protein function. Some of these modifications appear to be random, whereas others appear to be specific, following the same rules for site-specific chemical modification as discussed next. These reactions include oxidation, various reactions with nitric oxide, and glycation.

Oxidation of proteins occurs *in vivo* with generally unfavorable consequences. For example, oxidation of the active-site methionine residue in alpha-1-antitrypsin (alpha-1-antiprotease inhibitor) results in pulmonary damage.[139–141] Methionine is quite susceptible to oxidation, first to the sulfoxide, a reversible reaction, and then to the sulfone.[142,143] There is wide spectrum of oxidizing agents, including free radicals such as hydroxyl radical,[144,145] hypochlorite,[146] and organic peroxides such as lipid peroxides in oxidative stress.[147–149] Hypochlorites can be formed by the action of myeloperoxidase and might be responsible for protein oxidation in atherosclerosis and Alzheimer's disease.[150,151] It is also suggested that oxidative covalent modifications of proteins are signals for degradation.[152]

Nitric oxide is a potent physiological agent with diverse systemic effects.[153–156] Peroxynitrite, formed from nitric oxide by reaction with superoxide,[157–160] is a mediator of some of these physiological effects of nitric oxide. The reaction of proteins with peroxynitrite results primarily in the modification of tyrosine residues to form 3-nitrotyrosine.[161–163] Peroxynitrite can also modify tryptophanyl residues[164] and oxidize methionine residues.[165] Other oxidation pathways exist, resulting in carbonyl formation and dityrosine.[166–168] Carbon dioxide influences the reaction of peroxynitrite with proteins[169,170] by forming an adduct with peroxynitrite.[171] Peroxynitrite can also react with nucleic acids,[172] resulting in chain cleavage. Nitric oxide can react directly with sulfhydryl groups on proteins, but the reactions appear to be somewhat complicated, resulting in the formation of an *S*-nitroso derivative of cysteine.[173,174] The reaction of proteins with nitroxyl radical is somewhat less studied.[175,176]

Glycation is the term used to identify the reaction of reducing sugars with proteins. This involves the initial formation of a Schiff base followed by rearrangement in the Maillard reaction,[177,178] eventually resulting in advanced glycation end products (AGEs).[179] Reaction can occur at lysine and arginine residues, with resulting cross-link formation.[180,181] The glycation of proteins by methyl glyoxal has been of specific interest.[182,183]

This information is provided as an overview to the current use of site-specific chemical modification in protein chemistry. Specific information on individual modification reactions is presented in the following chapters. The material presented is intended to be comprehensive, but the reader is also referred to other reviews on this topic.[184–199] In addition, several volumes of *Methods in Enzymology*[200–210] are extremely useful. The reader is also referred to Appendix I for information on the availability of various reagents, addresses of suppliers, and a list of current journals active in publishing the results of site-specific chemical modification studies.

An additional purpose of this chapter is to briefly introduce the concept of site-specific chemical modification, including methods of characterizing the product of chemical modification reactions. Site-specific chemical modification is strictly defined as a process that yields a stoichiometrically altered protein with the quantitative covalent derivatization of a single, unique amino acid residue without either modification of any other amino acid residues or conformational change. This objective is rarely obtained with most reagents, as several factors confound this goal. First, few reagents are specific for the modification of a single functional group.

Most reagents react with nucleophiles on proteins and the nucleophilic character is, in part, dependent on the protonation state of the residue. Table 1.1 presents the acid dissociation constants for typical amino acid functional groups. The acid dissociation constant is dependent on the microenvironment surrounding the specific amino acid residues, which is discussed next.

The environments of the various amino acid residues in a protein are not identical. As a result of this lack of homogeneity, a variety of surface polarities surround the various functional groups. The physical and chemical properties of any given functional group are strongly influenced by the nature (e.g., polarity) of the local microenvironment. For example, consider the effect of the addition of an organic solvent, ethyl alcohol, on the pK_a of acetic acid. In 100% H_2O, acetic acid has a pK_a of 4.70. Addition of 80% ethyl alcohol results in an increase of the pK_a to 6.9. In 100% ethyl alcohol, the pK_a of acetic acid is 10.3. This is particularly important while considering the reactivity of nucleophilic groups such as amino groups, cysteine, carboxyl groups, and the phenolic hydroxyl group. When primary amines are present in a protein, these functional groups are not reactive except in the free base form. In other words, the proton present at neutral pH must be removed from the ε-amino group of lysine before this functional group can function as an effective nucleophile. Table 1.1 also gives a listing of the average pK_a values for the various functional groups present in protein. Many modification reactions take advantage of differences in pK_a values in similar chemical groups. The difference in pK_a values between an α-amino group and an ε-amino group makes it possible to selectively modify the γ-amino group in a protein (see Chapter 2). Another example is the selective modification of the β/γ-carboxyl groups on a protein without modification of the α-carboxyl groups on a protein, because the protonated form of the carboxylic acid is required for successful reaction (see Chapter 5).

Other factors that can influence the pK_a of a functional group in a protein include hydrogen binding with an adjacent functional group, the direct electrostatic effect of the presence of a charged group in the immediate vicinity of a potential nucleo phile, and direct steric effects on the availability of a given functional group. An excellent example of the effect of a neighboring group on the reaction of a specific amino acid residue is provided by the comparison of the rates of modification of the active-site cysteinyl residue by chloroacetic acid and chloroacetamide in papain.[211,212] A rigorous evaluation of the effect of pH and ionic strength on the reaction of papain with chloroacetic acid and chloroacetamide demonstrated the importance of a neighboring imidazolium group in enhancing the rate of reaction at low pH. Similar results had been reported earlier by Gerwin[213] for the essential cysteine residues in streptococcal proteinase. The essence of the experimental observations is that the plot of the pH dependence of the second-order rate constant for the reaction with chloroacetic acid is bell shaped, with an optimum at ca. pH 6.0, whereas that of chloroacetamide is S shaped, approaching maximal rate of reaction at pH 10.0. Gerwin demonstrated that the reaction of chloroacetic acid and chloroacetamide with reduced glutathione did not demonstrate this difference in pH dependence. Other excellent examples of the effect of neighboring functional groups are the reaction of 2,4-dinitrophenyl acetate with a lysine residue at the active site of phosphonoacetaldehyde hydrolase, in which the pK_a of the lysine residue is decreased to 9.3 as a result of a positively charged environment[214] and the effect of remote sites on the reactivity of histidine residues in ribonuclease A.[215] Schmidt and Westheimer[216] made the seminal observations on the reaction of 2,4-dinitrophenyl propionate with the active-site lysine of acetoacetate decarboxylase, demonstrating a pK_a of 5.9 for this residue. These examples clearly demonstrate the effect of electrostatic effects on the reactivity of amino acid residues in proteins.

Another consideration can in a sense be considered either a cause or consequence of microenvironmental polarity. This concerns the environment immediately around the residue modified. These are the factors that can cause a selective increase (or decrease) in reagent concentration in the vicinity of a potentially reactive species. The most clearly understood example of this is the process of affinity labeling.[217] Another consideration is the partitioning of a reagent such as tetranitromethane between the aqueous environment, which is polar, and the interior of the protein, which in nonpolar. Tetranitromethane is an organic compound and can, in principle, react equally well with exposed and buried tyrosyl residues.[218]

Establishing the stoichiometry of modification is a relatively straightforward process. First, the molar quantity of modified residue is established by analysis. This could be spectrophotometric as, for example, with trinitrophenylation of primary amino groups, nitration of tyrosine with tetranitromethane, or alkylation of tryptophan with 2-hydroxy-5-nitrobenzyl bromide, or by amino acid analysis to determine either the loss of a residue as, for example, in photooxidation of histidine and the oxidation of the indole ring of tryptophan with N-bromosuccinimide or the appearance of a modified residue such as with S-carboxymethylcysteine or N^1-or N^3-carboxymethylhistidine. In the situation wherein spectral change or radiolabel incorporation is used to establish stoichiometry, analysis must be performed to determine that there is no reaction with another amino acid. For example, the extent of oxidation of tryptophan by N-bromosuccinimide can be determined spectrophotometrically, but amino acid analysis or mass spectrometric analysis is required to determine whether modification has also occurred with another amino acid such as histidine or methionine. It is clear that the evolution of mass spectrometry from an esoteric, specialized laboratory resource to a technique that is as common in the protein chemistry as amino acid analysis has provided another tool for the evaluation of protein structure after chemical modification.[219–221] As a result of the availability of mass spectrometric analysis of proteins, there is increasing use of this technology to evaluate the chemical modification of proteins.[222–230]

In the case of site-specific chemical modification of a protein, it must be established that the modification of one residue mole per mole of protein (or functional subunit) has occurred without modification of another amino acid. (For example, modification has occurred only with lysine and not with tyrosine.) The reaction pattern of a given reagent with free amino acids or amino acid derivatives does not necessarily provide the basis for reaction with such amino acid residues in a protein. Furthermore, the reaction pattern of a given reagent with one protein cannot necessarily be extrapolated to all proteins. Results of a chemical modification can be markedly affected by reaction conditions (e.g., pH, temperature, solvent or buffer used, and degree of illumination). Establishment of stoichiometry does not necessarily mean that this modification has occurred at a unique residue (unique in terms of position in the linear peptide chain, not necessarily unique with respect to reactivity). It is useful if there is a change in biological activity (catalysis, substrate binding, ion binding, etc.) occurring concomitant with the chemical modification. Ideally, one would like to establish a direct relationship (i.e., 0.5 mol/mol of protein with 50% activity modification; 1.0 mol/mol of protein with 100% activity modification). More frequently, there is the situation in which there are several moles of a given residue modified per mole of protein, but there is reason to suspect stoichiometric chemical modification. In some of these situations, it is possible to fractionate the protein into uniquely modified species. The separation of carboxy-methyl-His-12-pancreatic ribonuclease from carboxy-methyl-His-119-pancreatic ribonuclease is a classic example of this type of situation.[231] More recently, it has been possible to separate various derivatives of lysozyme obtained from the modification of carboxyl groups.[232] Frequently, however, although there is good evidence that multiple modified species are obtained as a result of the reaction, it is not possible to separate apparently uniquely modified species. In the reaction of tetranitromethane with thrombin,[233] apparent stoichiometry of inactivation was obtained with equivalent modification of two separate tyrosine residues (Tyr-71 and Tyr-85 in the B chain), and it was not possible to separate these derivatives.

Assessing stoichiometry of modification from only the functional consequences of such modification is a far more difficult proposition. First, there must be a clear, unambiguous signal that can be effectively measured. In a situation in which there are clearly multiple sites of reaction that can be distinguished by analytical techniques, the approach advanced by Ray and Koshland[234] is useful. This analysis is based on establishing a relationship between the rate of loss of biological activity and the rate of modification of a single residue. A similar approach advanced by Tsou[235–237] is based on establishing a relationship between the number of residues modified and the change in biological activity. Horiike and McCormick[238] explored the approach of relating changes in activity to extent of chemical modification. They state that the original concepts that form the basis of this approach are sound, but extrapolation from a plot of activity remaining vs. residues modified is not necessarily sound. Such extrapolation is only valid if the nonessential residues react much slower (rate at least 10 times slower). Given a situation wherein all residues within a given group are equally reactive toward the reagent in question, the number of essential residues obtained from such a plot is correct only when the total number of residues is equal to the number of essential residues which is, in turn, equal to 1.0. However, it is important to emphasize that this approach is useful when there is a difference in the rate of reaction of an essential residue or residues and all other residues in that class, as is the example in the modification of histidyl residues with diethylpyrocarbonate in lactate dehydrogenase[239,240] and pyridoxamine-5'-phosphate oxidase.[241] A major advantage in relating changes in activity to a specific chemical modification is being able to demonstrate that the reversal of modification is directly associated with the reversal of the changes in biological activity. Demonstrating that the effects of a specific chemical modification are reversible lends support against the argument that such effects are a result of irreversible and nonspecific conformational change. The issue is complicated when more than one residue is modified in the course of the chemical reaction. Whether the residues are like or unlike amino acids, it is still difficult to assign the functional consequences to the modification of a single residue. The mathematical approaches described previously provide an approach to this specific situation. Another complication is provided when there is incomplete inactivation at the completion of the chemical modification.[233,242–244]

A key difference between the use of site-specific chemical modification and site-specific mutagenesis to study protein structure and function is the ability to measure the rate of reaction of a specific amino acid residue or residues. The reader is referred to a review by Rakitzis[245] for a discussion of the kinetics of protein chemical modification. There has been continuing use of this approach during the 20 years since the article was published.[246–251] An example of the use of reaction rate is provided from the study of the modification of an aminopeptidase by diethylpyrocarbonate.[252] It was demonstrated that the reaction of the aminopeptidase with diethylpyrocarbonate resulted in the modification of histidine residues. A difference of the reactivity of the two histidine residues modified by diethylpyrocarbonate in the presence and absence of calcium ions permitted the identification of one of the two histidine residues as critical for the binding of calcium ions. Careful analysis of the effect of pH on the reaction rate in the presence and absence of calcium ions allowed the assignment of pK_a value to the two residues.

Functional characterization of the modified protein can provide a significant challenge. The discussion of this problem is biased toward the study of enzymes, but the same general considerations are valid for receptors, protein ligands, structural proteins, and carrier proteins such as hemoglobin and transferrin. Functional characterization is relatively straightforward when activity is totally abolished, such as that which occurs when the active-site histidine in a serine protease is modified with a peptide chloromethylketone.[253] A more difficult problem is encountered for a modified protein with fractional activity.[233,242–244] The most critical aspect in the characterization of the modified protein is the method used to determine activity. The rigorous determination of binding constants and kinetic constants is absolutely essential; the reporting of percent change in activity is clearly inadequate. The reader is directed to several classic works[254,255] as well as more recent expositions[256–260] in this area. For the reader who, like the author, is somewhat challenged by physical biochemistry, consideration of some more basic information[261–263] will be useful. Finally, the reader is directed to an excellent review by Plapp.[264] The discussion is directed toward the use of site-specific mutagenesis for the study of enzymes, but much of the content is equally applicable to the characterization of chemically modified proteins. Particular consideration should be given to the section on kinetics, with emphasis on the importance of V/K (catalytic efficiency) to evaluate the effect of a modification on catalytic activity and the discussion on the importance of understanding that K_m is not necessarily a measure of affinity. Evaluation contribution of individual residues to the overall catalytic process is also discussed, for which the reader is referred to another review.[265] This type of analysis would markedly increase the value of studies in which several different reagents are used for the chemical modification of a protein.

Conformational change must be considered in the interpretation of the results of site-specific modification of a protein.[266–268] More frequently, though, site-specific chemical modification has been used to assess conformational change in proteins,[269–275] including some extremely elegant recent examples.[276–278] Salhany and co-workers[276] used modification of a carboxyl group with Woodward's reagent K to study conformational effects in a membrane transport protein. D'Ambrosio and co-workers[277] used a combination of chemical cross-linking, modification of lysine residues with acetic anhydride, and modification of tyrosine residues with tetranitro-methane to study the dimeric structure of porcine aminoacylase 1. Mass spectrometry was used for the analysis of various chemical modification procedures. Li and Bigelow[278] modified tyrosine with tetranitromethane for use as a fluorescence resonance energy transfer (FRET) receptor for the study of the interaction between the transmembrane and cytosolic domains of phospholamban.

Fortunately, a number of analytic techniques are available to evaluate conformation change, which have been applied to the characterization of chemically modified proteins. Mass spectrometry[279–281] is a useful method for evaluating changes in the surface topology of a protein when combined with cross-linking and chemical modification.[282–284] Raman spectroscopy[285,288] has been used to study sulfhydryl groups in myoinositol monophosphatase,[287–289] histidine residues in bacteriochlorophyll,[290] the modification of a sulfhydryl group in hemoglobin A,[291] and the oxidation of methionine in amyloid plaque.[151] Circular dichroism[286, 292,293] has been used to study the modification of arginine residues with methyl glyoxal in α-crystalline[294]; reduction and alkylation of disulfide bonds in the ABA-1 allergen of the nematode *Ascaris*[295]; reductive methylation or N-acetylation of collagen[296]; and modification of human serum albumin with diethyl pyrocarbonate (His), acetyl salicylic acid (Lys), *p*-nitrophenyl anthranilate (Trp) and glucose (Lys).[297] Intrinsic fluorescence[298–303] has been used to study the modification of human serum albumin with diverse reagents[297] and the reaction of bovine serum albumin with alkyl anhydride and sulfenyl chlorides,[301] the folding pathway of recombinant human macrophage-colony stimulating factor in combination with chemical trapping and mass spectrometric mapping,[284] the reaction of 2-hydroxy-5-nitrobenzyl bromide with the rat ovarian LH/hCG receptor,[302] and modification of cysteine residues in β-lactoglobulin with *N*-ethylmaleimide.[303] This last study was of particular interest in that the effect of high pressure on the reaction was studied; high pressure forces the exposure of a buried sulfhydryl group, which then reacts with *N*-ethylmaleimide. X-ray crystallographic analysis[304,305] has used to study the reductive methylation of lysine residues in pokeweed antiviral protein-II.[306] This study showed that reductive methylation improved the quality of the crystal structure. Other crystallographic studies include the modification of cysteine in actin with fluorescent dyes[307] and the modification of hemoglobin with alkyl isothiocyanates.[308] Nuclear magnetic resonance has also been useful.[309–311] As noted previously, spin-labels and electron paramagnetic resonance have also been useful to study protein conformation.[312,313]

Careful control of pH, temperature, and ionic strength is necessary for any study of the solution chemistry of proteins. While selecting buffers for control of pH, there are concerns beyond buffering capacity or efficiency (β). Tris buffers cannot be used with *N*-acetyl imidazole as Tris is a competing nucleophile; both the reagent and reaction product are unstable in the presence of Tris.[313a]

The real danger in the use of chemical reagents for the site-specific modification of proteins is to make assumptions based on behavior in other systems. For example, phenylmethylsulfonyl fluoride (PMSF) is best known for reacting with the active-site serine residue in serine proteases such as trypsin, chymotrypsin, and thrombin. Thus, it comes as a surprise to observe the specific reaction of PMSF with a tyrosine residue in a superoxide dismutase.[314] Means and Wu previous described the modification of a tyrosine residue in human serum albumin with diisopropylphosphorofluoridate.[315] Peptide chloromethylketones, developed as reagents for the modification of histidine residues at the active sites of serine proteases,[253] react with other amino acid residues in disparate proteins.[316] This unexpected broad spectrum of reactivity undoubtedly provides an explanation for the effect of peptide chloromethylketones in complex systems.[317,318]

Chemical reagents can react with substrates and have value for this ability.[319-321] However, this precludes the consideration of value of protection by substrate.

REFERENCES

1. Degenhardt, T.P., Thorpe, S.R., and Baynes, J.W., Chemical modification of proteins by methylglyoxal, *Chem. Mol. Biol.* 44, 1139, 1998.
2. Seidler, N.W., and Kowalewski, C., Methylglyoxal-induced glycation affects protein topography, *Arch. Biochem. Biophys.* 410, 149, 2003.
3. Sheng, J., and Preiss, J., Arginine[294] is essential for the inhibition of *Anabaena* PCC 7120 ADP-glucose pyrophosphorylase by phosphate, *Biochemistry* 36, 13077, 1997.
4. Wu, X. et al., Alteration of substrate selectivity through mutation of two arginine residues in the binding site of amadoriase II from *Aspergillus* sp., *Biochemistry* 41, 4453, 2002.
5. Zhang, W., Hu, Y., and Kabak, H.R., Site-directed sulfydryl labeling of helix IX in the lactose permease of *Escherichia coli*, *Biochemistry* 42, 4904, 2003.
6. Lin, T.-K. et al., Specific modification of mitochondrial protein thiols in response to oxidative stress: a proteomics approach, *J. Biol. Chem.* 277, 17048, 2002.
7. Qiu, H. et al., Activation of human acid sphingomyelinase through modification or deletion of C-terminal cysteine, *J. Biol. Chem.* 278, 32744, 2003.
8. Karim, S.B. et al., Cysteine reactivity and oligomeric structures of phospholamban and its mutants, *Biochemistry* 37, 12074, 1998.
9. Bahar, S. et al., Persistence of external chloride and DIDS binding after chemical modification of Glu-681 in human band 3, *Am. J. Physiol.* 277, C791, 1999.
10. Schmid, B. et al., Biochemical and structural characterization of the cross-linked complex of nitrogenase: comparison to the ADP-A1F4(−)-stabilized structure, *Biochemistry* 41, 15557, 2002.
11. Kalkum, M., Przybylski, M., and Glocker, M.O., Structure characterization of functional histidine residues and carboxyethylated derivatives in peptides and proteins by mass spectrometry, *Bioconj. Chem.* 9, 226, 1998.
12. Orth, J.H.C., Blöcker, D., and Aktories, K., His1205 and His1223 are essential for the activity of the mitogenic *Pasteurella multocida* toxin, *Biochemistry* 42, 4971, 2003.
13. Dage, J.L., Sun, H., and Halsall, H.B., Determination of diethylpyrocarbonate-modified amino residues in α_1-acid glycoprotein by high-performance liquid chromatography electrospray ionization-mass spectrometry and matrix-assisted laser desorption/ionization time-of-flight-mass spectrometry, *Anal. Biochem.* 257, 176, 1998.

14. Kusnezow, W. et al., Antibody microarrays: an evaluation of production parameters, *Proteomics* 3, 254, 2003.
15. Muller, E.C. et al., Covalently crosslinked complexes of bovine adrenodoxin with adrenodoxin reductase and cytochrome P450scc. Mass spectrometry and Edman degradation of complexes of the steroidogenic hydroxylase system, *Eur. J. Biochem.* 268, 1837, 2001.
16. Hopkins, C.E. et al., Chemical-modification rescue assessed by mass spectrometry demonstrates that gamma-thia-lysine yields the same activity as lysine in aldolase, *Protein Sci.* 11, 1591, 2002.
17. Rodriguez, J.C., Wong, L., and Jennings, P.A. (2003), The solvent in CNBr cleavage reactions determines the fragmentation efficiency of ketosteroid isomerase fusion proteins used in the production of recombinant peptides, *Protein Express. Purif.* 28, 224, 2003.
18. Beckingham, J.A. et al., Studies on a single immunoglobulin-binding domain of protein L from *Peptostreptococcus magnus*: the role of tyrosine-53 in the reaction with human IgG, *Biochem. J.* 353, 395, 2001.
19. Leite, J.F., and Cascio, M., Probing the topology of the glycine receptor by chemical modification coupled to mass spectrometry, *Biochemistry* 41, 6140, 2002.
20. Munagala, N., Basus, V.J., and Wang, C.C., Role of the flexible loop of hypoxanthine-guanine-xanthine phosphoribosyltransferase from *Tritrichomas foetus* in enzyme catalysis, *Biochemistry* 40, 4303, 2001.
21. Mann, M., and Jensen, O.N., Proteomic analysis of post-translational modifications, *Nature Biotechnol.* 21, 255, 2003.
22. Chalkley, R.J., and Burlingame, A.L., Identification of novel sites of O-N-acetylglucosamine modification of serum response factor using quadrapole time-of-flight mass spectrometry, *Mol. Cell. Proteom.* 2, 182, 2003.
23. Zhang, K. et al., Histone acetylation and deacetylation: identification of acetylation and methylation sites of HeLa histone H4 by mass spectrometry, *Mol. Cell. Proteom.* 2, 500, 2002.
24. Chen, S.L. et al., Mass spectrometry-based methods for phosphorylation site mapping of hyperphosphorylated proteins applied to Net1, a regulator of exits from mitosis in yeast, *Mol. Cell. Proteom.* 2, 186, 2002.
25. Taylor, C.F. et al., A systematic approach to modeling, capturing, and disseminated proteomics experimental data, *Nature Biotechnol.* 21, 247, 2003.
26. Patterson, S.D., Data analysis: the Achilles heel of proteomics, *Nature Biotechnol.* 21, 221, 2003.
27. Dutt, M.J., and Lee, K.H., Proteomic analysis, *Curr. Opin. Biotechnol.* 11, 176, 2000.
28. Seillier-Moiseiwitsch, F., Trost, D.C., and Moiseiwitsch, J., Statistical methods for proteomics, *Methods Mol. Biol.* 184, 51, 2002.
29. Persson, B., Bioinformatics in protein analysis. In *Proteomics in Functional Genomics. Protein Structure Analysis* (Jollès, P., and Jörnvall, H., Eds.) Birkhäuser Verlag, Basel, pp. 215–232, 2000.
30. Gygi, S.P. et al., Quantitative analysis of complex protein mixtures using isotope-coded affinity tags, *Nature Biotechnol.* 17, 994, 1999.
31. Griffin, T.J., and Aebersold, R., Advances in proteome analysis by mass spectrometry, *J. Biol. Chem.* 276, 45479, 2001.
32. Tao, W.A., and Aebersold, R., Advances in quantitative proteomics via stable isotope tagging and mass spectrometry, *Curr. Opin. Biotechnol.* 14, 110, 2003.
33. Smolka, M., Zhou, H., and Aebersold, R., Quantitative protein profiling using two-dimensional gel electrophoresis, isotope-coded affinity tag labeling, and mass spectrometry, *Mol. Cell. Proteom.* 1, 19, 2002.

34. Smolka, M.B. et al., Optimization of the isotope-coded affinity tag-labeling procedure for quantitative proteome analysis, *Anal. Biochem.* 297, 25, 2001.
35. Zhang, R. et al., Fractionation of isotopically labeled peptides in quantitative proteomics, *Anal. Chem.* 73, 5142, 2001.
36. Hansen, K.C. et al., Mass spectrometric analysis of protein mixtures at low levels using cleavable ^{13}C-isotope-coded affinity tag and multidimensional chromatography, *Mol. Cell. Proteom.* 2, 229, 2003.
37. Goshe, M.B. et al., Phosphoprotein isotope-coded affinity tag approach for isolating and quantitating phosphopeptides in proteome-wide analyses, *Anal. Chem.* 73, 2578, 2001.
38. Kuyama, H. et al., An approach to quantitative proteome analysis by labeling tryptophan residues, *Rapid Commn. Mass Spectrom.* 17, 1642, 2003.
39. Goshe, M.B., and Smith, R.D., Stable isotope-coded proteomic mass spectrometry, *Curr. Opin. Biotechnol.* 14, 101, 2003.
40. Qiu, Y. et al., Acid-labile isotope-coded extractants: a class of reagents for quantitative mass spectrometric analysis of complex protein mixtures, *Anal. Chem.* 74, 4969, 2002.
41. Ong, S.-E. et al., Stable isotope labeling by amino acids in cell culture, SILAC, as a simple and accurate approach to expression proteomics, *Mol. Cell. Proteom.* 1, 376, 2002.
42. Michels, D.A. et al., Fully automated two-dimensional capillary electrophoresis for high sensitivity protein analysis, *Mol. Cell. Proteom.* 1, 69, 2002.
43. Kwon, S.W. et al., Selective enrichment of thiophosphorylated polypeptides as a tool for the analysis of protein phosphorylation, *Mol. Cell. Proteom.* 2, 242, 2003.
44. Gevaert, K. et al., Chromatographic isolation of methionine-containing peptides for gel-free proteome analysis, *Mol. Cell. Proteom.* 1, 896, 2002.
45. Spande, T.F. et al., Selective cleavage and modification of peptides and proteins. *Adv. Protein Chem.* 24, 97, 1970.
46. Brown, J.R., and Hartley, B.S., Location of disulphide bridges by diagonal paper electrophoresis. The disulphide bridges of bovine chymotrypsinogen A, *Biochem. J.* 101, 214, 1966.
47. Tang, J., and Hartley, B.S., A diagonal electrophoretic method for selective purification of methionine peptides, *Biochem. J.* 102, 593, 1967.
48. Greenbaum, D. et al., Chemical approaches for functionally probing the proteome, *Mol. Cell. Proteom.* 1, 60, 2002.
49. Adam, G.C., Sorenson, E.J., and Cravatt, B.F., Trifunctional chemical probes for the consolidated detection and identification of enzyme activities from complex proteomes, *Mol. Cell. Proteom.* 1, 828, 2002.
50. Ruoppolo, M. et al., Analysis of transglutaminase protein substrates by functional proteomics, *Protein Sci.* 12, 1290, 2003.
51. Nikov, G. et al., Analysis of nitrated proteins by nitrotyrosine-specific affinity probes and mass spectrometry, *Anal. Biochem.* 320, 214, 2003.
52. Sokolovsky, M, Riordan, J.F., and Vallee, B.L., Conversion of 3-nitrotyrosine to 3-aminotyrosine in peptides and proteins, *Biochem. Biophys. Res. Commn.* 6, 358, 1967.
53. Scherrer, P., and Stoeckenius, W., Selective nitration of tyrosines-26 and -64 in bacteriorhodopsin with tetranitromethane, *Biochemistry* 23, 6195, 1984.
54. Biomolecular Technologies, Inc., http://www.bmtusa.com.
55. Stark, G.R., Recent developments in chemical modification and sequential degradation of proteins, *Adv. Protein Chem.* 24, 261, 1970.

56. Shen, M. et al., Isolation and isotope labeling of cysteine- and methionine-containing tryptic peptides, *Mol. Cell. Proteom.* 2, 315, 2003.
57. Fontana, A., and Scoffone, E., Sulfenyl halides as modifying reactions for polypeptides and proteins, *Methods Enzymol.* 25, 482, 1972.
58. Fung, E.T. et al., Protein chips for differential profiling, *Curr. Opin. Biotechnol.* 12, 65, 2001.
59. Metzger, S.W., Lochhead, M.J., and Grainger, D.W., Improving performance in protein-based microarrays, *IVD Technol.* 39, June 2002.
60. Schweitzer, B., and Kingsmore, S., Measuring proteins on microarrays, *Curr. Opin. Biotechnol.* 13, 14, 2002.
61. Cutler, P., Protein arrays: the current state-of-the art, *Proteomics* 3, 3, 2003.
62. Zhu, H., and Snyder, M., Protein chip technology, *Curr. Opin. Biotechnol.* 7, 55, 2003.
63. Kusnezow, W. et al., Antibody microarrays: an evaluation of production parameters, *Proteomics* 3, 254, 2003.
64. Mills, J.S. et al., Biologically active fluorescent derivatives of spinach calmodulin that report calmodulin target binding, *Biochemistry* 27, 991, 1988.
65. Lee, K.-N. et al., Protein patterning by virtual mask photolithography using a micromirror array, *J. Micromech. Microeng.* 13, 18, 2003.
66. Peluso, P. et al., Optimizing antibody immobilization strategies for the construction of protein microarrays, *Anal. Biochem.* 312, 113, 2003.
67. Johnson, C.P. et al., Engineering protein A for the orientation control of immobilized proteins, *Bioconj. Chem.* 14, 974, 2003.
68. Zhang, Z. et al., A new strategy for the site-specific modification of proteins *in vivo*, *Biochemistry* 42, 6735, 2003.
69. Smith, R.A. et al., Chemical derivatization of therapeutic proteins, *Trends Biotechnol.* 11, 297, 1993.
70. Pozansky, M.J., Soluble enzyme-albumin conjugates: new possibilities for enzyme replacement therapy, *Methods Enzymol.* 137, 566, 1988.
71. Sharifi, J. et al., Improving monoclonal antibody pharmacokinetics via chemical modification, *J. Nucl. Med.* 42, 242, 1998.
72. Awwad, M. et al., Modification of monoclonal antibody carbohydrates by oxidation, conjugation, or deoxymannojirimycin does not interfere with antibody effector functions, *Cancer Immunol. Immunother.* 38, 23, 1994.
73. Brader, M.L., et al., Hybrid insulin cocrystals for controlled release delivery, *Nature Biotechnol.* 20, 800, 2002.
74. Lundblad, R.L., and Bradshaw, R.A., Application of site-specific chemical modification in the manufacture of biopharmaceuticals. I. An overview, *Biotechnol. Appl. Biochem.* 26, 143, 1997.
75. Kozlowski, A., and Harris, J.M., Improvements in protein PEGylation: pegylated interferons for treatment of hepatitis C, *J. Control. Release* 72, 217, 2001.
76. Harrington, K.J., Mabasher, H.S., and Palu, G., Polyethylene glycol in the design of tumor-targeting radiolabelled macromolecules: lessons from liposomes and monoclonal antibodies, *Q. J. Nucl. Med.* 46, 171, 2002.
77. Chapman, A.P., Therapeutic antibody fragments with prolonged *in vivo* half-lives, *Nat. Biotechnol.* 54, 531, 2002.
78. Veronese, F.M., and Harris, J.M., Introduction and overview of peptide and protein PEGylation, *Adv. Drug Deliv. Rev.* 54, 453, 2002.
79. Crawford, J., Clinical uses of pegylated pharmaceuticals in oncology, *Cancer Treat. Rev.* 28 (Suppl. A), 7, 2002.

80. Harris, J.M., and Chess, R.B., Effect of PEGylation on pharmaceuticals, *Nat. Rev. Drug Discov.* 2, 214, 2003.

81. Wylie, D.C. et al., Carboxyalkylated histidine is a pH-dependent product of PEGylation with SC-PEG, *Pharm. Res.* 18, 1354, 2001.

82. Tsutsumi, Y. et al., Site-specific chemical modification with polyethylene glycol of recombinant immunotoxin anti-Tac(Fv)-PE38 (LMB-2) improves antitumor activity and reduces animal toxicity and immunogenicity, *Proc. Natl. Acad. Sci. USA* 97, 8548, 2000.

83. Kerwin, B.A. et al., Interactions between PEG and type I soluble tumor necrosis factor receptor: modulation by pH and by PEGylation at the N terminus, *Protein Sci.* 11, 1825, 2002.

84. Lee, H. et al., N-terminal site-specific mono-PEGylation of epidermal growth factor, *Pharm. Res.* 20, 818, 2003.

85. Greenwald, R.B. et al., Controlled release of proteins from their poly(ethylene glycol) conjugates: drug delivery systems employing 1,6-elimination, *Bioconj. Chem.* 14, 395, 2003.

86. Kim, J.-G. et al., Development and characterization of a glucagon-like peptide 1-albumin conjugate. The ability to activate the glucagon-like peptide 1 receptor *in vivo*, *Diabetes* 52, 751, 2003.

87. Poznonsky, M.J., and Antohe, F., Enzyme-albumin conjugates: advantages and problems with targeting. In *Pharmaceutial Enzymes* (Sawyers, A., and Sharpé, S., Eds.), Marcel Dekker, New York, 1997, Chap. 3.

88. Wilbur, D.S. et al., Development of new biotin/streptavidin reagents for pretargeting, *Biomol. Eng.* 16, 113, 1999.

89. Doronina, S.O. et al., Development of potent monoclonal antibody auristatin conjugates for cancer therapy, *Nat. Biotechnol.* 21, 778, 2002.

90. Gladysheva, I.P. et al., Immunoconjugates of soybean Bowman-Birk protease inhibitor as targeted antitumor polymeric agents, *J. Drug Target.* 9, 303, 2001.

91. Ruppert, C. et al., Chemical crosslinking of urokinase to pulmonary surfactant protein B for targeting alveolar fibrin, *Thromb. Haemostas.* 89, 53, 2003.

92. Raiche, A.T., and Puleo, D.A., Triphasic release model for multilayered gelatin coatings that can recreate growth factor profiles during wound healing, *J. Drug Target.* 9, 449, 2001.

93. Ozeki, M. et al., Controlled release of hepatocyte growth factor from gelatin hydrogels based on hydrogel degradation, *J. Drug Target.* 9, 461, 2001.

94. Stevens, K.R. et al., *In vivo* biocompatibility of gelatin-based hydrogels and interpenetrating networks, *J. Biomater. Sci. Polym. Ed.*, 13, 1353, 2002.

95. Friess, W. et al., Characterization of absorbable collagen sponges as recombinant human bone morphogenetic protein-2 carrier, *Int. J. Pharm.* 185, 51, 1999.

96. Freiss, W. et al., Characterization of absorbable collagen sponges as rhBMP-2 carriers, *Int. J. Pharm.* 187, 91, 1999.

97. Mistrello, G. et al., Monomeric chemically modified allergens: immunologic and physiochemical characterization, *Allergy* 51, 8, 1999.

98. Bagnasco, M. et al., Pharmacokinetics of an allergen and a monomeric allergoid for oromuscosal immunotherapy in allergic volunteers, *Clin. Exp. Allergy* 31, 54, 2001.

99. Akdis, C.A., and Blaser, K., Regulation of specific immune responses by chemical and structural modifications of allergens, *Int. Arch. Allergy Immunol.* 121, 261, 2000.

100. Müller, U.R., New developments in the diagnosis and treatment of hymenoptera venom allergy, *Int. Arch. Allergy Immunol.* 124, 447, 2001.

101. Kelleher, N.L. et al., Top down versus bottom up protein characterization by tandem high-resolution mass spectrometry, *J. Am. Chem. Soc.* 121, 806, 1999.

102. Ge, Y. et al., Detection of four oxidation sites in viral prolyl-4-hydroxylase by top-down mass spectrometry, *Protein Sci.* 12, 2320, 2003.

103. Kadlik, V., Strohalm, M., and Kodicek, M., Citraconylation: a simple method for high protein sequence coverage in MALDI-TOF mass spectrometry, *Biochem. Biophys. Res. Commn.* 305, 1091, 2003.

104. Pál, G. et al., The first semi-synthetic serine protease made by native chemical ligation, *Protein Express. Purif.* 29, 185, 2003.

105. Dawson, P.E. et al., Synthesis of proteins by native chemical ligation, *Science* 266, 776, 1994.

106. Kuliopulos, A., and Walsh, C.T., Production, purification, and cleavage of tandem repeats of recombinant peptides. *J. Am. Chem. Soc.* 116, 4599, 1994.

107. Karamloo, F. et al., Pyr c 1, the major allergen from pear (*Pyrus communis*), is a new member of the Bet v 1 allergen family, *J. Chromatogr.* 756, 281, 2001.

108. Fairle, W. et al., A fusion protein system for the recombinant production of short disulfide-containing peptides, *Protein Express. Purif.* 26, 171, 2002.

109. Rodríguez, J.C. et al., The solvent in CNBr cleavage reactions determines the fragmentation efficiency of ketosteroid isomerase fusion proteins used in the production of recombinant peptides, *Protein Express. Purif.* 28, 224, 2003.

110. Jenny, R.J., Mann, K.G., and Lundblad, R.L., A critical review of the methods for cleavage of fusion proteins with thrombin and factor Xa, *Protein. Express. Purif.* 31, 1, 2003.

111. Kirtley, M.E., and Koshland, D.E., Jr., The introduction of a "reporter" group at the active site of glyceraldehyde-3-phosphate dehydrogenase, *Biochem. Biophys. Res. Commn.* 23, 810, 1966.

112. Loudon, G.M., and Koshland, D.E., Jr., The chemistry of a reporter group: 2-hydroxy-5-nitrobenzyl bromide, *J. Biol. Chem.* 245, 2247, 1970.

113. Riordan, J.R., Sokolovsky, M., and Vallee, B.L., Environmentally sensitive tyrosyl residues. Nitration with tetranitromethane, *Biochemistry* 6, 358, 1967.

114. Herzyk, E., Owen, J.S., and Chapman, D., The secondary structure of apolipoproteins in human HDL3 particles after chemical modification of their tyrosine, lysine, cysteine or arginine residues. A Fourier transform infrared spectroscopy study, *Biochim. Biophys. Acta* 962, 131, 1988.

114a. Skawinski, W.J., Adebodun, F., Cheng, J.T., Jordan, F., and Mendelsohn, R., Labeling of tyrosines in proteins with [N-15] tetranitromethane, a new NMR reporter for nitrotyrosine, *Biochim. Biophys. Acta* 1162, 297, 1993.

114b. Tcherkasskaya, O. and Ptitsyn, O. B., Direct energy transfer to study the 3D structure of non-native proteins: AGH complex in molten globule state of apomyoglobin, *Protein Eng.* 12, 485, 1999.

114c. Tschenko, V.M., Zav'yalova, G.A., Bliznyukov, O.P., and Zav'yalov, V.P., Thermodynamic, conformational and functional properties of the human C1q globular heads in the intact C1q molecule in solution, Mol.Immunol. 40, 1225, 2004.

115. Waggoner, A., Covalent labeling of proteins and nucleic acids with fluorophores, *Methods Enzymol.* 246, 362, 1995.

116. Bech, L.M. et al., Introduction of a free cysteine residue at position 68 in the subtilisin Savinase, based on homology with proteinase K, *FEBS Lett.* 297, 164, 1992.

117. Kunkel, T.A., Rapid and efficient site-specific mutagenesis without phenotypic selection, *Proc. Natl. Acad. Sci. USA* 82, 488, 1985.

118. Hogrefe, H.H. et al., Creating randomized amino acid libraries with the QuikChange® multi site-directed mutagenesis kit, *BioTechniques* 33, 1158, 2002.

119. Karlin, A., and Akabas, M.H., Substituted-cysteine accessibility method, *Methods Enzymol.* 293, 123, 1998.

120. Ratner, V. et al., A general strategy for site-specific double labeling of globular proteins for kinetic FRET studies, *Bioconjug. Chem.* 13, 1163, 2002.

121. Heyduk, T., Measuring protein conformational changes by FRET/LRET, *Curr. Opin. Biotechnol.* 13, 292, 2002.

122. Watrob, H.M., Pan, C.P., and Barkley, M.D., Two-step FRET as a structural tool, *J. Am. Chem. Soc.* 125, 7336, 2003.

123. Rhoades, E., Gussakovsky, E., and Haran, G., Watching proteins fold one molecule at a time, *Proc. Natl. Acad. Sci. USA* 100, 7418, 2003.

124. Buschmann, V., Weston, K.D., and Sauer, M., Spectroscopic study and evaluation of red-absorbing fluorescent dyes, *Bioconj. Chem.* 14, 195, 2003.

125. Kondo, T. et al., Application of sensitive fluorescent dyes in linkage of laser microdissection and two-dimensional gel electrophoresis as a cancer proteomic study tool, *Proteomics* 3, 1758, 2003.

126. Berliner, L.J. (Ed.), *Spin Labeling: Theory and Applications*, Academic Press, New York, 1975.

127. Berliner, L.J., and Wong, S.S., Evidence against two "pH locked" conformations of phosphorylated trypsin, *J. Biol. Chem.* 248, 1118, 1973.

128. Berliner, L. J., and Wong, S. S., Spin-labeled sulfonyl fluorides as active site probes of protease structure. I. Comparison of the active site environments in α-chymotrypsin and trypsin, *J. Biol. Chem.* 249, 1668, 1974.

129. Wong, S.S. et al., Spin-labeled sulfonyl fluorides as active site probes of protease structure. II. Spin label synthese and enzyme inhibition, *J. Biol. Chem.* 249, 1678, 1974.

130. Berliner, L.J., and Shen, Y.Y., Probing active site structure by spin label (ESR) and fluorescence methods. In *Chemistry and Biology of Thrombin* (Lundblad, R.L., Fenton, J.W., II, and Mann, K.G., Eds.), Ann Arbor Science, Ann Arbor, MI, 1977, p. 197.

131. Berliner, L.J. et al., Active site topography of human coagulant(α) and noncoagulant(β) thrombins, *Biochemistry* 20, 1831, 1981.

132. Nienaber, V.L., and Berliner, L.J., Atomic structures of two nitroxide spin labels complexed with human thrombin: comparison with solution studies, *J. Protein Chem.* 19, 129, 2000.

133. Twining, S.S., Sealy, R.C., and Glick, D.M., Preparation and activation of spin-labelled pepsinogen, *Biochemistry* 20, 1267, 1981.

134. Taylor, J.C., and Markham, G.D., Conformational dynamics of the active site loop of S-adenosylmethionine synthetase illuminated by site-directed spin labeleing, *Arch. Biochem. Biophys.* 415, 164, 2003.

135. Morozzo della Rocca, B. et al., The mitochondrial oxoglutarate carrier: structural and dynamic properties of transmembrane segment IV studied by site-directed mutagenesis, *Biochemistry* 42, 5493, 2003.

136. Chen, C.A., and Manning, D.R., Regulation of G proteins by covalent modification, *Oncogene* 20, 1643, 2001.

137. Bradshaw, R.A., and Yi, E., Methionine aminopeptidases and angiogenesis, *Essays Biochem.* 38, 65, 2002.

138. de Ruijter, A.J. et al., Histone deactylases (HCACs): characterization of the classical HDAC family, *Biochem. J.* 370, 737, 2003.

139. Johnson, D., and Travis, J., The oxidative inactivation of human alpha-1-proteinase inhibitor. Further evidence for methionine at the reactive center, *J. Biol. Chem.* 254, 4022, 1979.

140. Matheson, N.R., and Travis, J., Differential effects of oxidizing agents on human plasma alpha 1 proteinase inhibitor and human neutrophils myeloperoxidase, *Biochemistry* 24, 1941, 1985.

141. Ueda, M., Mashiba, S., and Uchida, K., Evaluation of oxidized alpha-1-antitrypsin in blood as an oxidative stress marker using anti-oxidative alpha1-AT monoclonal antibody, *Clin. Chim. Acta* 317, 125, 2002.

142. Vogt, W., Oxidation of methionyl residues in proteins: tools, targets, and reversal, *Free Radical Biol. Med.* 18, 93, 1995.

143. Maleknia, S.D., Brenowitz, M., and Chance, M.R., Millisecond radiolytic modification of peptides by synchrotron X-rays identified by mass spectrometry, *Anal. Chem.* 71, 3965, 1999.

144. Heyduk, E., and Heyduk, T., Mapping protein domains involved in macromolecular interactions: a novel protein footprinting approach, *Biochemistry* 33, 9643, 1994.

145. Rashidzadeh, H. et al., Solution structure and interdomain interactions of the *Saccharomyces cerevisiae* "TATA binding protein" (TBP) probed by radiolytic protein footprinting, *Biochemistry* 42, 3655, 2003.

145a. Kiselar, J.G. et al., Structural analysis of gelsolin using synchrotron protein footprinting, *Mol. Cell. Proteom.*, 2, 1120, 2003.

146. Hawkins, C.L., and Davies, M.J., Hypochlorite-induced oxidation of proteins in plasma: formation of chloramines and nitrogen-centered radicals and their role in protein fragmentation, *Biochem. J.* 340, 539, 1999.

147. Gay, C.A., and Gebicki, J.M., Measurement of protein and lipid hydroperoxides in biological systems by the ferric-xylenol orange method, *Anal. Biochem.* 315, 29, 2003.

148. Refsgaard, H.H., Tsai, L., and Stadtman, E.R., Modifications of proteins by polyunsaturated fatty acid peroxidation products, *Proc. Natl. Acad. Sci. USA*, 97, 611, 2000.

149. Hidalgo, F.J., Alaiz, M., and Zamora, R., A spectrophotometric method for the determination of proteins damaged by oxidized lipids, *Anal. Biochem.* 262, 129, 1998.

150. Woods, A.A., Linton, S.M., and Davies, M.J., Detection of HOCl-mediated protein oxidation products in the extracellular matrix of human atherosclerotic plaques, *Biochem. J.* 370, 729, 2003.

151. Dong, J. et al., Metal binding and oxidation of amyloid- within isolated senile plaque cores: Raman microscopic evidence, *Biochemistry* 42, 2768, 2003.

152. Stadtman, E.R., Covalent modification reactions are marking steps in protein turnover, *Biochemistry* 29, 6323, 1990.

153. Koppenol, W.H., and Trayham, J.G., Say NO to nitric oxide: nomenclature for nitrogen- and oxygen-containing compounds, *Methods Enzymol.* 268, 3, 1996.

153. Schwentker, A., and Billiar, T.R., Nitric oxide and wound repair, *Surg. Clin. N. Am.* 83, 521, 2003.

154. Mamas, M.A., Reynard, J.M., and Brading, A.F., Nitric oxide and the lower urinary tract: current concepts, future prospects, *Urology* 61, 1079, 2003.

155. von Haehling, S., Ander, S.D., and Bassenge, E., Statins and the role of nitric oxide in chronic heart failure, *Heart Fail. Rev.* 8, 99, 2003.

156. Ricciardolo, F.L., Multiple roles of nitric oxide in the airways, *Thorax* 58, 175, 2003.

157. Freels, J.L. et al., Enhanced activity of human IL-10 after nitration in reducing human IL-1 production by stimulated peripheral blood mononuclear cells, *J. Immunol.* 169, 4568, 2002.

158. Galiñanes, M., and Matata, B.M., Protein nitration is predominantly mediated by a peroxynitrite-dependent pathway in cultured human leukocytes, *Biochem. J.* 367, 467, 2002.

159. Pryor, W.A. et al., A practical method for preparing peroxynitrite solutions of low ionic strength and free of hydrogen peroxide, *Free Radical Biol. Med.* 18, 75, 1995.

160. Bartlett, D. et al., The kinetics of the oxidation of L-ascorbic acid by peroxynitrite, *Free Radical Biol. Med.* 18, 85, 1995.

161. Kong, S.-K. et al., Peroxynitrite disables the tyrosine phosphorylation regulatory mechanism: lymphocyte-specific tyrosine kinase fails to phosphorylate nitrated $CDC2(6-20)NH_2$ peptide, *Proc. Natl. Acad. Sci. USA* 93, 3377, 1996.

162. MacMillan-Crow, L.A. et al., Nitration and inactivation of manganese superoxide dismutase in chronic rejection of human renal allografts, *Proc. Natl. Acad. Sci. USA* 93, 11853, 1999.

163. Jiao, K. et al., Site-selective nitration of tyrosine in human serum albumin by peroxynitrite, *Anal. Biochem.* 293, 43, 2003.

164. Ischiropoulos, H., Biological selectivity and functional aspects of protein tyrosine nitration, *Biochem. Biophys. Res. Commn.* 305, 776, 2003.

165. Pryor, W.A., Jin, X., and Squadrito, G.L., One- and two-electron oxidations of methionine by peroxynitrite, *Proc. Natl. Acad. Sci. USA* 91, 11173, 1994.

166. Beckman, J.S. et al., Oxidative chemistry of peroxynitrite. *Methods Enzymol.* 233, 229, 1994.

167. Morton, L.W., Puddey, I.B., and Croft, K.D., Comparison of nitration and oxidation of tyrosine in advanced human carotid plaque proteins, *Biochem. J.* 370, 339, 2003.

168. Norris, E.H. et al., Effects of oxidative and nitrative challenges on α-synuclein fibrillogenesis involve distinct mechanisms of protein modification, *J. Biol. Chem.* 278, 27230, 2003.

169. Berlett, B.S., Levine, R.L., and Stadtman, E.R., Carbon dioxide stimulates peroxynitrite-mediated nitration of tyrosine residues and inhibits oxidation of methionine residues of glutamine synthetase: Both modifications mimic effects of adenylation, *Proc. Natl. Acad. Sci. USA* 95, 2784, 1998.

170. Tien, M. et al., Peroxynitrite-mediated modification of proteins at physiological carbon dioxide concentration: pH dependence of carbonyl formation, tyrosine nitration, and methionine oxidation, *Proc. Natl. Acad. Sci. USA* 96, 7809, 1999.

171. Lymar, S.V., and Hurst, K., Rapid reaction between peroxynitrite ion and carbon dioxide. Implications for biological activity, *J. Am. Chem. Soc.* 117, 8867, 1995.

172. Saigo, M.G. et al., Peroxynitrite causes DNA nicks in plasmid pBR232, *Biochem. Biophys. Res. Commn.* 210, 1025, 1995.

173. Nedospasov, A. et al., An autocatalytic mechanism of protein nitrosylation, *Proc. Natl. Acad. Sci. USA* 97, 13543, 2000.

174. Tao, L., and English, A.M., Mechanism of *S*-nitrosation of recombinant human brain calbindin D_{28K}, *Biochemistry* 42, 3326, 2000.

175. Cook, N.M. et al., Nitroxyl-mediated disruption of thiol proteins: inhibition of the yeast transcription factor Ace1, *Arch. Biochem. Biophys.* 410, 89, 2003.

176. Houk, K.N. et al., Nitroxyl disulfides, novel intermediates in transnitrosation reactions, *J. Am. Chem. Soc.* 125, 6972, 2003.

177. Maillard, L.C., Action des acides amines sur les sucres: formation des melanoidines par voie methodique, *C. R. Hebd. Seanes Acad. Sci.*, 154, 66, 1912.

178. Ulrich, P., and Cerami, A., Protein glycation, diabetes, and aging, *Recent Prog. Horm. Res.* 56, 1, 2001.

179. Biemel, K.M., and Lederer, M.O., Site-specific quantitative evaluation of the protein glycation product N(6) 0 (2,3-dihydroxy-5,6-dioxohexyl-1-lysinate by LC-(ESI) MS peptide mapping: evidence for its key role in AGE formation, *Bioconj. Chem.* 14, 619, 2003.

180. Tessier, F.J. et al., Triosidines: novel Maillard reaction products and cross-links from the reaction of triose sugars with lysine and arginine residues, *Biochem. J.* 369, 705, 2003.

181. Miller, A.G., Meade, S.J., and Gerrard, J.A., New insights into protein crosslinking via the Maillard reaction: structural requirements, the effects on enzyme function, and predicted efficacy of crosslinking inhibitors as anti-aging therapeutics, *Bioorg. Med. Chem.* 11, 843, 2003.

182. Degenhardt, T.P., Thorpe, S.R., and Baynes, J.W., Chemical modification of proteins by methylglyoxal, *Cell. Mol. Biol.* 44, 1139, 1998.

183. Seidler, N.W., and Kowalewski, C., Methylglyoxal-induced glycation affects protein topography, *Arch. Biochem. Biophys.* 410, 149, 2003.

184. Lundblad, R.L., and Noyes, C.M., *Chemical Reagents for Protein Modification,* CRC Press, Boca Raton, FL, 1984.

185. Lundblad, R.L., *Chemical Reagents for Protein Modification,* 2nd ed., CRC Press, Boca Raton, FL, 1991.

186. Walker, J.M. (Ed.), *Methods in Molecular Biology, Vol. 1, Proteins,* Humana Press, Clifton, NJ, 1984.

187. Shivley, J.E. (Ed.), *Methods of Protein Microcharacterization. A Practical Handbook,* Humana Press, Clifton, NJ, 1986.

188. Oxender, D.L., and Fox, C.F. (Eds.), *Protein Engineering,* Alan R. Liss, New York, 1987.

189. Fasman, G.D., *Practical Handbook of Biochemistry and Molecular Biology,* CRC Press, Boca Raton, FL, 1989.

190. Fasman, G.D., *Handbook of Biochemistry and Molecular Biology, Vol. 2, Proteins,* 3rd. ed., CRC Press, Boca Raton, FL, 1976.

191. Walker, J.M. (Ed.), *Methods in Molecular Biology, Vol. 3, New Protein Techniques,* Humana Press, Clifton, NJ, 1988.

192. Lundblad, R.L., *Techniques in Protein Modification,* CRC Press, Boca Raton, FL, 1994.

193. Means, G.E., Zhang, H., and Le, M., The chemistry of protein functional groups. In *Proteins. A Comprehensive Treatise,* Vol. 2 (Allen, G., Ed.), JAI Press, Stamford, CT, 1999, Chap. 2.

194. DeSantis, G., and Jones, J.B., Chemical modification of enzymes for enhanced functionality. *Curr. Opin. Biotechnol.* 10, 324, 1999.

195. Stark, G.R., Recent developments in chemical modification and sequential degradation of proteins, *Adv. Protein Chem.* 24, 261, 1970.

196. Offord, R.E., Chemical approaches to protein engineering. In *Protein Engineering: A Practical Approach* (Sternberg, M.J.E., and Wetzel, R., Eds.), IRL Press at Oxford University Press, Oxford, 1992, Chap. 10.

197. Li-Chan, E.C.Y., Methods to monitor process-induced changes in food proteins. In *Process-Induced Changes in Food* (Shahadi, F., and Ho, C.-T., Eds.), Plenum Press, New York, 1998, Chap. 2.

198. Spande, T.F. et al., Selective cleavage and modification of peptides and proteins, *Adv. Protein Chem.* 24, 97, 1970.

199. Davis, B.G., Chemical modification of biocatalysts, *Curr. Opin. Biotechnol.* 14, 379, 2003.

199a. Price, N.C. (Ed.), Protein LabFax, Bios Scientific Publishers, Academic Press, Oxford, 1996.
199b. Engel, P.C. (Ed.), Enzymology LabFax, Bios Scientific Publishers, Academic Press, Oxford, 1996.
200. Hirs, C.H.W. (Ed.), Enzyme structure, *Methods Enzymol.* 11, 1967.
201. Hirs, C.H.W., and Timasheff, S.N. (Eds.), Enzyme structure, *Methods Enzymol.* 25, 1971.
202. Hirs, C.H.W., and Timasheff, S.N. (Eds.), Enzyme structure, *Methods Enzymol.* 47, 1976.
203. Hirs, C.H.W., and Timasheff, S.N. (Eds.), Enzyme structure, *Methods Enzymol.* 91, 1983.
204. Packer, L. (Ed.), Oxygen radicals in biological systems, *Methods Ezymol.* 233, 1994.
205. Lee, Y.C., and Lee, R.T., Neoglycoconjugates, Part A. Synthesis, *Methods Enzymol.* 242, 1994.
206. Spears, K., Biochemical spectroscopy, *Methods Enzymol.* 246, 1995.
207. Spears, K., Enzyme kinetics, *Methods Enzymol.* 249, 1995.
208. Packer, L., Biothiols. Part A. Monothiols and dithiols, protein thiols, and thiyl radicals, *Methods Enzymol.* 251, 1995.
209. Packer, L., Nitric oxide. Part A. Sources and detection of NO; NO synthase, *Methods Enzymol.* 268, 1996.
210. Marriott, G., and Parker, I. (Eds.), Biophotonics, Part A, *Methods Enzymol.* 360, 2003.
211. Chaiken, I.M., and Smith, E.L., Reaction of chloroacetamide with the sulfydryl groups of papain, *J. Biol. Chem.* 244, 5087, 1969.
212. Chaiken, I.M., and Smith, E.L., Reaction of the sulfydryl group of papain with chloroacetic acid, *J. Biol. Chem.* 244, 5095, 1969.
213. Gerwin, B.I., Properties of the single sulfydryl group of streptococcal proteinase. A comparison of the rates of alkylation by chloroacetic acid and chloroacetamide, *J. Biol. Chem.* 242, 451, 1967.
214. Zhang, G., Mazurkie, A.S., Dunaway-Mariano, D., and Allen, K.N., Kinetic evidence for a substrate-induced fit in phosphonoacetaldehyde hydrolase catalysis, *Biochemistry* 41, 13370, 2002.
215. Fisher, B.M., Schultz, L.W., and Raines, R.T., Coulombic effects of remote subsites on the active site of ribonuclease A, *Biochemistry* 37, 17386, 1998.
216. Schmidt, D.E., Jr., and Westheimer, F.H., pK of the lysine amino group at the active site of acetoacetate decarboxylase, *Biochemistry* 10, 1249, 1971.
217. Plapp, B.V., Application of affinity labeling for studying structure and function in enzymes, *Methods Enzymol.* 87, 469, 1982.
218. Myers, B.H., and Glazer, A.N., Spectroscopic studies of the exposure of tyrosine residues in proteins with special references to the subtilisins, *J. Biol. Chem.* 26, 412, 1971.
218a. Skov, K., Hofmann, T., and Williams, G.R., The nitration of cytochrome c, *Can. J. Biochem.* 47, 750, 1969.
219. Huang, L. et al., Functional assignment of the 20 S proteasome from *Trypanosoma soma* using mass spectrometry and new bioinformatics approaches, *J. Biol. Chem.* 276, 28327, 2001.
220. Glish, G.L., and Vachet, R.W., The basics of mass spectrometry in the twenty-first century, *Nature Rev. Drug Discov.* 2, 140, 2003.
221. Lin, D., Tabb, D.L., and Yates, J.R., III, Large-scale protein identification using mass spectrometry, *Biochim. Biophys. Acta* 1646, 1, 2003.

222. Bennett, K.L. et al., Rapid characterization of chemically-modified proteins by electrospray mass spectrometry, *Bioconj. Chem.* 7, 16, 1996

223. Kelleher, N.L. et al., Localization of labile posttranslational modifications by electron capture dissociation: the case of γ-carboxyglutamic acid, *Anal. Chem.* 71, 4250, 1999.

224. Fligge, T.A. et al., Direct monitoring of protein-chemical reactions by utilizing nanoelectrospray mass spectrometry, *J. Am. Soc. Mass Spectrom.* 10, 112, 1999.

225. Jahn, O. et al., The use of multiple ion chromatograms in on-line HPLC-MS for the characterization of post-translational and chemical modifications of proteins, *Int. J. Mass Spectrom.* 214, 37, 2002.

226. Leite, J.F., and Cascio, M., Probing the topology of the glycine receptor by chemical modification coupled to mass spectrometry, *Biochemistry* 41, 6140, 2002.

227. Fenaille, F., Guy, P.A., and Tabet, J.C., Study of protein modification by 4-hydroxy-2-nonenol and other short chain aldehydes analyzed by electrospray ionization tandem mass spectrometry, *J. Am. Soc. Mass Spectrom.* 14, 215, 2003.

228. Hakansson, S., Viljanen, J., and Broo, K.S., Programmed delivery of novel functional groups to the alpha class glutathione transferases, *Biochemistry* 42, 10260, 2003.

229. Standing, K.G., Peptide and protein de novo sequencing by mass spectrometry, *Curr. Opin. Struct. Biol.* 13, 595, 2003.

230. Fenaille, F., Tabet, J.C., and Guy, P.A., Study of peptides containing modified lysine residues by tandem mass spectrometry: precursor ion scanning of hexanal-modified peptides, *Rapid Commun. Mass Spectrom.* 18, 67, 2004.

230a. Sharp, J.S., Becker, J.M., and Hettich, R.L., Analysis of protein solvent accessible surfaces by photochemical oxidation and mass spectrometry, *Anal. Chem.* 76, 672, 2004.

230b. Novak, P., Kruppa, G.H., Young, M.M., and Schoeniger, J., A top-down method for the determination of residue-specific solvent accessibility in proteins, *J. Mass Spectrom.* 39, 322, 2004.

231. Crestfield, A.M., Stein, W.H., and Moore, S., Alkylation and identification of the histidine residues at the active site of ribonucleae, *J. Biol. Chem.* 238, 2418, 1963.

232. Yamada, A., Imoto, T., Fujita, K., Okasaki, K., and Motomura, M., Selective modification of aspartic acid-101 in lysozyme by carbodiimide reaction, *Biochemistry* 20, 4836, 1981.

233. Lundblad, R.L., Noyes, C.M., Featherstone, G.L., Harrison, J.H., and Jenzano, J.W., The reaction of bovine alpha-thrombin with tetranitromethane. Characterization of the modified protein, *J. Biol. Chem.* 263, 3729, 1988.

234. Ray, W.J., Jr., and Koshland, D.E., Jr., A method for characterizing the type and numbers of groups involved in enzyme action. *J. Biol. Chem.* 236, 1973, 1961.

235. Tsou, C.-L., Relation between modification of functional groups of proteins and their biological activity. I. A graphical method for the determination of the number and type of essential groups, *Sci. Sinica* 11, 1535, 1962.

236. Tsou, C.-L., Kinetics of substrate reaction during irreversible modification of enzyme activity, *Adv. Enzymol.* 61, 381, 1988.

237. Zhou, J.-M., Liu, C., and Tsou, C.-L., Kinetics of trypsin inhibition by its specific inhibitors, *Biochemistry* 28, 1070, 1989.

238. Horiike, K., and McCormick, D.B., Correlations between biological activity and the number of functional groups chemically modified, *J. Theor. Biol.* 79, 403, 1979.

239. Holbrook, J.J., and Ingram, V.A., Ionic properties of an essential histidine residue in pig heart lactate dehydrogenase, *Biochem. J.* 131, 729, 1973.

240. Bloxham, D.P., The chemical reactivity of the histidine-195 residue in lactate dehydrogenase thiomethylated at the cysteine-165 residue, *Biochem. J.* 193, 93, 1981.

241. Horiike, K., Tsuge, H., and McCormick, D.B., Evidence for an essential histidyl residue at the active site of pyridoxamine (pyridoxine)-5′-phosphate oxidase from rabbit liver, *J. Biol. Chem.* 254, 6638, 1979.
242. Levy, H.M., Leber, P.D., and Ryan, E.M., Inactivation of myosin by 2,4-dinitrophenol and protection by adenosine triphosphate and other phosphate compounds, *J. Biol. Chem.* 238, 3654, 1963.
243. Grouselle, M., and Pudlis, J., Chemical studies on yeast hexokinase. Specific modification of a single tyrosyl residue with 1-ethyl-3-(3-dimethylaminopropyl) carbodiimide, *Eur. J. Biochem.* 74, 471, 1977.
244. Makinen, K.K. et al., Chemical modification of *Aeromonas* aminopeptidase. Evidence for the involvement of tyrosyl and carboxyl groups in the activity of the enzyme, *Eur. J. Biochem.* 128, 257, 1982.
245. Rakitzis, E.T., Kinetics of protein modification reactions, *Biochem. J.* 217, 341, 1984.
246. Rakitzis, E.T., Kinetic analysis of regeneration by dilution of a covalently modified protein, *Biochem. J.* 268, 669, 1990.
247. Page, M.G.P., The reaction of cephalosporins with penicillin-binding protein 1b from *Escherichia coli*, *Biochim. Biophys. Acta* 1205, 1994.
248. Dubus, A., Normark, S., Kania, M., and Page, M.G.P., Role of asparagine 152 in catalysis of -lactam hydrolysis by *Escherichia coli* ampC -lactamase studied by site-directed mutagenesis, *Biochemistry* 34, 7757, 1995.
249. Yang, S.J. et al., Involvement of tyrosine residue in the inhibition of plant vacuolar H^+-pyrophosphatase by tetranitromethane, *Biochim. Biophys. Acta* 1294, 89, 1996.
250. Chu, C.L. et al., Inhibition of plant vacuolar H^+-ATPase by diethylpyrocarbonate, *Biochim. Biophys. Acta* 1506, 12, 2001.
251. Hsiao, Y.Y. et al., Diethylpyrocarbonate inhibition of vacuolar H^+-pyrophosphatase possibly involves a histidine residue, *J. Protein Chem.* 21, 51, 2002.
252. Yang, S.-H., Wu, C.-H., and Lin, W.-Y., Chemical modification of aminopeptidase isolated from Pronase, *Biochem. J.* 302, 595, 1994.
253. Powers, J.C., Reaction of serine proteases with halomethyl ketones, *Meth. Enzymol.* 46, 197, 1977.
253a. Collen, D., Lijnen, H.R., De Cock, F., Durieux, J.P., and Loffet, A., Kinetic properties of tripeptide lysine chloromethyl ketones and lysyl-*p*-nitroanilide derivatives toward trypsin-like serine proteinases, *Biochim. Biophys. Acta* 615, 158, 1980.
253b. Kettner, C. and Shaw, E., Inactivation of trypsin-like enzymes with peptides of arginine chloromethyl ketones, *Meth. Enzymol.* 80, 826, 1981.
254. Dixon, M., and Webb, E.C., *Enzymes*, 3rd ed., Academic Press, New York, 1979.
255. Siegal, I.H., *Enzyme Kinetics. Behavior and Analysis of Rapid Equilibrium and Steady-State Enzyme Systems*, Wiley-Interscience, New York, 1975.
256. Purich, D.L. (Ed.), *Contemporary Enzyme Kinetics and Mechanism*, Academic Press, New York, 1983.
257. Northrup, D.B., Rethinking fundamentals of enzyme action, *Adv. Enzymol.* 73, 25, 1999.
258. Wang, J. et al., A graphical method of analyzing pH dependence of enzyme activity, *Biochim. Biophys. Acta* 1435, 177.
259. Raguin, O., Gruaz-Guyon, A., and Barbet, J., Equilibrium expert: an add-in to Microsoft Excel for multiple binding equilibrium simulations and parameter estimations, *Anal. Biochem.* 310, 1, 2002.
260. Liao, F. et al., Kinetic substrate quantification by fitting the enzyme reaction curve to the integrated Michaelis-Menten equation, *Anal. Bioanal. Chem.* 375, 756, 2003.

261. Brey, W.S., *Physical Chemistry and Its Biological Applications*, Academic Press, New York, 1978.
262. Sheehan, D., *Physical Biochemistry: Principles and Applications*, John Wiley & Sons, New York, 2000.
263. Price, N.C. et al., *Physical Chemistry for Biochemists*, 3rd ed., Oxford University Press, Oxford, 2001.
264. Plapp, B.V., Site-directed mutagenesis: a tool for studying enzyme catalysis, *Methods Enzymol.* 249, 91, 1995.
265. Peracchi, A., Enzyme catalysis: removing chemically 'essential' residues by site-directed mutagenesis, *Trends Biochem. Sci.* 26, 497, 2001.
266. Rao, A.G., and Neet, K.E., Tryptophan residues of the gamma subunit of 7S nerve growth factor: intrinsic fluorescence, solute quenching, and *N*-bromosuccinimide oxidation, *Biochemistry* 21, 6843, 1982.
267. Davies, K.J.A., Protein damage and degradation by oxygen radicals III. Modification of secondary and tertiary structure, *J. Biol. Chem.* 262, 9908, 1987.
268. Okajima, T., Kawata, Y., and Hamaguchi, K., Chemical modification of tryptophan residues and stability changes in proteins, *Biochemistry* 29, 9168, 1990.
269. Suckau, D., Mak, M., and Przybylski, M., Protein surface topology-probing by selective chemical modification and mass spectrometric peptide mapping. *Proc. Natl. Acad. Sci. USA* 89, 5630, 1992.
270. Gettins, P.G.W., Fan, B., Crews, B.C., Turko, I.V., Olson, S.T., and Streusand, V.J., Transmission of conformational change from the heparin binding site to the reactive center of antithrombin, *Biochemistry* 32, 8385, 1993.
271. Buechler, J.A., Vedvick, T.A., and Taylor, S.S., Differential labeling of the catalytic subunit of cAMP-dependent protein kinase with acetic anhydride: substrate-induced conformational changes, *Biochemistry* 28, 3018, 1989.
272. Mykkanen, H.M., and Wasserman, R.H., Reactivity of sulfhydryl groups in the brush-border membranes of chick duodena is increased by 1,25-dihydroxycholecalciferol, *Biochim. Biophys. Acta* 1033, 282, 1990.
273. Landfear, S.M., Evans, D.R., and Lipscomb, W.N., Elimination of cooperativity in aspartate transcarbamylase by nitration of a single tyrosine residue, *Proc. Natl. Acad. Sci. USA* 75, 2654, 1978.
274. Kumar, G.K., Beegen, N., and Wood, H.G., Involvement of tryptophans at the catalytic site and subunit-binding domains of transcarboxylase, *Biochemistry* 27, 5972, 1988.
275. Winkler, M.A., Fried, V.A., Merat, D.L., and Cheung, W.Y., Differential reactivities of lysines in calmodulin complexed to phosphatase, *J. Biol. Chem.* 262, 15466, 1987.
276. Salhany, J.M., Sloan, R.L., and Cordes, K.S., The carboxyl side chain of glutamate 681 interacts with a chloride binding modifier site that allosterically modulates the dimeric conformational state of Band 3 (AE1). Implications for the mechanism of anion/proton cotransport. *Biochemistry* 42, 1589, 2003.
277. D'Ambrosio, C. et al., Probing the dimeric structure of porcine aminoacylase 1 by mass spectrometric and modeling procedures. *Biochemistry* 42, 4430, 2003.
278. Li, J., and Bigelow, D.J., Phosphorylation by cAMP-dependent protein kinase modulates the structural coupling between the transmembrane and cytosolic domains of phospholamban. *Biochemistry* 42, 10674, 2003.
279. Cristoni, S. and Bernardi, L.R., Development of new methodologies for the mass spectrometric study of bioorganic macromolecules, *Mass Spectrom. Rev.* 22, 369, 2003.
280. Baldwin, M.A., Protein identification by mass spectrometry: issues to be considered, *Mol. Cell. Proteom.* 3, 1, 2004.

281. Wu, Y. and Engen, J.R., What mass spectrometry can reveal about protein function, *Analyst* 129, 290, 2004.
282. Bennett, K.L., Matthiesen, T., and Roepstorff, P., Probing protein surface topology by chemical surface labeling, crosslinking, and mass spectrometry. *Methods Mol. Biol.* 146, 113, 2000.
283. Happersberger, H.P., Przybylski, M., and Glocker, M.O., Selective bridging of bis-cysteinyl residues by arsonous acid derivatives as an approach to the characterization of protein tertiary structures and folding pathways by mass spectrometry, *Anal. Biochem.* 264, 237, 1998.
284. Happersberger, H.P., Cowgill, C., and Glocker, M.O., Structural characterization of monomeric folding intermediates of recombinant human macrophage-colony stimu-lating factor (rhM-CSF) by chemical trapping, chromatographic separation and mass spectrometric peptide mapping. *J. Chromatogr. B* 782, 393, 2002.
285. Parker, F.S., *Applications of Infrared, Raman, and Resonance Raman Spectroscopy in Biochemistry*, Plenum Press, New York, 1983.
286. Wright, J.R. et al., *Physical Methods for Inorganic Biochemistry*, Plenum Press, New York, 1986.
287. Carey, P.R. and Dong, J., Following ligand binding and ligand reactions in protein via Raman crystallography, *Biochemistry* 43, 8885–8893, 2004.
288. DuBois, J.L. and Klinman, J.P., Methods for characterizing TPQ-containing proteins, *Methods Enzymol.* 378, 17–31, 2004.
289. Ganzhorn, A.J., Vincendon, P., and Pelton, J.T., Structural characterization of myo-inositol monophosphatase from bovine brain by secondary structure prediction, flu-orescence, circular dichroism and Raman spectroscopy. *Biochim. Biophys. Acta* 1161, 303, 1993.
290. Buche, A., Ellis, G., and Ramirez, J.M., Probing the binding site of 800-nm bacteri-ochlorophyll in the membrane-linked LH2 protein of *Rhodobacter capsulatus* by local unfolding and chemical modification: evidence for the involvement of a His20 residue, *Eur. J. Biochem.* 268, 2792. 2001.
291. Juszczak, L.J., Manjula, B., Bbonaventura, C., Acharya, S.A., and Friedman, J.M., UV resonance Raman study of 93-modified hemoglobin A: chemical modifier-specific effects and added influences of attached poly(ethylene glycol) chains, *Biochemistry* 41, 376, 2002.
292. Coleman, J.E. and Oakley, J.L., Physical chemical studies of the structure and function of DNA binding (helix-stabilizing) proteins, *CRC Crit. Rev. Biochem.* 7, 247, 1980.
293. Miller, G.J. and Ball, E.H., Conformational change in the vinculin C-terminal depends on a critical histidine residue (His-906), *J. Biol. Chem.* 276, 28829, 2000.
294. Nagaraj, R.H. et al., Enhancement of chaperone function of α-crystalline by meth-ylglyoxal modification, *Biochemistry* 42, 10746, 2003.
295. McDermott, L. et al., Mutagenic and chemical modification of the ABA-1 allergen of the nematode *Ascaris*: consequences for structure and lipid binding properties, *Biochemistry* 40, 9918, 2001.
296. Giudici, C. et al., Molecular stability of chemically modified collagen triple helices, *FEBS Lett.* 547, 170, 2003.
297. Romanini, D., Müller, G., and Picó, G., Use of amphotericin B as optical probe to study conformational changes and thermodynamic stability in human serum albumin, *J. Protein Chem.* 21, 505, 2002.
298. Penzer, G., Molecular emission spectroscopy (fluoresence and phosphorescence). In *An Introduction to Spectroscopy for Biochemists* (Brown, S.B., Ed.), Academic Press, London, 1980, Chap. 3.

299. Engelborghs, Y., The analysis of time resolved protein fluorescence in multi-tryptophan proteins, *Spectrochim. Acta A Mol. Biomol. Spectrosc.* 57, 2255, 2001.
300. Persson, D. et al., Application of a novel analysis to measure the binding of the membrane-translocating peptide penetratin to negatively charged liposomes, *Biochemistry* 42, 421, 2003.
301. Gerbanowski, A. et al., Grafting of aliphatic and aromatic probes on bovine serum albumin: influence on its structural and physicochemical characteristics, *J. Protein Chem.* 18, 325, 1999.
302. Scsukova, S. et al., Fluorescence quenching studies of the rat ovarian LH/hCG receptor, *Gen. Physiol. Biophys.* 15, 451, 1996.
303. Tanaka, N. et al., Modification of the single unpaired sulfydryl group of beta-lactoglobulin under high pressure and the role of intermolecular S-S exchange in the pressure denaturation (single SH of beta lactoglobulin and pressure denaturation), *Int. J. Biol. Macromol.* 19, 63, 1996.
304. McRee, D.E., *Practical Protein Crystallography*, 2nd ed., Academic Press, San Diego, 1993.
305. Foadi, J. et al., A flexible and efficient procedure for the solution and phase refinement of protein structures, *Acta Crystallogr. D. Biol. Crystallogr.* 56, 1137, 2000.
306. Kurinov, I.V. et al., X-ray crystallographic analysis of pokeweed protein-II after reductive methylation of lysine residues, *Biochem. Biophys. Res. Commn.* 275, 549, 2000.
307. Yasunaga, T., and Wakabayashi, T., Relocation of Cys374 of actin induced by labeling with fluorescent dyes. *J. Biochem.* 129, 201, 2001.
308. Park, S. et al., Regioselective covalent modification of hemoglobin in search of antisickling agents, *J. Med. Chem.* 46, 936, 2003.
309. Claiborne, A. et al., Structural, redox, and mechanistic parameters for cysteine-sulfenic acid function in catalysis and regulation, *Adv. Protein Chem.* 58, 215, 2001.
310. Geyer, M. et al., Glucosylation of Ras by *Clostridium sordellii* lethal toxin: consequences for effector loop conformations observed by NMR spectroscopy, *Biochemistry* 42, 11951, 2003.
311. Dvoretsky, A., Gaponenko, V., and Rosevear, P.R., Derivation of structural restraints using a thiol-reactive chelator, *FEBS Lett.* 528, 189, 2002.
312. Cammack, R., and Shergill, J.K., EPR spectroscopy in enzymology. In *Enzymology LabFax* (Engel, P.C., Ed.), Bios Scientific Publishers, Oxford, 1996, Chap. 8.
313. della Rocca, B.M. et al., The mitochondrial oxoglutarate carrier: structural and dynamic properties of transmembrane segment IV studied by site-directed spin labeling, *Biochemistry* 42, 5493, 2003.
313a. Riordan, J.F., and Vallee, B.L., *O*-Acetyltyrosine, *Methods Enzymol.* 25, 500, 1972.
314. De Vendittis, E. et al., Phenylmethylsulfonyl fluoride inactivates an archael superoxide dismutase by chemical modification of a specific tyrosine residue. Cloning, sequencing and expression of the gene coding for *Sulfolobus solfataricus*, *Eur. J. Biochem.* 268, 1794, 2001.
315. Means, G.E., and Wu, H.L., The reactive tyrosine residues of human serum albumin: characterization of its reaction with diisopropylfluorophosphate, *Arch. Biochem. Biophys.* 194, 526, 1979.
316. Frey, J., Kordel, W., and Schneider, F., The reaction of aminoacylase with chloromethylketone analogues of amino acids, *Zeit. Naturforsch.* 32, 769, 1977.
317. Abate, A., Oberle, S., and Schröder, H., Lipopolysaccharide-induced expression of cyclooxygenase-2 in mouse macrophages is inhibited by chloromethylketones and a direct inhibitor of NF-B translocation, *Prostaglandins Other Lipid Mediat.* 56, 277, 1998.

318. Lee, M.J., and Goldsworthy, G.J., Chloromethyl ketones are insulin-like stimulators of lipid synthesis in locust fat body. *Arch. Insect Biochem. Physiol.* 39, 9, 1998.

319. Mahotra, O.P., and Dwivedi, U.N., Formation of enzyme-bound carbanion intermediate in the isocitrate lysase-catalyzed reaction: Enzymatic behavior on tetranitromethane with substrates and its dependence on effector, pH, and metal ions, *Arch. Biochem. Biophys.* 250, 238, 1986.

320. Medda, R., Padiglia, A., Pedersen, J.Z., and Floris, G., Evidence of α-proton abstraction and carbanion formation involving a functional histidine residue in lentil seedling amine oxidase, *Biochem. Biophys. Res. Commn.* 196, 1349, 1993.

321. Rùa, J., Soler, J., Busto, F., and de Arriaga, D., The pH dependence and modification by diethylpyrocarbonate of isocitrate lyase from *Phycomyces blakesleeanus*, *Eur. J. Biochem.* 232, 381, 1995.

2 The Modification of Amino Groups

It is difficult to obtain site-specific modification of lysine or α-amino groups in proteins. Amino groups must be unprotonated to function as satisfactory nucleophiles; therefore, alkaline pH conditions are required. As shown in Table 1.1, the pK_a for the ε-amino group of lysine is 10.79 and 9.68 for an α-amino group. It is normal practice for modification reactions to be performed at pH 8.0 or higher, although it has been possible to obtain site-specific modification at a lower pH. As with other functional groups in proteins, amino-group reactivity is subject to microenvironmental influences. The issue of microenvironmental influences on functional group reactivity is discussed in Chapter 1 and other chapters. Schmidt and Westheimer[1] reported the acylation of the amino group at the active site of acetoacetate decarboxylase by 2,4-dinitrophenyl propionate. The observed pK_a for this amino group was 5.9, which, as noted by the authors, is ca. 4 pK units less than that of an ordinary ε-amino group of lysine. Notwithstanding the difficulties in the interpretation of such data,[2] it is clear that the pH dependence of the reaction is consistent with a difference in lysine group reactivity. Differences in lysine group reactivity can be further studied by competitive labeling, as discussed later. Table 2.1 presents a listing of reagents commonly used to modify lysine residues.

Acylation of amino groups in proteins with acid anhydrides (Figure 2.1) or less frequently with acid chlorides is an extensively used modification procedure. Modification most often occurs at the ε-amino group of lysine, somewhat less frequently at α-amino groups, and amino sugars, if present.[3] Reaction can also occur at other nucleophilic functional groups, including sulfhydryl, phenolic hydroxyl, histidine imidazole nitrogens, and aspartic or glutamic via mixed acid anhydride formation. Most of these modifications are either exceedingly transient or labile under conditions (mild base) in which N-acyl groups are stable. Acylation can be performed with a variety of acid anhydrides, including citraconic anhydride,[4,5] maleic anhydride,[6–8] succinic anhydride,[9–11] 3-hydroxyphthalic anhydride,[12] trimellitic anhydride,[12] methyltetrahydrophthalic anhydride,[13,14] cis-aconitic anhydride,[10,11] fatty acid anhydrides,[15] hexahydrophthalic anhydride,[14,16] and phthalic anhydride.[17] Some of these modifications are discussed in greater detail later.

Succinic anhydride (Figure 2.1) has also proved useful in the modification of lysine.[18] Modification of lysine residues with succinic anhydride results in charge reversal. Reaction with succinic anhydride frequently results in the dissociation of multimeric proteins and has also been used to solubilize insoluble proteins. Meighen and coworkers[19] have produced a variant form of bacterial luciferase through a reaction with succinic anhydride. The succinylated protein retained the dimeric subunit structure of the native enzyme. By complementation experiments involving

TABLE 2.1
Some Reagents Used to Obtain Site-Specific Modification of Lysyl Residues in Proteins

Reagent	Other AA Modified	Ref.
Acetic anhydride	α-Amino groups, tyrosine, histidine, cysteine	1–13
Methyl acetyl phosphate	α-Amino groups	4–21
Pyridoxal-5′-phosphate		22–29
Reductive alkylation		30–40
2,4,6-Trinitrobenzenesulfonic acid	α-Amino groups	41–50

References for Table 2.1

1. Shen, S., and Strobel, H.W., Role of lysine and arginine residues of cytochrome P450 in the interaction between cytochrome P4502B1 and NADPH-cytochrome P450 reductase, *Arch. Biochem. Biophys.* 304, 257, 1993.
2. Suckau, D., Mak, M., and Przybylski, M., Protein surface topology-probing by selective chemical modification and mass spectrometric peptide mapping, *Proc. Natl. Acad. Sci. USA* 89, 5630, 1992.
3. Cervenansky, C., Engstrom, A., and Karlsson, E., Study of structure-activity relationship of fasciculin by acetylation of amino groups, *Biochim. Biophys. Acta* 1199, 1, 1994.
4. Ohguro, H. et al., Topographical study of arrestin using differential chemical modifications and hydrogen deuterium exchange, *Protein Sci.* 3, 2428, 1994.
5. Zappacosta, F. et al., Surface topology of Minibody by selective chemical modification and mass spectrometry, *Protein Sci.* 6, 1901, 1997.
6. Hochleitner, E.O. et al., Characterization of a discontinuous epitope of the human immunodeficiency virus (HIV) core protein p24 by epitope excision and differential chemical modification followed by mass spectrometric peptide mapping analysis, *Protein Sci.* 9, 487, 2000.
7. Hlavica, P., Schulze, J., and Lewis, D.F.V., Function interaction of cytochrome P450 with its redox partners: a critical assessment and update of the topology of the predicted contact regions, *J. Inorg. Biochem.* 96, 279, 2003.
8. Calvete, J.J. et al., Characterisation of the conformational and quaternary structure-dependent heparin-binding region of bovine seminar plasma protein PDC-109, *FEBS Lett.* 444, 260164, 1999.
9. Smith, C.M. et al., Mass spectrometric quantification of acetylation at specific lysines within the amino-terminal tail of histone H4. *Anal. Biochem.* 316, 23, 2003.
10. Yadev, S.P., Brew, K., and Puett, D., Holoprotein formation of human chorionic gonadotropin: differential labeling with acetic anhydride, *Mol. Endocrinol.* 8, 1547, 1994.
11. Gao, J. et al., Determination of the effective charge of a protein in solution by capillary electrophoresis, *Proc. Natl. Acad. Sci. USA* 91, 12027, 1994.
12. Gao, J. et al., Using capillary electrophoresis to follow the acetylation of the amino groups of insulin and to estimate their basicities, *Anal. Chem.* 68, 3093, 1995.
13. Taralp, A., and Kaplan, H., Chemical modification of lysophilized proteins in non-aqueous environments, *J. Protein Chem.* 16, 183, 1997.
14. Kluger, R., and Tsui, W.-C., Methyl acetyl phosphate. A small anionic acetylating agent, *J. Org. Chem.* 45, 2723, 1980.
15. Ueno, H., Pospischil, M.A., Manning, J.M., and Kluger, R., Site-specific modification of hemoglobin by methyl acetyl phosphate, *Arch. Biochem. Biophys.* 244, 795, 1986.
16. Ueno, H., Pospischil, M.A., and Manning, J.M., Methyl acetyl phosphate as a covalent probe for anion-binding sites in human and bovine hemoglobins, *J. Biol. Chem.* 264, 12344, 1989.
17. Ueno, H., and Manning, J.M., The functional, oxygen-linked chloride binding sites of hemoglobin are contiguous within a channel in the central cavity, *J. Protein Chem.* 11, 177, 1992.

TABLE 2.1 (continued)
Some Reagents Used to Obtain Site-Specific Modification of Lysyl Residues in Proteins

18. Ueno, H., Popowicz, A.M., and Manning, J.M., Random chemical modification of the oxygen-linked chloride-binding sites of hemoglobin: those in the central dyan axis may influence the transition between deoxy- and oxy-hemoglobin, *J. Protein Chem.* 12, 561, 1993.
19. Manning J.M., Preparation of hemoglobin derivatives selectively or randomly modified at amino groups, *Methods Enzymol.* 231, 225, 1994.
20. Kataoka, K. et al., Identification of active site lysyl residues of phenylalanine dehydrogenases by chemical modification with methyl acetylphosphate combined with site-directed mutagenesis. *J. Biochem.* 116, 1370, 1994.
21. Xu, A.S., Labotka, R.J., and London, R.E., Acetylation by human hemoglobin by methyl acetylphosphate; evidence of broad region-selectivity revealed by NMR studies. *J. Biol. Chem.* 274, 26629, 1999.
22. Lo Bello, M. et al., Chemical modification of human placental glutathione transferase by pyridoxal-5'-phosphate, *Biochim. Biophys. Acta,* 1121, 167, 1992.
23. Paine, L.J. et al., The identification of a lysine residue reactive to pyridoxal-5-phosphate in the glycerol dehydrogenase from the thermophile *Bacillus stearothermophilus, Biochem. Biophys. Acta* 1202, 235, 1993.
24. Basu, S., Basu, A., and Modak, M.J., Pyridoxal 5'-phosphate mediated inactivation of *Escherichia coli* DNA polymerase I: identification of lysine-635 as an essential residue for the processive mode of DNA synthesis, *Biochemistry* 27, 6710, 1988.
25. Hale, J.A., and Maloney, P.C., Pyridoxal-5-phosphate inhibition of substrate selectivity mutants of UhpT, the sugar 6-phosphate carrier of *Escherichia coli, J. Bacteriol.* 184, 3756, 2002.
26. Celestina, F., and Suryananarayana, T., Biochemical characterization and helix stabilizing properties of HSNP-C' from the thermoacidophilic archeon *Sulfolobus acidocaldarius, Biochem. Biophys. Res. Commn.* 267, 614, 2000.
27. Talarico, T.L., Guise, K.J., and Stacey, C.J., Chemical characterization of pyridoxylated hemoglobin polyoxyethylene conjugate, *Biochim. Biophys. Acta* 1476, 53, 2000.
28. Lapko, A.G., and Ruckpaul, K., Discrimination between conformational states of mitochondrial cytochrome P-450scc by selective modification with pyridoxal-5-phosphate, *Biochemistry (Moscow)* 63, 568, 1998.
29. Anderberg, S.J., Topological disposition of lysine 943 in native $Na^+/K(+)$-transporter ATPase, *Biochemistry* 34, 9508, 1995.
30. Brown, E.M. et al., Accessibility and mobility of lysine residues in β-lactoglobulin, *Biochemistry* 27, 5601, 1988.
31. Fujita, Y., and Noda, Y., Effect of alkylation with different sized substituents on thermal stability of lysozyme, *Int. J. Pept. Protein Res.* 40, 103, 1992.
32. Rypniewski, W.R., Holden, H.M., and Rayment, I., Structural consequences of reductive methylation of lysine residues in hen egg white lysozyme: an X-ray analysis at 1.8 Å resolution, *Biochemistry* 32, 9851, 1993.
33. Kim, T.K., and Burgess, D.J., Pharmacokinetic characterization of ^{14}C-vascular endothelial growth factor controlled release microspheres using a rat model. *J. Pharm. Pharmacol.* 54, 897, 2002.
34. Kinesetter, O. et al., Mono-*N*-terminal poly(ethylene glycol)-protein conjugates. *Adv. Drug Deliv. Rev.* 54, 4770485, 2002.
35. Yammani, R.R., Seetharam, S., and Seetharam, B., Identification and characterization of two distinct ligand binding regions of cubilin, *J. Biol. Chem.* 276, 44777, 2001.

(continued)

TABLE 2.1 (continued)
Some Reagents Used to Obtain Site-Specific Modification of Lysyl
Residues in Proteins

36. Zhang, M., Thulin, E., and Vogel, H.J., Reductive methylation and pK_a determination of the lysine side chains in calbindin D9k. *J. Protein Chem.* 13, 527, 1994.
37. Means, G.E., and Feeney, F.E., Reductive alkylation of proteins, *Anal. Biochem.* 224, 1, 1995.
38. Rayment, I., Reductive alkylation of lysine residues to alter crystallization properties of proteins, *Methods Enzymol.* 276, 171, 1997.
39. Kurinov, I.V. et al., X-ray crystallographic analysis of pokeweed antiviral protein-II after reductive methylation of lysine residues, *Biochem. Biophys. Res. Commn.* 275, 549, 2000.
40. Iwata, K., Eyles, S.J., and Lee, J.P., Exposing asymmetry between monomers in Alzheimer's amyloid fibrils via reductive alkylation of lysine residues, *J. Am. Chem. Soc.* 123, 6728, 2001.
41. Coffee, C.J. et al., Identification of the sites of modification of bovine liver glutamate dehydrogenase reacted with trinitrobenzenesulfonate, *Biochemistry* 10, 3516, 1971.
42. Means, G.E., Congdon, W.I., and Bender, M.L., Reactions of 2,4,6-trinitrobenzenesulfonate ion with amines and hydroxide ion, *Biochemistry* 11, 3564, 1972.
43. Kurono, Y. et al., Kinetic study on rapid reaction of trinitrobenzenesulfonate with human serum albumin, *J. Pharm. Sci.* 70, 1297, 1981.
44. Cahalan, M.D., and Pappone, P.A., Chemical modification of potassium channel gating in frog myelinated nerve by trinitrobenzene sulfonic acid. *J. Physiol.* 342, 119, 1983.
45. Whitson, P.A., Burgumn, A.A., and Matthews, K.S., Trinitrobenzenesulfonate modification of the lysine residues in lactose repressor protein, *Biochemistry* 23, 6046, 1984.
46. Sams, C.F., and Matthews, K.S., Diethyl pyrocarbonate reaction with the lactose repressor protein affects both inducer and DNA binding, *Biochemistry* 27, 2277, 1988.
47. Haniu, M. et al., Structure-function relationship of NAD(P)H:quinone reductase: characterization of NH_2-terminal blocking group and essential tyrosine and lysine residues, *Biochemistry* 27, 6877, 1988.
48. Xia, C. et al., Chemical modification of GSH transferase P1-1 confirms the presence of Arg-13, Lys-44 and one carboxylate group in the GSH-binding domain of the active site, *Biochem. J.* 293, 357, 1993.
49. Cayot, P., and Tainturier, G., The quantitation of protein amino groups by the trinitrobenzenesulfonic acid method: a reexamination, *Anal. Biochem.* 249, 184, 1997.
50. Cayot, P., Roullier, L., and Tainturier, G., Electrochemical modifications of proteins. 1. Glycitolation, *J. Agric. Food Chem.* 47, 1915, 1999.

the mixing/hybridization of the modified and native enzyme, it was determined that succinylation of bacterial luciferase resulted in the inactivation of the β-subunit without markedly affecting the function of the α-subunit. Shetty and Rao[20] studied the reaction of succinic anhydride with arachin. In this study, reaction of the protein was performed in 0.1 M sodium phosphate, pH 7.8, with the pH maintained over the course of the reaction by adding 2.0 M NaOH. The extent of modification was determined by reaction of the unmodified primary amino groups on the protein with trinitrobenzenesulfonic acid (see later). With a 200:1 molar excess of succinic anhydride, 82% of the available amino groups were succinylated, with concomitant dissociation of the subunits of this protein. The reaction of chymotrypsinogen with succinic anhydride has been studied.[21] In these experiments, the reaction was performed under ambient conditions in 0.05 M sodium phosphate, pH 7.5. During the course of the reaction, the pH was maintained at 7.5 by the addition of 1.0 M NaOH.

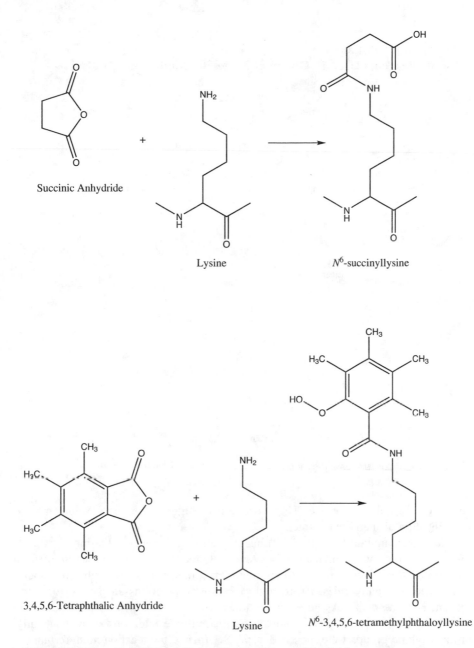

FIGURE 2.1 Reaction of some alkyl acid anhydrides with lysine in proteins.

Chymotrypsinogen (1 g) was dissolved in the sodium phosphate buffer, and 50 mg of succinic anhydride was added over a 30-min period. Under these conditions, 8 of the 14 lysine residues were modified. A related reaction involves the trimesylation of amino groups in proteins.[22] This reaction involves modification of the protein with di(trimethysilyethyl)trimesic acid. Removal of the blocking groups results in

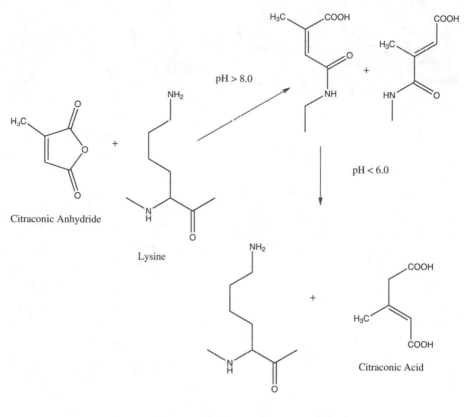

FIGURE 2.2 Reversible modification of lysine with citraconic anhydride.

an extremely polar derivative. The procedure is suggested to have value in the solubilization of membrane proteins.

Citraconic anhydride (Figure 2.2) has proven useful, because the modification of lysine residues with this reagent is a reversible reaction. Reaction conditions for the modification of lysine residues in proteins are similar to those described for other carboxylic acid anhydrides. Atassi and Habeeb[23] have discussed the use of this reagent in some detail. As an example, the reaction of egg-white lysozyme with citraconic anhydride has been studied.[24] With multiple additions of reagent, all primary amino groups were modified at pH 8.2 (pH of the reaction mixture maintained with a pH-stat). The product of the reaction was heterogeneous as judged by polyacrylamide gel electrophoresis. All citraconyl groups could be removed by treatment with 1.0 M hydroxylamine, pH 10.0. This treatment also resulted in an electrophoretically homogeneous species. Complete removal of the citraconyl groups could also be achieved by incubation at pH 4.2 for 3 h at 40°C.

Reaction with citraconic anhydride has been used to dissociate nucleoprotein complexes.[25] Modification of the lysine residues with citraconic anhydride (pH 8.0

to 9.0 maintained with pH-stat) resulted in a marked change in the charge relationship between the α-amino groups of lysine and the phosphate backbone of the nucleic acid, allowing subsequent separation of protein from nucleic acid. The citraconyl groups were subsequently removed from this protein by incubation at pH 3.0 to 4.0 at 30°C for 3 h.

Mahley and co-workers[26,27] prepared the acetoacetyl derivatives of lipoproteins by reaction with diketene in 0.3 borate, pH 8.5. The modification of tyrosyl and seryl residues also can occur under these conditions, but the O-acetoacetyl groups can be removed by dialysis against a mild alkaline buffer such as bicarbonate. The modification at lysyl residues can be reversed by 0.5 M hydroxylamine, pH 7.0, at 37°C. A 0.06 M solution of diketene was prepared by taking 50 μl diketene into 10 ml of 0.1 M sodium borate, pH 8.5. The extent of modification was determined by subsequent titration with fluorodinitrobenzene. The effect of the modification of lysine residues on the *in vivo* clearance of lipoproteins in rats has been investigated.[27]

Urabe and co-workers[28] prepared various mixed carboxylic acid anhydrides of tetradecanoic acid and oxa derivatives that varied in their hydrophobicity. This represented an attempt to change the surface properties of the enzyme molecule, in this case, thermolysin. The carboxylic acid anhydrides were formed *in situ* from the corresponding acid and ethylchloroformate in dioxane with triethylamine. The modification reaction was performed in 0.013 M barbital, 0.013 M $CaCl_2$, pH 8.5, containing 39% (v/v) dioxane and was terminated with neutral hydroxylamine, which also served to remove O-acyl derivatives. The extent of reaction was determined by titration with trinitrobenzenesulfonic acid. Derivatives obtained with tetradecanoic acid and 4-oxatetradecanoic acid were insoluble. Derivatives obtained with 4,7,10-trioxatertradecanoic acid and 4,7,10,13-tetraoxotetradecanoic acid both had approximately seven amino groups modified per mole of enzyme, showed little if any loss in either proteinase or esterase activity, and possessed enhanced thermal stability. Pool and Thompson[15,29] selectively attached long-chain (C_{12}–C_{16}) fatty acids to the amino-terminal amino acid of chemically modified (guanidinated) bovine pancreatic trypsin inhibitor. Fatty acid anhydrides were used to selectively acylate the α-amino group [1.5 μMol guanidinated bovine pancreatic trypsin inhibitor was dissolved in 1.0 ml 100 mM phosphate, pH 7.8, and 100 μMol acid anhydride in 1.0 ml tetrahydrofuran added with stirring at 20 to 50°C (temperature depended on the physical nature of the anhydride-solid or liquid) for 1 h. The modified protein was precipitated by the addition of tetrahydrofuran and purified by gel filtration followed by reverse-phase HPLC (C_3). Fatty acid chlorides can also be used for the acylation reaction.[30]

Howlett and Wardrop[31] were able to dissociate the components of human erythrocyte membrane by the use of 3,4,5,6-tetrahydrophthalic anhydride (Figure 2.1). The reaction was performed in 0.02 M Tricine, pH 8.5. The 3,4,5,6-tetrahydrophthalic anhydride was introduced into the reaction mixture as a dioxane solution (a maximum of 0.10 ml/5 ml reaction mixture). The pH was maintained at 8.0 to 9.0 with 1.0 M NaOH. The reaction was considered complete when no further change in pH was observed. The extent of modification was determined by titration with trinitrobenzenesulfonic acid. The reaction could be reversed by incubation for 24 to 48 h at ambient temperature following the addition of an equal volume of 0.1 M potassium phosphate, pH 5.4. (Final pH of the reaction mixture was 6.0.)

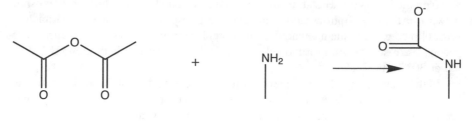

FIGURE 2.3 Reaction of acetic anhydride with protein amino groups.

Modification of amino groups with acetic anhydride (Figure 2.3) is one of the oldest[32] and still most extensively used approaches for the chemical modification of proteins. As originally described in glacial acetic acid or later at alkaline pH in either a pH-stat or in saturated sodium acetate,[33] extensive modification of amino groups (and other nucleophilic groups) occurs. Performing the modification reaction in conditions of saturated sodium acetate results in increased specificity, because O-acetyl tyrosine is unstable in sodium acetate. It was suggested that reaction occurred with groups on the surface of the protein, because lysine is a charged amino acid residue.[33] As with other reagents, the issue of accessibility is more a question of microenvironment and reaction rate rather than an all-or-none phenomenon.

Acetylation has been used to study calcitonin[34] and a bacterial cytochrome.[35] Acetic and maleic anhydrides have been used to study porcine pancreatic elastase.[36] In these studies, the reaction was carried out in a pH-stat or in various buffers. Inactivation of elastase at neutral pH (7.6, pH stat) with acetic anhydride was reversible and activity recovered when acetic anhydride was no longer added to the reaction mixture. At pH 9.5, reactivation was complete within 1 min. It was established that reaction occurred at both lysyl and tyrosyl residues. It is relatively easy to differentiate between the two sites of modification, because O-acyl tyrosyl residues are unstable at pH 9.0. The inactivation by acetic anhydride was irreversible at pH 10.5 or at 9.0 to 9.5 in 2 and 4 M urea. The amino-terminal valine was not available for modification at pH 7.4, but could be modified at pH 11.0. Modification of this residue could be achieved in the presence of urea (4.0 M) at a lower pH (9.0). It was suggested that the reversible inactivation was likely due to the transient acetylation of a histidine residue (see Chapter 3) and that the irreversible inactivation was a reflection of the modification of the N-terminal valine residue.

In a study that surprisingly has been largely ignored, Smit[37] used native electrophoresis to study the modification of human interleukin-3 with acetic anhydride. Changes in electrophoretic mobility correlated well with the amino group modification as measured by reaction with trinitrobenzenesulfonic acid.[38] This study is also quite useful in that it provides a study of the effect of pH on the modification of amino groups with acetic anhydride. Human interleukin-3 has eight amino groups for potential modification. Approximately two groups were available for modification at pH 5.0 (acetate), three at pH 6.0 (MES), and approximately eight at pH 9.0 (borate).

The use of acetic anhydride to modify amino groups in proteins has evolved into a much more sophisticated approach over the past decade. This is strictly a reflection of advances in analytical technology. In particular, mass spectrometry has become extremely useful for the study of chemical modification in proteins.[39–42] A number of studies on the use of acetic anhydride have appeared in the past 10 years and are listed in Table 2.1. The following sections describe certain of these studies in some detail.

Przybylski and colleagues[43] used modification with acetic anhydride to study the surface topology of hen egg-white lysozyme. Modification was performed in 0.5 M $(NH_4)_2CO_3$, pH 7.0 (maintained by the addition of NH_4OH [30% NH_c] with a 10 to 10,000 molar excess [to amino groups]) or acetic anhydride for 30 min at 20°C. Analysis by mass spectrometry before and after tryptic hydrolysis permitted the identification of modified residues and the assignment of relative reactivity of the individual amino groups. This is an excellent paper that has been cited extensively since its publication in 1992 and continues to be of interest to investigators.[44]

Palczewski and colleagues[45] used the reaction with acetic anhydride to study conformational changes in arrestin as well as the association of arrestin with P-Rho. The effect of light/dark cycles was evaluated on reactivity with acetic anhydride as well as on the interaction of arrestin with P-Rho. The initial modification was performed with low levels (1 to 10 mM) of deuterated acetic anhydride at 0°C in 100 mM sodium borate, pH 8.5. This was followed by modification with higher concentrations of acetic anhydride (20 mM) in 100 mM sodium borate, pH 9.0, containing 6.0 M guanidine hydrochloride. The ratio of deuterated to protiated modification permitted the identification of residues protected from modification by interaction with P-Rho as well three lysine residues that were more reactive as a result of the interaction. Pucci and co-workers[46] used a similar approach to study the interaction of thyroid transcription factor 1 homeodomain with DNA. A 1- to 10-fold molar excess of acetic anhydride was added to free or DNA-complexed thyroid transcription factor 1 homeodomain in 20 mM Tris-HCl, 75 mM KCl, pH 7.5, at 25°C for 10 min. The acetylated samples were subjected to cyanogen bromide cleavage in 70% formic acid at room temperature or 18 h in the dark. The fractions were separated by HPLC and analyzed by mass spectrometry. Scaloni and co-workers[47] have also used this differential modification approach to study the dimeric structure of porcine aminoacylase 1. Reaction with acetic anhydride was performed in 10 mM NH_4CO_3, 1 mM DTT, 1 mM $ZnCl_2$, pH 7.5, at 25°C for 10 min using a 100- to 5000-fold molar excess of reagent. Modification occurred readily at the amino terminus and at 8 of the 17 lysine residues.

Calvete and co-workers[48] used a clever approach to identify the heparin-binding domain of bovine seminal plasma protein PDC-109. The PDC-109 protein was bound to heparin-agarose in 16.6 mM Tris, 50 mM NaCl, 1.6 mM EDTA, 0.025% NaN_3, pH 7.4. After washing the column to remove protein not bound to the matrix, the column was recycled at room temperature with the application buffer containing acetic anhydride (25- to 1600-fold molar excess over protein lysine). A similar experiment was performed with 1,2-cyclohexanedione to study arginine modification. Six basic residues were protected from modification by binding to the heparin matrix.

Pucci and co-workers[49] measured amino reactivity on Minobody, a small *de novo*-designed β-protein, by reaction with a low concentration of acetic anhydride. The modification reaction was performed in 50 mM NH$_4$CO$_3$, pH 7.5, at 25°C for 10 min with a twofold molar excess (to amino groups) of acetic anhydride. Two lysine residues were highly reactive, one lysine and the α-amino group were less reactive, and one lysine residue was not modified. Note that the results are qualitative; however, it is still a clever, well-performed study that provides considerable information.

Taralp and Kaplan[50] examined the reaction of acetic anhydride with lyophilized α-chymotrypsin *en vacuo*. α-Chymotrypsin was lyophilized from an unbuffered solution at pH 9.0 in one chamber in a reaction vessel. ³H-acetic anhydride was added to another compartment in the reaction vessel. The reaction vessel was evacuated and placed in an oven at 75°C. Several reaction vessels were used and removed at various time intervals for analysis. The proteins were then modified with ¹⁴C-acetic anhydride, and the ratio of ³H to ¹⁴C was used to determine the extent of modification. Complete modification of amino groups is achieved at pH 9.0 in aqueous solution, but in the nonaqueous system only 25% of the ε-amino groups and 90% of the α-groups were modified. It also appeared that mixed anhydrides formed with carboxyl groups on the protein surface.

Smith and co-workers[51] used the reaction with acetic anhydride to determine the extent of posttranslational acetylation in histone H4. Histone H4 was modified with deuterated acetic anhydride. (Dried histone samples were suspended in deuterated glacial acid acetic and deuterated acetic anhydride was added; the reaction mixture was allowed to stand for 6 h at ambient temperature.) The modified samples were subjected to mass spectrometric analysis. The extent of endogenous acetylation was determined from the ratio of protiated to deuterated fragmentation ions.

Tomer and co-workers[52] used the reaction with acetic anhydride to define a discontinuous epitope in human immunodeficiency virus core protein p24. The protein was bound to the immobilized antibody and digested with endoprotease lys-C. Modification was then accomplished with acetic anhydride (10,000-fold molar excess of reagent, pH 7.8, 50 mM NH$_4$CO$_3$, 20 min), followed by a 100,000-fold excess of the hexadeuterated reagents under the same conditions; pH was maintained by the addition of 10% NH$_4$OH.

Competitive labeling (trace labeling) is a technique for determining the ionization state or constant and intrinsic reactivity of individual amino groups in a protein.[53] The method is based on the hypothesis that individual amino groups will compete for a trace amount of radiolabeled reagent. (The reagent is selected on the basis of nonselective reactivity with amino groups; with most studies, acetic anhydride has been the reagent of choice.) The extent of radiolabel incorporation into the protein at a given site will then be a function of the pK_a, microenvironment, and inherent nucleophilicity of that particular amino group.[53] After the reaction with the radiolabeled reagent is complete, the protein is denatured and complete modification at each amino group is achieved by the addition of an excess of unlabeled reagent. A reproducible digestion method (i.e., tryptic or chymotryptic hydrolysis) is used to obtain peptides from the completely modified protein. The peptides are separated by a chromatographic technique, and the extent of radiolabel at each site is determined. The extent of radiolabel incorporation at a given site is a function of the reactivity of that individual amino group under the reaction conditions used at the radiolabel step. An alternative approach[54,55] involves a trace labeling step with tritiated acetic anhydride, followed by complete modification with unlabeled acetic anhydride under denaturing conditions. This modified protein is then mixed with a preparation of the same protein that has been uniformly labeled with the ^{14}C-labeled acetic anhydride. Digestion and separation of the peptide are performed by conventional techniques (see previously), and the extent of radiolabeling is determined. The ratio of ^{3}H to ^{14}C in peptides containing amino groups is an indication of functional group reactivity. This method is somewhat more sensitive than the original method. Reductive methylation has also been used.[56]

Although this is a laborious technique, the data obtained are excellent and provide considerable insight into the solution structure of proteins. This technique has been consistently used for the study of troponin-T,[57] troponin-C,[58] troponin-I,[59] calmodulin,[60-62] and tropomyosin.[63] In particular, studies[62,63] that have used this technique to assess conformational change in solution have been particularly rewarding.

Trifluoroacetylated derivatives have been of interest in the study of protein structure. In these studies, ethylthiotrifluoroacetate was used to modify cytochrome c in 0.14 M sodium phosphate, pH 8.0.[64,65] The pH was maintained at 8.0 by a pH-stat. Singly substituted derivatives of cytochrome c can be separated by chromatography on an anion-exchange resin (Bio Rex 70) and carboxymethylcellulose. It is critical to avoid lyophilization during the preparation of the various derivatives. These derivatives have been subjected to further investigation,[66,67] including the use of ^{19}F containing derivatives for nuclear magnetic resonance (NMR) probes.[68]

Lysine residues can be modified by reaction with α-ketoalkyl halides such as iodoacetic acid.[69] Acylation can occur at pH >7.0, but the rate of reaction is much slower than that with cysteinyl residues. Both mono- and disubstituted derivatives have been reported. The monosubstituted derivative migrates close to methionine on amino acid analysis, whereas the disubstituted derivative migrates near aspartic acid. The reaction with α-ketoalkyl halides is not considered particularly useful for the modification of primary amino groups. This reaction can be a possible side reaction occurring during the reduction and carboxymethylation of proteins. The reactivity of a given lysyl residue is affected by the nature of surrounding amino residues.

Both fluoronitrobenzene and fluorodinitrobenzene have been of considerable value in protein chemistry since Sanger and Tuppy's work on the structure of insulin.[70] Carty and Hirs[71] developed the use of 4-sulfonyl-2-nitrofluorobenzene to modify amino groups in pancreatic ribonuclease. This reagent also is more stable than, for example, fluorodinitrobenzene under alkaline conditions, permitting more accurate measurement at pH > 9.0. The lysine residue at position 41 is the site of major substitution, which is a reflection of the lower pK_a for the ε-amino group of this residue. Use of this compound did not present the solubility and reactivity problems posed by fluoronitrobenzene compounds. It was possible to qualitatively determine the classes of amino groups in ribonuclease; these were the α-amino group, nine "normal" amino groups, and lysine 41. The reactivity of Lys-41 was influenced by neighboring functional groups. This effect was lost at pH >11 or on thermal denaturation of the protein. The reaction of 1-dimethylaminonaphthalene-5-sulfonyl chloride (dansyl chloride) has been useful both in the structural analysis and amino group modification with proteins. In one study,[72] dansyl chloride (in acetone) was added to a solution of trypsin in 0.1 M phosphate, pH 8.0. The reaction was terminated after 24 h at 25°C by acidification to pH 3.0 with 1.0 M HCl. Insoluble material was removed by centrifugation and the supernatant fraction placed in dialysis. These investigators reported modification of the amino-terminal isoleucine and one lysine residue. The extent of modification was determined by absorbance at 336 nm ($\varepsilon_m = 3.4 \times 10^4$ M^{-1} cm^{-1}). The reaction of dansyl chloride with phosphoenolpyruvate carboxylase was used to introduce a fluorescent probe into this protein.[73] A somewhat specific modification of one of the eight lysine residues was achieved. The extent of modification was determined by spectral analysis at 355 nm, using an extinction coefficient of 3400 M^{-1} cm^{-1}.

The reaction of 2-carboxy-4,6-dinitrochlorobenzene with proteins has been explored.[74,75] This reagent reacts with amino, sulfhydryl, and amino groups. This reagent has recently been used to modify specific lysine residues in cytochrome c.[76,77] The modification reaction (ca. sixfold molar excess of reagent) was performed in 0.2 M sodium bicarbonate, pH 9.0, at ambient temperature for 24 h. The extent of modification was determined as described by Brautigan and co-workers.[6] The absorbance maximum of derivatives formed with various alkylamines was 436 nm, with an extinction coefficient of 6.9×10^3 M^{-1} cm^{-1}. Chromatographic fractionation of the modified protein (sulfoethyl-cellulose) yielded six fractions with major lysine group modification. Hiratsuka and Uchida[78] examined the reaction of N-methyl-2-anilino-6-naphthalenesulfonyl chloride with lysyl residues in cardiac myosin. There was a difference in the nature of the reaction in the presence and absence of a divalent cation. N-methyl-2-anilino-6-naphthalenesulfonyl chloride has been suggested for use as a fluorescent probe for hydrophobic regions of protein molecules.[79–81] The extent of incorporation of the N-methyl-2-anilino-6-naphthalenesulfonyl moiety into protein can be determined by spectral analysis at 327 nm ($\Delta\varepsilon = 2.0 \times 10^4$ M^{-1} cm^{-1}).[79,80] Modification of protein amino groups with isothiocyanate derivatives of various dyes has proved to be an effective means of introducing structural probes into proteins at specific sites.[81] Eosin isothiocyanate has been used to modify the lysyl residues in phosphoenolpyruvate carboxylase.[73] The reagent was dissolved in dimethylsulfoxide/50 mM HEPES, pH 8.0 (50:50), immediately before use and added to the protein (in 50 mM HEPES, pH 8.0). The modified derivatives were used to determine the spatial proximity of the modified lysine residues by using resonance energy transfer. Fluorescein isothiocyanate has been used to modify cytochrome P-450 (reaction performed in 30 mM Tris, pH 8.0, containing 0.1% Tween 80; 2 h at 0°C in the dark),[82] actin (2 mM borate, pH 8.5; 3 h at ambient temperature and then at 4°C for 16 h),[83] and ricin (pH 8.1; 6°C for 4 h).[84] The extent of modification with fluorescein isothiocyanate can be determined by spectroscopy, using an extinction coefficient of 80,000 M^{-1} cm^{-1} at 495 nm (1% SDS with 0.1 M NaOH)[82] or 74,500 M^{-1} cm^{-1} (0.1 M Tris, pH 8.0).[83] Antibodies labeled with fluorescein have been used as targeted phototoxic agents.[85] In this approach, the fluorescein moiety is iodinated, resulting in a photodynamic sensitizer. The reader is referred to an elegant study on the effect of microenvironment on the fluorescence of arylaminophthalenesulfonates.[86]

Welches and Baldwin[87] examined the reaction of bacterial luciferase with 2,4-dinitrofluorobenzene. Modification was associated with inactivation at the rate of

157 M^{-1} cm^{-1}, pH 7.0 (0.05 M phosphate). Both lysyl and cysteinyl residues can be modified under the experimental conditions (0.05 M phosphate, pH 7.0 at 25°C) used in these studies. To assess the significance of reaction at primary amino groups, the cysteinyl residues were blocked with methyl methanethiosulfonate. Reaction of luciferase with methyl methanethiosulfonate resulted in more than 95% loss of catalytic activity (twofold molar excess of methyl methanethiosulfonate in 0.02 M phosphate, pH 7.0 at 25°C). The loss of activity was completely reversed with 2-mercaptoethanol (97 mM). The small amount of residual activity present after treatment with methyl methanethiosulfonate was further reduced on treatment with 2,4-dinitrofluorobenzene, and the recovery of activity subsequent to 2-mercapto-ethanol was greatly reduced. Quantitative analysis was not performed, but qualitative analysis suggested that the modification occurred at the α-amino group of methionine on the α- or -subunits, or both. A combination of site-specific mutagenesis and site-specific chemical modification with 2,4-dinitroflurobenzene was used to study lysine residues in angiogenin.[88]

The reaction of primary amino groups in proteins with cyanate (Figure 2.4) has been a useful procedure for several decades[89] and a technical challenge in more recent years.[90–92] Stark and co-workers[89] pursued the observation that ribonuclease was inactivated by urea in a time-dependent reaction. It was established that this inactivation was a reflection of the content of cyanate in the urea preparation. This observation was subsequently developed into a method for the quantitative determination of amino-terminal residues in peptides and proteins.[93] Stark[94] reported the reaction of cyanate with amino acid residues. The ε-amino group of lysine is less reactive ($k = 2.0 \times 10^3$ M^{-1} min^{-1}) than the α-amino group of glycylglycine ($k = 1.4 \times 10^1$ M^{-1} min^{-1}). The carbamyl derivative of histidine is quite unstable as is the corresponding derivative of cysteine. Importance should be given to reaction at residues other than amines. For example, the reaction of chymotrypsin with cyanate results in loss of catalytic activity associated with the carbamylation of the active site serine residue.[95]

Manning and co-workers[96–99] established that the modification of sickle cell hemoglobin with cyanate increased the oxygen affinity of this protein. It has been established that the amino-terminal value of hemoglobin is more reactive to cyanate in deoxygenated blood than in partially deoxygenated blood. At pH 7.4, the amino-terminal valyl residues of oxyhemoglobin S are carbamylated 50 to 100 times faster than lysyl residues.[99] The same laboratory has examined the carbamylation of α- and β-chains in some detail.[98] With the deoxy protein, the ratio of radiolabel from ^{14}C-cyanate on the α-chain to that on the β-chain is 1.7:1.0, whereas it is 1:1 with the oxy protein. Carbamylation of the amino-terminal valine residues of hemoglobin is ca. 2.5-fold higher in partially deoxygenated media as compared to fully oxygenated media. Thus, it appears that the reactivity of the amino-terminal valine is a sensitive index of conformational change.[99] It is also of interest that removal of Arg-141 (α) with carboxypeptidase B abolishes the enhancement of carbamylation observed with the removal of oxygen from hemoglobin.

Weisgraber and co-workers[26] used carbamylation to explore the role of lysyl residues in the binding of plasma lipoprotein to fibroblasts. The reaction was performed in 0.3 M sodium borate, pH 8.0. The extent of modification was determined

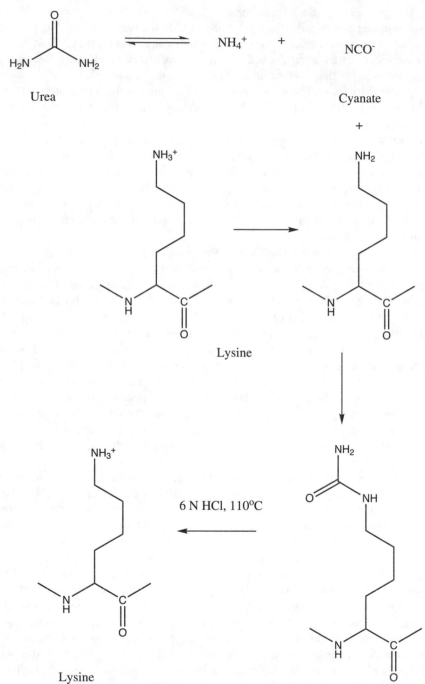

FIGURE 2.4 Scheme for the formation of cyanate from urea and subsequent carbamylation of lysine.

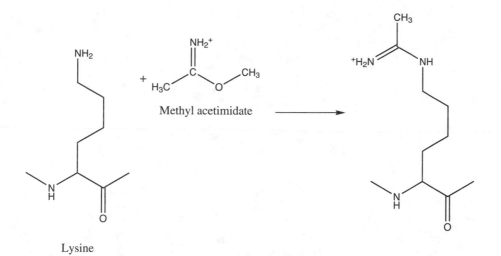

FIGURE 2.5 Reaction of methyl acetimidate with lysine.

in two ways. In the first, the modified protein was subjected to acid hydrolysis. The amount of homocitrulline, the product of the reaction of the ε-amino group of lysine with cyanate, was considered equivalent to the number of lysine residues modified. However, homocitrulline is partially degraded on acid hydrolysis to produce lysine (17 to 30%). To obviate this difficulty, these investigators removed a portion of the modified protein and reacted in under denaturing conditions with 2,4-dinitrifluo-robenzene, yielding an acid-stable derivative. The number of lysine residues modified was therefore the sum of free lysine and homocitrulline obtained on amino acid analysis following acid hydrolysis. In an elegant study by Plapp and co-workers,[100] the modification of lysyl residues in bovine pancreatic deoxyribonuclease A by several different reagents, including cyanate, was studied in great detail. The modification with cyanate was performed at 37°C in 1.0 M triethanolamine hydrochloride, pH 8.0. The extent of modification was determined by analysis for homocitrulline following acid hydrolysis. A time course of hydrolysis was used to provide for the accurate determination of homocitrulline, because this amino acid slowly decomposes to form lysine during acid hydrolysis (see previously).

The reaction of imidoesters with the primary amino groups of proteins (Figure 2.5) has been the subject of considerable investigation. The most extensive use of this class of reagents has been the covalent cross-linking of proteins. These reagents have the particular advantage that the charge of the lysine residue is maintained during the modification. The reaction was performed with a 1000-fold molar excess of reagent in 0.02 M sodium borate, 0.15 M NaCl, pH 8.5. Amino acid analysis indicated that ca. 80% of the lysyl residues were modified under these conditions. Sekiguchi and co-workers[101] studied the modification of a glutamine synthetase from *Bacillus stearothermophilus* with ethyl acetimidate. The modification was performed at pH 9.5 with 0.2 M phosphate for 1 h at 35°C and terminated by dialysis at pH 7.2. The extent of modification was determined by titration of the modified protein

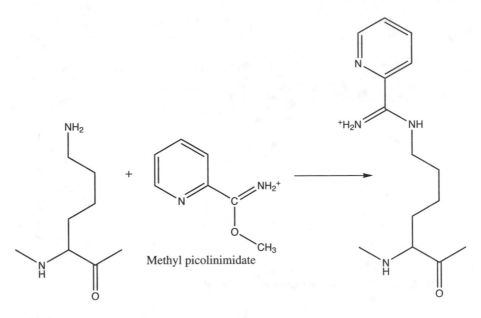

Methyl picolinimidate

Lysine

FIGURE 2.6 Reaction of methyl picolinimidate with lysine.

with trinitrobenzenesulfonic acid (see later). As these investigators suggest, consideration must be given to the possibility of cross-linking occurring with this reagent under the conditions used.[102] Monneron and d'Alayer[103] examined the reaction of either methyl acetimidate or dimethyl suberimidate with particulate adenylate cyclase. The reaction was performed in 0.05 M triethanolamine, 10% (w/v) sucrose, 0.005 M MgCl$_2$, pH 8.1. Plapp and co-workers[100] examined the reaction of methyl picolinimidate with pancreatic deoxyribonuclease. Methyl picolinimidate (Figure 2.6) is an imidoester that reacts with the primary amino groups in proteins. The reaction was performed in 0.5 M triethanolamine hydrochloride, pH 8.0, containing 1 mM CaCl$_2$ with 0.1 M methyl picolinimidate for 22 h at 25°C, then with 0.2 M methyl picolinimidate for an additional 8 h. The extent of modification of a protein by methyl picolinimidate can be determined by spectral analysis. Under these conditions, essentially all the primary amino groups in deoxyribonuclease (nine lysine and one amino-terminal amino group) were modified, but there was no change in biological activity. Plapp also studied the reaction of methyl picolinimidate with horse liver alcohol dehydrogenase.[104] This study was somewhat unique in that modification of the enzyme resulted in enhanced catalytic activity reflecting more rapid dissociation of the enzyme–coenzyme complex. Note that the derivatized lysine reverts to lysine (60% yield) under the normal conditions of acid hydrolysis.

A number of investigators have used pyridoxal-5′-phosphate (Figure 2.7) to modify lysyl residues in proteins. Pyridoxal-5′-phosphate is the cofactor form of vitamin B$_6$ and plays an important role in biological catalysis.[105] Pyridoxal phosphate is useful for the modification of lysine because of selectivity of reaction, spectral

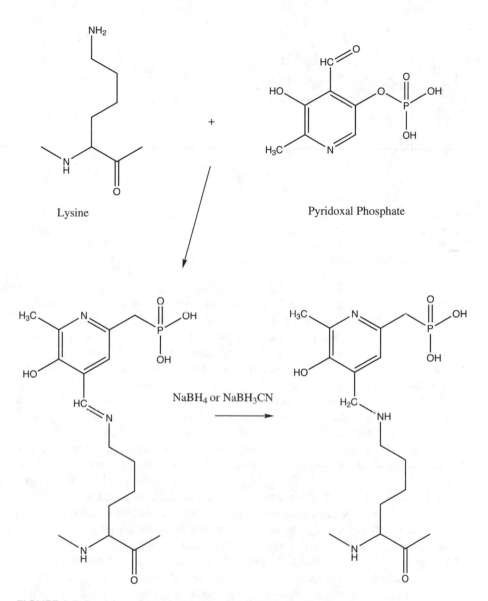

FIGURE 2.7 Reaction of pyridoxal phosphate with lysine and subsequent reduction of the Schiff base with borohydride.

properties of the modified residue, reversibility of reaction, and establishment of stereochemistry by use of radiolabeled sodium borohydride (sodium borotritiide) to reduce the Schiff base initially formed on the reaction of pyridoxal phosphate with a primary amine. Pyridoxal phosphate reacts with all primary amines (both ε-amino groups of lysine and the amino-terminal α-amino function) in a protein. In general, pyridoxal-5′-phosphate is far more reactive than pyridoxal because of intramolecular

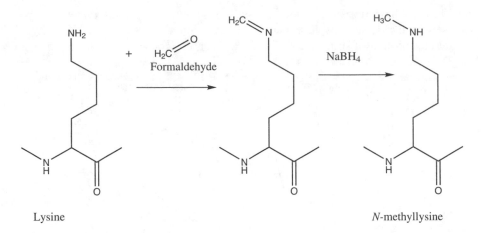

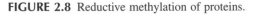

Lysine N-methyllysine

FIGURE 2.8 Reductive methylation of proteins.

hemiacetal formation and the neighboring group effect of the phosphate moiety. Shapiro and co-workers investigated the reaction of pyridoxal phosphate with rabbit muscle aldolase.[106] The initial reaction produced a species with an absorbance maximum at 430 to 435 nm, reflecting the protonated Schiff base form of the pyridoxal phosphate–protein complex. After reduction with sodium borohydride, the absorbance maximum was at 325 nm, which is characteristic of the reduced Schiff base. This is quite a useful study in that the difference in reactivity between pyridoxal and pyridoxal-5′-phosphate is demonstrated as is the reversible nature of the initial complex. Schnackerz and Noltmann[107] compared the reaction of pyridoxal-5′-phosphate and other aldehydes in reaction with rabbit muscle phosphoglucose at pH 8.0. Pyridoxal-5′-phosphate (0.19 mM) resulted in 82% inactivation, whereas the following results were obtained with other aldehydes: pyridoxal (8.4 mM), 16% inactivation; acetaldehyde (75 mM), 75% inactivation; and acetone (75 mM), 31% inactivation. This last reaction is of interest as many investigators are unaware that acetone can react with amino groups in proteins. The reaction of acetone with primary amino groups has been known for some time[108] and is discussed in further detail later under reductive alkylation. Paech and Tolbert[109] studied the reaction of ribulose 1,5-bis-phosphate carboxylase/oxygenase with pyridoxal-5′-phosphate. Pyridoxal-5′-phosphate inactivated the enzyme with or without reduction with NaBH$_4$. This reaction was performed in 0.1 M Bicine (N,N-(2-hydroxyethyl) glycine), 0.010 M MgCl$_2$, 0.2 mM EDTA, and 0.001 M dithiothreitol. The reaction demonstrated an optimum at pH 8.4. Spectral studies showed the formation of a species absorbing at 432 nm. As is characteristic for the Schiff base derivative, this peak disappears on reduction to yield a species with an optimum at 325 nm ($\varepsilon = 4800$ M^{-1} cm^{-1}). This supports the suggestion that the loss of activity observed on reaction with pyridoxal-5′-phosphate is because of the formation of a Schiff base, which can be reduced with NaBH$_4$ to form a stable derivative, as opposed to the formation of a 2-azolidine ring with a second nucleophile as has been observed by other investigators.[110–112] Jones and Priest[113] have investigated the modification of apo-serine hydroxymethyltrans-ferase with pyridoxal phosphate and the subsequent use of the enzyme-bound pyri-doxal phosphate as a structural probe. Cortijo and co-workers[114] suggested the use of the ratio of absorbance at 415 to that at 335 nm of enzyme-bound pyridoxal phosphate as an indication of the polarity of the medium. Cake and co-workers[115] demonstrated that modification of activated hepatic glucorticoid receptor with pyri-doxal-5′-phosphate obviated the binding of the receptor to DNA. Greatly reduced inhibition was seen with pyridoxamine-5′-phosphate, pyridoxamine, or pyridoxine. Inhibition could be reversed by gel filtration or treatment with dithiothreitol, whereas treatment with $NaBH_4$ resulted in irreversible inhibition of DNA binding. These investigators used 0.2 M borate, 0.25 M sucrose, and 0.003 M $MgCl_2$ (pH 8.0) as the solvent for reaction with pyridoxal-5′-phosphate. Slebe and Martinez-Carrion[116] introduced the use of phosphopyridoxal trifluoroethyl amine as a probe for pyridoxal phosphate binding sites in enzymes. Nishigori and Toft[117] explored the reaction of pyridoxal-5′-phosphate with the avian progesterone receptor. Reaction with pyri-doxal-5′-phosphate was performed in 0.02 M barbital, 10% (v/v) glycerol, 0.005 mM dithiothreitol, and 0.010 M KCl, pH 8.0. The modification was stabilized by $NaBH_4$. These investigators noted that the modification was readily reversed in Tris buffer unless stabilized by $NaBH_4$. Sugiyama and Mukohata[118] observed that modification with pyridoxal-5′-phosphate of the lysine residue in chloroplast coupling factor using 0.020 M Tricine, 0.001 M EDTA, and 0.010 M $MgCl_2$, pH 8.0, resulted in complete inactivation of the ATPase activity. Peters and co-workers[119] reported the inactivation of ATPase activity in a bacterial coupling factor by reaction with pyridoxal-5′-phosphate. The modification was performed in 0.050 M morpholinosulfonic acid, pH 7.5. The inhibition was readily reversed by dilution or by 0.01 M lysine and was, as expected, stabilized by $NaBH_4$. Gould and Engel[120] reported the reaction of mouse testicular lactate dehydrogenase with pyridoxal-5′-phosphate in 0.050 M sodium pyrophosphate, pH 8.7, at 25°C. This reaction resulted in inactivation of the dehydrogenase activity. The inactivation was reversed by cysteine and stabilized by $NaBH_4$. These investigators reported that the observed absorption coefficient at 325 nm might be decreased by as much as 50% with protein-bound pyridoxal phosphate. Thus, estimation of the number of lysine residues modified using the absorption coefficient obtained with model compounds might provide a minimum value only. Ogawa and Fujioka[121] studied the reaction of pyridoxal-5′-phosphate with saccharopine dehydrogenase in 0.1 M potassium phosphate, pH 6.8, at ambient temperature in the dark. Both spectral analysis and tritium incorporation from sodium borohydride reduction were consistent with the modification of one lysine residue per mole of enzyme being responsible for the loss of enzyme activity. A value of 1×10^4 M^{-1} cm^{-1} for the extinction coefficient at 325 nm[122] was used in this study. The concentrations of pyridoxal and pyridoxal-5′-phosphate were determined spectro-photometrically in 0.1 M NaOH, using an extinction coefficient of 5.8×10^3 M^{-1} cm^{-1} at 300 nm and 6.6×10^3 M^{-1} cm^{-1} at 388 nm, respectively.[123] Amine compounds have the potential to interfere in the reaction of pyridoxal-5′-phosphate with proteins. Moldoon and Cidlowski[124] demonstrate that 0.1 M Tris, pH 7.4, markedly interfered with the modification of rat uterine estrogen receptor with pyridoxal-5′-phosphate.

These investigators also noted that, as in the other studies, 0.05 M lysine blocked the modification reaction and could also reverse the modification if the Schiff base had not been reduced. Stock solutions of pyridoxal phosphate were prepared in 0.01 M NaOH to avoid acid decomposition. Ohsawa and Gualerzi[125] have emphasized the importance of local environmental factors in the specificity of modification by pyridoxal phosphate. These investigators examined the modification of *Escherichia coli* initiation factor by pyridoxal phosphate in 0.020 M triethanolamine and 0.03 M KCl, pH 7.8. In the course of the studies, it was observed that pyridoxal phosphate did not react with poly(AUG). These investigators also reported the preparation of N^6-pyridoxal lysine by reaction of pyridoxal phosphate with polylysine in 0.01 M sodium phosphate, pH 7.2, at 37°C, followed by reduction with NaBH$_4$. The reduction was terminated by the addition of acetic acid. Acid hydrolysis (6 N HCl, 110°C, 22 h) yielded N^6-pyridoxal-L-lysine. Bürger and Görisch[126] reported the inactivation of histidinol dehydrogenase on reaction with pyridoxal phosphate in 0.02 M Tris, pH 7.6. This modification could be reversed by dialysis unless the putative Schiff base was stabilized by reduction with NaH$_4$ (n-octyl alcohol added to prevent foaming). These investigators used a ε for ε-amino pyridoxal lysine of 1×10^4 M^{-1} cm^{-1} at 325 nm. Other applications of pyridoxal-5'-phosphate modification have been used to study hydroxymethylbilane synthetase,[127] DNA polymerase I,[128] and rabbit glycogen synthase isozymes.[129] A novel affinity label (pyridoxal-5'-diphospho-5'-adenosine) that uses pyridoxal-5'-phosphate chemistry has been used to study the adenine nucleotide binding sites in yeast hexokinase.[130]

A substantial portion of the specificity of pyridoxal-5'-phosphate in protein modification arises from electrostatic interactions via the phosphate group with positively charged groups (i.e., arginine) on the protein surface. A conceptually related compound is methyl acetyl phosphate. The reagent was originally developed as an affinity label for D-3-hydroxybutyrate dehydrogenase.[131] Manning and co-workers examined the chemistry of the reaction of methyl acetyl phosphate with hemoglobin in some detail.[132,133] It appears to be the affinity label for the 2,3-diphosphoglycerate binding site.[132] More recent work suggests that this reagent might be a useful probe for other anion binding sites in proteins.[134]

The modification of primary amines in proteins by reductive alkylation (Figure 2.8) has proved to be a useful reaction. This reaction has the advantage of the basic charge properties of the modified residue being preserved. Means[134] reviewed the early work on this modification. Both monosubstituted and disubstituted derivatives can be prepared, depending on reaction conditions and nature of the carbonyl compound.

Rice and co-workers[135] reported the stabilization of trypsin by reductive methylation. This reaction used formaldehyde/sodium borohydride in 0.2 M sodium borate, pH 9.2, in the cold. Unsubstituted amino groups were present after the reaction, as demonstrated by titration with trinitrobenzenesulfonic acid. The amino-terminal isoleucine residue was not modified under these conditions. Morris and co-workers[136] investigated the reductive methylation of monellin. The modification was performed in 0.2 M sodium borate, pH 8.0, with 1 mM monellin (11 mM) with respect to primary amino groups in the cold. Sodium borohydride was added to give a final concentration of 0.5 mg ml^{-1}, and 1 to 5 ml of 6 to 8 M formaldehyde was added per milliliter of solution. Tritiated formaldehyde was used to establish the extent of modification. One of the problems with the use of formaldehyde in this reaction is the presence of paraformaldehyde. Chen and Benoiton[137] obviated this difficulty by the *in situ* generation of formaldehyde from methanol.

The introduction of sodium cyanoborohydride as a reducing agent for this reaction represented a real advance. Sodium cyanoborohydride is stable in aqueous solution at pH 7.0. Unlike sodium borohydride, which can reduce aldehydes and disulfide bonds, sodium cyanoborohydride only reduces the Schiff base formed in the initial process of reductive alkylation. Radiolabeling of proteins by using ^{14}C-formaldehyde and sodium cyanoborohydride has been reported.[138] The modification was performed in 0.04 M phosphate, pH 7.0, at 25°C. The modification can be performed equally well at 0°C, but, as expected, it takes a longer period of time; there is no effect on the extent of modification. In this regard, these authors estimated that the same extent of modification obtained in 1 h at 37°C could be achieved in 4 to 6 h at 25°C or 24 h at 0°C. Although the majority of experiments in this study were performed in phosphate buffer at pH 7.0, equivalent results can be obtained in Tris or HEPES buffer at pH 7.0. A greater extent of modification was observed with sodium cyanoborohydride at pH 7.0 than with sodium borohydride at pH 9.0. Jentoft and Dearborn[139] studied the use of sodium cyanoborohydride in some detail. In particular, the preparation of sodium cyanoborohydride is critical, and most, if not all, commercial preparations require recrystallization before use. This reflects the presence of impurities, which limit the extent of the reductive alkylation (see later). Recrystallization is accomplished by dissolving 11 g of sodium cyanoborohydride in 25 ml acetonitrile. Insoluble material is removed by centrifugation. Crystallization is accomplished by adding 150 ml methylene chloride and allowing it to stand overnight at 4°C. The recrystallized sodium cyanoborohydride is collected by filtration and stored in a vacuum desiccator. A fresh solution of reagent is prepared daily. Using [^{14}C]-formaldehyde and sodium cyanoborohydride, the major product is ε-methylated lysine, with minor incorporation of radiolabel into arginine and histidine. Optimal reductive methylation was obtained at pH values higher than 8.0 during a short-term (10 min) incubation. The effect of pH is much less pronounced at longer periods of incubation (1 to 2 h), with optimal reductive methylation occurring between pH 7.0 and 8.0. These investigators also noted that Tris, 2-mercaptoethanol, dithiothreitol, ammonium ions (as ammonium sulfate), and guanidine (5 M) inhibited the reductive alkylation of albumin by formaldehyde and sodium cyanoborohydride in 0.1 M HEPES, pH 7.5.

Reductive methylation with ^{3}C-enriched formaldehyde has been used to introduce an NMR probe to study protein conformation.[140] A similar approach has been developed by using deuterated acetone.[141]

Fretheim and co-workers[142] examined the effect of carbonyl compounds of different sizes on the extent of reductive alkylation. The extent of modification is more a reflection of the type of alkylating agent and reaction conditions than an intrinsic property of the protein under study. For example, nearly 100% disubstitution can be obtained with formaldehyde and ca. 35% disubstitution with *n*-butanol, whereas only monosubstitution can be obtained with acetone, cyclopentanone, cyclohexanone, and benzaldehyde. Whereas most of the products of reductive alkylation retained solubility, the reaction products obtained with cyclohexanone and benzaldehyde tended to precipitate. Examination of the reductive alkylation of ovomucoid, lysozyme, and ovotransferrin with different aldehydes suggests that such modification occurs without major conformational change, as judged by circular dichroism measurements.[143] The same study also examined the stability of the modified proteins by scanning differential calorimetry. The extensive modification of amino groups decreases thermal stability. The destabilization effect increases with increasing size (and hydrophobicity) of the modifying aldehyde.

In another study, the reversible reductive alkylation of proteins was examined.[144] Both glycolaldehyde and acetol react with the primary amino groups in proteins to yield derivatives that can be cleaved with periodate under mild basic condition to yield the free amine. Sodium cyanoborohydride is much more effective in the pH range of 6.0 to 8.0, whereas sodium borohydride is more effective under more alkaline conditions. Treatment of 30.0 mg lysozyme in 6.0 ml 0.2 M sodium borate, pH 9.0, with 60 mg glycoaldehyde and 10 mg sodium borohydride at ambient temperature resulted in 60% 2-hydroxyethylation. Treatment of 20 mg ovomucoid in 2.0 ml 0.2 M sodium borate, pH 9.0, with 10% acetol and 30 mg sodium borohydride (added in portions) resulted in 55% hydroxyisopropylation. In both situations, the reaction was terminated by adjusting the pH to 5 with glacial acetic acid. The extent of modification was determined either by titration with trinitrobenzenesulfonic acid or by amino acid analysis after acid hydrolysis. Periodate oxidation could be accomplished with 0.015 M sodium periodate, pH 7.9, for 30 min at ambient temperature.

The use of [^{13}C]-formaldehyde in the reductive alkylation of ribonuclease has been reported.[145] In a subsequent study,[146] Jentoft and Dearborn characterized the inhibition by cyanide of reductive alkylation with sodium cyanoborohydride. This is of some importance, because cyanide is a product of reductive alkylation with sodium cyanoborohydride. Inhibition by cyanide can be blocked by nickel (II) or cobalt (III). The observation that nickel (II) can preclude the inhibition of reductive alkylation by cyanide was shown to obviate the previously observed necessity for recrystallization of the sodium cyanoborohydride. Additional studies on the development of reagents alternative to sodium borohydride have been reported from other laboratories. Geoghegan and co-workers[147] compared sodium cyanoborohydride, dimethylamine borane, and trimethylamine borane with respect to effectiveness in reductive alkylation. Reduction at disulfide bonds was not observed with any of the three reagents. Dimethylamine borane was only slightly less effective than sodium cyanoborohydride, whereas trimethylamine borane was much less effective. This decrease in effectiveness in reductive alkylation is balanced by the absence of toxic byproducts, such as cyanide, evolving during the reaction. Quantitative reductive methylation (equal to or greater than one methyl group per lysyl residue) is achieved at 10 mM formaldehyde with dimethylamine borane and at 50 mM formaldehyde with trimethylamine borane. Note that a similar extent of modification is obtained with 5 mM formaldehyde by using sodium cyanoborohydride. In a subsequent study,[148] this laboratory reported the successful use of pyridine borane in the reductive alkylation of proteins. Wu and Means[149] used reductive alkylation with a nonpolar aldehyde (dodecylaldehyde) to subsequently prepare insoluble proteins by binding the modified protein to octyl-Sepharose.

The reaction of glyceraldehyde with carbonmonoxyhemoglobin S has been explored by Acharya and Manning.[150] This reaction was performed with 0.010 M glyceraldehyde in phosphate-buffered saline, pH 7.4, and the resultant Schiff bases were stabilized by reduction with sodium borohydride. Using radiolabeled glyceraldehyde, these investigators were able to obtain support for the concept that there is selectivity in the reaction of sugar aldehydes with hemoglobin. The reaction product between glyceraldehyde and hemoglobin S did have stability properties without reduction that were not consistent with only Schiff base products. These investigators suggested that the glyceraldehyde-hemoglobin Schiff base could undergo an Amadori rearrangement to form a stable ketoamine adduct, which could be reduced with sodium borohydride to form a product identical to that obtained by direct reduction of the Schiff base. In a subsequent study, these investigators did show that the glyceraldehyde-hemoglobin S Schiff base could rearrange to form a ketamine via an Amadori rearrangement.[151] These investigators were able to use reaction with phenylhydrazine to detect the protein-bound ketamine adduct.

Another class of aldehydes that reacts with protein to give interesting products are simple monosaccharides, which exist in solution in enol and keto forms. Wilson[152] showed that bovine pancreatic ribonuclease dimer would react with lactose in the presence of sodium cyanoborohydride to yield an active derivative that shows selectivity in uptake by the liver during in vivo experiments. The modification of ribonuclease dimer was performed in 0.2 M potassium phosphate, pH 7.4 (phosphate buffer was used to protect Lys-41 from modification) at 37°C for 5 days with lactose and sodium cyanoborohydride. Under these conditions, 80% of the amino groups were modified. Bunn and Higgins[153] explored the reaction of monosaccharides with protein amino groups in the presence of sodium cyanoborohydride in some detail. These investigators studied the reaction of hemoglobin with various monosaccharides in Krebs–Ringer phosphate buffer, pH 7.3. The extent of modification was determined by using tritiated sodium cyanoborohydride. The rate of modification was demonstrated to be a direct function of the amount of each sugar in the carbonyl (or keto) form. Thus, the k_1 ($\times 10^1$ mM^{-1} h^{-1}) for D-glucose is 0.6 with 0.002% in the carbonyl form, whereas the k_1 ($\times 10^3$ mM^{-1} h^{-1}) for D-ribose is 10.0 with 0.05% in the carbonyl form.

The reaction of 2,4,6-trinitrobenzenesulfonic acid (TNBS) with amino groups (Figure 2.9) has been of value in studying the function and reactivity of amino groups in proteins.[154–156] The modification of amino groups with TNBS is easy to monitor by spectral analysis. In the presence of an excess of sulfite, absorbance at 420 nm is the most sensitive index, having $\varepsilon = 2.0 \times 10^4$ M^{-1} cm^{-1}. Absorbance at 420 nm is dependent on the ability of the reaction product to form a complex with sulfite. It has proved convenient in our laboratories to use the fact that the spectrum of a trinitrobenzyl amino compound has an isosbestic point at 367 nm with $\varepsilon = 1.05 \times 10^4$ M^{-1} cm^{-1}. As suggested by Fields,[157] we recrystallize TNBS from 2.0 M HCl before use. We generally perform the modifications in phosphate buffer (pH 6.0 to 9.0). The derivatives of α- and ε-amino groups have similar spectra, with the exception that α-amino derivatives have a slightly higher extinction coefficient at 420 nm ($\varepsilon = 2.20 \times 10^4$ M^{-1} cm^{-1}) than ε-amino groups ($\varepsilon = 1.92 \times 10^4$ M^{-1} cm^{-1}). Both these derivatives have much higher extinction coefficients than those of

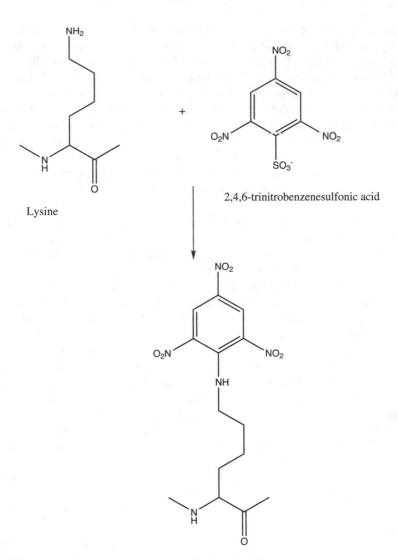

FIGURE 2.9 Reaction of trinitrobenzenesulfonic acid with lysine.

the derivative obtained by reaction of TNBS with cysteinyl residues ($\varepsilon = 2.25 \times 10^3 \ M^{-1} \ cm^{-1}$). α- and ε-amino derivatives can be differentiated by their stability to acid or base hydrolysis. α-amino derivatives are unstable to acid hydrolysis (8 h at 110°C) or base hydrolysis.[158]

More recently, Cayot and Tainturier[159] extended our understanding of the reaction of TNBS with proteins, with particular emphasis on the study of glycated proteins. These investigators carefully examined the conditions for the use of TNBS to determine amino groups in proteins. They observed that raising the temperature of the reaction only slightly increases the rate of reaction with amino groups, but the rate of hydrolysis of the reagent substantially increased. An increase in the pH of the reaction to 10 increases the rates of reaction, establishing the importance of the nucleophilic characters of the amino group in this reaction. It is also observed that, as expected, accessibility of the functional groups is important for reaction with TNBS. They recommend reaction at pH 10 (0.1 M borate) with a 100-fold molar excess (to protein amino groups). Reaction is usually complete in 15 min and can be measured by absorbance at 420 nm.

Frieden and co-workers[160,161] explored the reaction of trinitrobenzenesulfonic acid with bovine liver glutamate dehydrogenase. In these studies, the modification was performed in 0.04 M potassium phosphate, pH 8.0. Under these reaction conditions, the cysteinyl residues were not modified. The preparative reactions were terminated by reaction with 2-mercaptoethanol. It is of interest that under certain conditions (with reduced coenzyme), glutamate dehydrogenase catalyzed the conversion of TNBS to trinitrobenzene.[162]

Means and co-workers[163] studied in some detail the reaction of TNBS with simple amines and hydroxide ions. The reaction of TNBS with hydroxide is first order with respect to both trinitrobenzenesulfonate and hydroxide ions. The reaction with amines was considered in some detail. In general, reactivity of trinitrobenzenesulfonate with amines increases with increasing basicity, except that secondary amines and t-alkylamines are comparatively unreactive. The specific binding of trinitrobenzenesulfonate to proteins must be considered in the study of the reaction of this compound with proteins. Only amines with a pK_a > 8.7 follow a simple rate law. These investigators presented the following considerations regarding the reaction of trinitrobenzenesulfonic acid with proteins. Reactivity is a sensitive measure of the basicity of an amino group. Adjacent charged groups have an influence on the rate of reaction, with an increase observed with a positively charged group and a decrease with a negatively charged group. Proximity to surface hydrophobic regions that can bind TNBS can increase the observed reactivity of a particular amino group. The reaction of TNBS with ammonium has also been investigated by Whitaker and co-workers.[164] This reaction was performed in tetraborate buffer and 1 mM sulfite. The rate of the reaction was determined by following the increase in absorbance at 420 nm ($\varepsilon = 2.02 \times 10^4 M^{-1}$ cm^{-1}). The rate of reaction with ammonium ($k = 0.128$ min^{-1}) was slower than that with the average amine in a protein ($k = 0.907$ min^{-1} for enterotoxin). The reaction with ammonium does, however, provide a sensitive assay for ammonia (as low as 6 nmol), with a precision of 1 to 2%.

Flügge and Heldt explored the labeling of a specific membrane component with TNBS[165] and pyridoxal-5′-phosphate.[166] The modification of the phosphate translocation protein in spinach chloroplasts with TNBS was performed in 0.050 M HEPES, 0.33 M sorbitol, 0.001 M MgCl$_2$, 0.001 M MnCl$_2$, and 0.002 M EDTA, pH 7.6, at 4°C for periods of time up to 15 min, at which point tritiated sodium borohydride was added to both terminate the reaction and radiolabel the trinitrophenyl derivatives.[167] It is possible to label components on the surface of membranes with TNBS, as the sulfonate moiety does not permit membrane penetration. The same is true for pyridoxal-5′-phosphate.

Salem and co-workers[168] explored the use of trinitrobenzenesulfonate in the selective modification of membrane surface components. This study involved the modification of intact cells with the TNBS (dissolved in methyl alcohol) diluted to a 1% methanolic solution. As mentioned previously, trinitrobenzenesulfonate does not pass across (or into) membranes, being more hydrophilic than, for example, fluorodinitrobenzene.

Haniu and co-workers[169] examined the reaction of lysine residues in NAD(P)H:quinone reductase with TNBS as compared to the reaction of tyrosine residues with *p*-nitrobenzenesulfonyl fluoride. Isolation and characterization of the peptides containing the modified residues showed that the modified tyrosyl residues are in hydrophobic regions of the protein whereas the modified lysine residues are in hydrophilic regions.

With the exception of the Boulton–Hunter reagent,[170] the use of *N*-hydroxy-succinimide ester derivatives to modify lysine residues has been somewhat restricted to cross-linking reagents. However, the specificity demonstrated by this chemistry provides considerable potential for the introduction of structural probes and other unique functional groups into proteins. Yem and co-workers[171] have used *N*-hydroxy-succinimide chemistry to introduce biotin into recombinant interleukin-1-α. This is a fascinating technology with substantial promise.[172,173]

It is possible to selectively modify the α-amino groups of proteins by chemical transamination with glyoxylate at a slightly acid pH.[174,175] This modification has been applied to *Euglena* cytochrome C-552. This reaction was performed in 2.0 *M* sodium acetate, 0.10 *M* acetic acid, 0.005 *M* nickel sulfate, and 0.2 *M* sodium glyoxylate and resulted in the complete loss of the amino-terminal residue. Snake venom phospholipase A$_2$ has been subjected to chemical transamination.[175] This reaction was performed in 2.0 *M* sodium acetate, 0.4 *M* acetic acid, 0.010 *M* cupric ions, and 0.1 *M* glyoxylic acid, pH 5.5.

REFERENCES

1. Schmidt, D.F., Jr., and Westheimer, F.H., pK$_a$ of the lysine amino group of acetoacetate decarboxylase, *Biochemistry* 10, 1249, 1971.
2. Knowles, J.R., The intrinsic pK$_a$ values of functional groups in enzymes: improper deductions from the pH-dependence of steady-state parameters, *CRC Crit. Rev. Biochem.* 4, 165, 1976.
3. Kellam, B., De Bank, P.A., and Shakesheff, K.M., Chemical modification of mammalian cell surfaces, *Chem. Soc. Rev.* 32, 327, 2003.
4. Gould, A.R., and Norton, R.S., Chemical modification of cationic groups in the polypeptide cardiac stimula anthopleurin-A, *Toxicon* 33, 187, 1995.
5. Becker, L. et al., Identification of a critical lysine residue in apolipoprotein B-100 that mediates noncovalent interaction with apolipoprotein (a), *J. Biol. Chem.* 276, 36155, 2001.
6. Wink, M.R. et al., Effect of protein-modifying reagents on ecto-apyrase from rat brain, *Int. J. Biochem. Cell. Biol.* 32, 105, 2000.
7. Ehrhard, B. et al., Chemical modification of recombinant HIV-1 capsid protein p24 leads to the release of a hidden epitope prior to changes of the overall folding of the protein, *Biochemistry* 35, 9097, 1996.
8. Paetzel, M. et al., Use of site-directed chemical modification to study an essential lysine in *Escherichia coli* leader peptidase, *J. Biol.Chem.* 272, 994, 1997.

9. Alcalde, M. et al., Succinylation of cyclodextrin glucosyltransferase from *Thermoanaerobacter* s501 enhances its transferase activity using starch as a donor, *J. Biotechnol.* 86, 71, 2001.

10. Swart, P.J. et al., Antiviral effects of milk proteins: acylation results in polyanionic compounds with potent activity against human immunodeficiency virus types 1 and 2 *in vitro*, *AIDS Res. Hum. Retroviruses* 12, 769, 1996.

11. Swart, P.J. et al., Lactoferin. Antiviral activity of lactoferrin, *Adv. Exp. Med. Biol.* 443, 205, 1998.

12. Neurath, A.R. et al., Blocking of CD4 cell receptors for the human immunodeficiency virus type 1 (HIV-1) by chemically modified milk proteins: potential for AIDS prophylaxis, *J. Mol. Recog.* 8, 204, 1995.

13. Jonsson, B.A. et al., Lysine adducts between methyltetrahydrophthalic anhydride and collagen in guinea pig lung, *Toxicol. Appl. Pharmacol.* 135, 156, 1995.

14. Lindh, C.H., and Jonsson, B.A., Human hemoglobin adducts following exposure to hexhydrophthalic anhydride and methylhexahydrophthalic anhydride, *Toxicol. Appl. Pharmacol.* 153, 152, 1998.

15. Pool, C.T., and Thompson, T.E., Methods for dual, site-specific derivatization of bovine pancreatic trypsin inhibitor: trypsin protection of lysine-15 and attachment of fatty acids or hydrophobic peptides at the N-terminus. *Bioconj. Chem.* 10, 221, 1999.

16. Kristiansson, M.H., Jonsson, B.A., and Lindh, C.H., Mass spectrometric characterization of human hemoglobin adducts formed *in vivo* by hexahydrophthalic anhydride, *Chem. Res. Toxicol.* 15, 562, 2002.

17. O'Brien, A.M., Smith, H.T., and O''Fagain, C., Effects of phthalic anhydride modification on horse radish peroxidase stability and activity, *Biotechnol. Bioeng.* 81, 233, 2003.

18. Klotz, I.M., Succinylation, *Methods Enzymol.* 11, 576, 1967.

19. Meighen, E.A., Nicolim, M.Z., and Hustings, J.W., Hybridization of bacterial luciferase with a variant produced by chemical modification, *Biochemistry* 10, 4062, 1971.

20. Shetty, K.J., and Rao, M.S.N., Effect of succinylation on the oligomeric structure of arachin, *Int. J. Pept. Protein Res.*, 11, 305, 1978.

21. Shiao, D.D.F., Lumry, R., and Rajender, S., Modification of protein properties by change in charge. Succinylated chymotrypsinogen, *Eur. J. Biochem.* 29, 377, 1972.

22. Morton, R.C., and Gerber, G.E., Water solubilization of membrane proteins. Extensive derivatization with a novel polar derivatizing reagent, *J. Biol. Chem.* 263, 7989, 1988.

23. Atassi, M.Z., and Habeeb, A.F.S.A., Reaction of protein with citraconic anhydride, *Methods Enzymol.* 25, 546, 1972.

24. Habeeb, A.F.S.A., and Atassi, M.Z., Enzymic and immunochemical properties of lysozyme. Evaluation of several amino group reversible blocking reagents, *Biochemistry* 9, 4939, 1970.

25. Shetty, J.K., and Kinsella, J.E., Ready separation of proteins from nucleoprotein complexes by reversible modification of lysine residues, *Biochem. J.* 191, 269, 1980.

26. Weisgraber, K.H., Innerarity, T.L., and Mahley, R.W., Role of the lysine residues of plasma lipoproteins in high affinity binding to cell surface receptors on human fibroblasts, *J. Biol. Chem.* 253, 9053, 1978.

27. Mahley, R.W. et al., Accelerated clearance of low-density and high-density lipoproteins and retarded clearance of E apoprotein-containing lipoproteins from the plasma of rats after modification of lysine residues, *Proc. Natl. Acad. Sci. USA*, 76, 1746, 1979.

28. Urabe, I., Yamamoto, M., Yamada, Y., and Okada, H., Effect of hydrophobicity of acyl groups on the activity and stability of acylated thermolysin, *Biochim. Biophys. Acta* 524, 435, 1978.
29. Pool, C.T., and Thompson, T.E., Chain length and temperature dependence of the reversible association of model acylated proteins with lipid bilayers, *Biochemistry* 37, 10246, 1998.
30. Béven, L. et al., Ca^{2+}-myristoyl switch and membrane binding of chemical acylated neurocalcins, *Biochemistry* 40, 8152, 2001.
31. Howlett, G.J., and Wardrop, A.J., Dissociation and reconstitution of human erythrocyte membrane proteins using 3,4,5,6-tetrahydrophthalic anhydride, *Arch. Biochem. Biophys.* 188, 429, 1978.
32. Fraenkel-Conrat, H., Methods for investigating the essential groups for enzyme activity, *Methods Enzymol.* 4, 247, 1957.
33. Riordan, J.F., and Vallee, B.L., Acetylation, *Methods Enzymol.* 11, 565, 1967.
34. Merle, M. et al., Acylation of porcine and bovine calcitonin: effects on hypocalcemic activity in the rat, *Biochem. Biophys. Res. Commn.* 79, 1071, 1977.
35. Aviram, I., The role of lysines in *Euglena* cytochrome C-552. Chemical modification studies, *Arch. Biochem. Biophys.* 181, 199, 1977.
36. Karibian, D. et al., On the reaction of acetic and maleic anhydrides with elastase. Evidence for a role of the NH$_2$-terminal valine, *Biochemistry* 13, 2891, 1974.
37. Smit, V., Native electrophoresis to monitor chemical modification of human interleukin-3, *Electrophoresis* 15, 251, 1994.
38. Claasen, E., Kors, N., and Van, R.N., Influence of carriers on the development and localization of anti-trinitrophenyl antibody forming cells in the murine spleen, *Eur. J. Immunol.* 16, 271, 1986.
39. Glish, G.L., and Vacher, R.W., The basics of mass spectrometry in the twenty-first century, *Nature Rev. Drug Discov.* 2, 140, 2003.
40. Fligge, T.A. et al., Direct monitoring of protein-chemical reactions utilizing nano-electrospray mass spectrometry, *J. Am. Soc. Mass Spectrom.* 10, 112, 1999.
41. Bennett, K.L. et al., Rapid characterization of chemically-modified proteins by electrospray mass spectrometry, *Bioconjug. Chem.* 7, 16, 1996.
42. Jahn, O. et al., The use of multiple ion chromatograms in on-line HPLC-MS for the characterization of post-translational and chemical modifications of proteins. *Int. J. Mass Spectrom.* 214, 37, 2002.
43. Suckau, D., Mak, M., and Przybylski, M., Protein surface topology probing by selective chemical modification and mass spectrometric peptide mapping, *Proc. Natl. Acad. Sci.USA* 89, 5630, 1992.
44. Yeboah, F.K. et al., Effect of limited solid-state glycation on the conformation of lysozyme by ESI-MSMS peptide mapping and molecular modeling, *Bioconj. Chem.* 15, 27, 2004.
45. Ohguro, H. et al., Topographic study of arrestin using differential chemical modification and hydrogen/deuterium exchange, *Protein Sci.* 3, 2428, 1994.
46. Scaloni, A. et al., Topology of the thyroid transcription factor 1 homeodomain-DNA complex. *Biochemistry* 38, 64, 1999.
47. D'Ambrosio, C. et al., Probing the dimeric structure of porcine aminoacylase 1 by mass spectrometry and modeling procedures, *Biochemistry* 42, 4430, 2003.
48. Calvete, J.J. et al., Characterisation of the conformation and quaternary structure-dependent heparin-binding region of bovine seminal plasma protein PDC-109, *FEBS Lett.* 444, 260, 1999.

49. Zappacosta, F. et al., Surface topology of Minibody by selective chemical modification and mass spectrometry, *Protein Sci.* 6, 1901, 1997.
50. Taralp, A., and Kaplan, H., Chemical modification of lyophilized proteins in non-aqueous environments, *J. Protein Chem.* 16, 183, 1997.
51. Smith, C.M. et al., Mass spectrometric quantification of acetylation at specific lysines within the amino-terminal tail of histone H4, *Anal. Biochem.* 316, 23, 2003.
52. Hochleitner, E.O., Characterization of a discontinuous epitope of the human immuno-deficiency virus (HIV) core protein p24 by epitope excision and differential chemical modification followed by mass spectrometric peptide mapping analysis, *Protein Sci.* 9, 487, 2000.
53. Kaplan, H., Stevenson, K.J., and Hartley, B.S., Competitive labeling, a method for determining the reactivity of individual groups in proteins. The amino groups of porcine elastin, *Biochem. J.* 124, 289, 1971.
54. Bosshard, H.R., Koch, G.L.E., and Hartley, B.S., The aminoacyl tRNA synthetase-tRNA complex: detection by differential labeling of lysine residues involved in complex formation, *J. Mol. Biol.* 119, 125, 1978.
55. Richardson, R.H., and Brew, K., Lactose synthase. An investigation of the interaction site of alpha-lactalbumin for galactosyltransferase by differential kinetic labeling, *J. Biol. Chem.,* 255, 3377, 1980.
56. Rieder, R., and Bosshard, H.R., The cytochrome c oxidase binding site on cytochrome c. Differential chemical modification of lysine residues in free and oxidase-bound cytochrome c, *J. Biol. Chem.* 253, 6045, 1978.
57. Hitchcock, S.E., Zimmerman, C.J., and Smalley, C., Study of the structure of troponin-T by measuring the relative reactivities of lysines with acetic anhydride, *J. Mol. Biol.* 147, 125, 1981.
58. Hitchcock, S.E., Study of the structure of troponin-C by measuring the relative reactivities of lysines with acetic anhydride, *J. Mol. Biol.* 147, 153, 1981.
59. Hitchcock-De Gregori, S.E., Study of the structure of troponin-I by measuring the relative reactivities of lysine with acetic anhydride, *J. Biol. Chem.* 257, 7372, 1982.
60. Giedroc, D.P. et al., Differential trace labeling of calmodulin: investigation of binding sites and conformational states by individual lysine reactivities. Effects of beta endorphin, trifluoroperazine, and ethylene glycol bis(beta-aminoethyl ether)-*N,N,N'N*-tetraacetic acid, *J. Biol. Chem.* 260, 13406, 1985.
61. Wei, Q. et al., Effects of interactions of with calcineurin of the reactivities of calmodulin lysines, *J. Biol. Chem.* 263, 19541, 1988.
62. Winkler, M.A. et al., Differential reactivities of lysines in calmodulin complexed to phosphatase, *J. Biol. Chem.* 262, 15466, 1987.
63. Hitchcock-De Gregori, S.E. et al., Lysine reactivities of tropomyosin complexed with troponin, *Arch. Biochem. Biophys.* 264, 410, 1988.
64. Staudenmayer, N. et al., An enzyme kinetics and 19F nuclear magnetic resonance study of selectively trifluoroacetylated cytochrome c derivatives, *Biochemistry* 15, 3198, 1976.
65. Smith, M.B. et al., Use of specific trifluoroacetylation of lysine residues in cytochrome C to study the reaction with cytochrome b5, cytochrome, c1 and cytochrome oxidase, *Biochim. Biophys. Acta* 592, 303, 1980.
66. Web, M., Stonehuerner, J., and Millett, F., The use of specific lysine modifications to locate the reaction site of cytochrome C with sulfite oxidase, *Biochim. Biophys. Acta* 593, 290, 1980.
67. Ahmed, A.J., and Millett, F., Use of specific lysine modifications to identify the site of reaction between cytochrome C and ferricyanide, *J. Biol. Chem.* 256, 1611, 1981.

68. Smith, M.B., and Millett, F., A 19F nuclear magnetic resonance study of the interaction between cytochrome C and cytochrome C peroxidase, *Biochim. Biophys. Acta* 626, 64, 1980.
69. Gurd, F.R.N., Carboxymethylation, *Methods Enzymol.* 11, 532, 1967.
70. Sanger, F., and Tuppy, H., The amino acid sequence in the phenylalanyl chain of insulin. I. The identification of lower peptides from partial hydrolysates, *Biochem. J.* 49, 463, 1951.
71. Carty, R.P., and Hirs, C.H.W., Modification of bovine pancreatic ribonuclease A with 4-sulfonyloxy-2-nitrofluorobenzene, *J. Biol. Chem.* 243, 5254, 1968.
72. Franklin, J.G., and Leslie, J., Some enzymatic properties of trypsin after reaction with 1-dimethylaminonaphthalene-5-sulfonyl chloride, *Can. J. Biochem.* 49, 516, 1971.
73. Wagner, R., Podestá, F.E., González, D.H., and Andreo, C.S., Proximity between fluorescent probes attached to four essential lysyl residues in phosphoenolpyruvate carboxylase: a resonance energy transfer study, *Eur. J. Biochem.* 173, 561, 1988.
74. Brautigan, D.L., Ferguson-Miller, S., and Margoliuash, E., Definition of cytochrome c binding domains by chemical modification. I. Reaction with 4-chloro-3,5-dinitrobenzoate and chromatographic separation of singly substituted derivatives, *J. Biol. Chem.* 253, 130, 1978.
75. Bello, J., Iijima, H., and Kartha, G., A new arylating agent, 2-carboxy-4,6-dinitrochlorobenzene. Reaction with model compounds and bovine pancreatic ribonuclease, *Int. J. Pept. Protein Res.* 14, 199, 1979.
76. Hall, J. et al., Role of specific lysine residues in the reaction of *Rhodobacter sphaeroides* cytochrome c2 with the cytochrome bc1 complex, *Biochemistry* 28, 2568, 1989.
77. Long, J.E. et al., Role of specific lysine residues in binding cytochrome c2 to the *Rhodobacter sphaeroides* reaction center in optimal orientation for rapid electron transfer, *Biochemistry* 28, 6970, 1989.
78. Hiratsuka, T., and Uchida, K., Lysyl residues of cardiac myosin accessible to labeling with a fluorescent reagent, *N*-methyl-2-anilino-6-naphthalenesulfonyl chloride, *J. Biochem.* 88, 1437, 1980.
79. Onondera, M., Shiokawa, H., and Takagi, T., Flourescent probes for antibody active sites. I. Production of antibodies specific to the *N*-methyl-2-anilinonaphthalene-6-sulfonate group in rabbits and some fluorescent properties of the hapten bound to the antibodies, *J. Biochem.* 79, 195, 1976.
80. Cory, R.P., Becker, R.R., Rosenbluth, R., and Isenberg, I., Synthesis and fluorescent properties of some *N*-methyl-2-anilino-6-naphthalensulfonyl derivatives, *J. Am. Chem. Soc.* 90, 1643, 1968.
81. Haugland, R.P., Molecular probes. In *Handbook of Fluorescent Probes and Research Chemicals*, Molecular Probes, Eugene, OR, 1989, p. 37.
82. Tuls, J., Geren, L., and Millett, F., Fluorescein isothiocyanate specifically modifies lysine 338 of cytochrome P-450scc and inhibits adrenodoxin binding, *J. Biol. Chem.* 264, 16421, 1989.
83. Miki, M., Interaction of Lys-61 labeled actin with myosin subfragment-1 and the regulatory proteins, *J. Biochem. (Tokyo)* 106, 651, 1989.
84. Bellelli, A. et al., Binding and internalization of ricin labelled with fluorescein isothiocyanate, *Biochem. Biophys. Res. Commn.* 169, 602, 1990.
85. Devanathan, S. et al., Readily available fluorescein isothiocyanate-conjugated antibodies can be easily converted into targeted phototoxic agents for antibacterial, antiviral, and anticancer therapy, *Proc. Natl. Acad. Sci. USA* 87, 2980, 1990.
86. Turner, D.C., and Brand, L., Quantitative estimation of protein binding site polarity. Fluorescence of *N*-arylaminonaphthalenesulfonates, *Biochemistry* 7, 3381, 1968.

87. Welches, W.R., and Baldwin, T.O., Active center studies on bacterial luciferase: modification of the enzyme with 2,4-dinitrofluorobenzene, *Biochemistry* 20, 512, 1981.

88. Shapiro, R., Fox, E.A., and Riordan, J.F., Role of lysines in human angiogenin: chemical modification and site-directed mutagenesis, *Biochemistry* 28, 1726, 1989.

89. Stark, G.R., Stein, W.H., and Moore, S., Reaction of the cyanate present in aqueous urea with amino acids and proteins, *J. Biol. Chem.* 235, 3177, 1960.

90. Lippincott, J. and Apostol, I., Carbamylation of cysteine: a potential artifact in peptide mapping of hemoglobins in the presence of urea, *Anal. Biochem.* 267, 57-64.

91. McCarthy, J. et al., Carbamylation of proteins in 2-electrophoresis – myth or reality?, *J. Proteome Res.* 2, 239-242.

92. Lin, M.F. et al., Ion chromatographic quantification of cyanate in urea solutions: estimation of the efficiency of cyanate scavengers for use in recombinant protein manufacturing, *J. Chromatog. B.* 803, 353-362.

93. Stark, G.R., and Smyth, D.G., The use of cyanate for the determination of NH_2-terminal residues in proteins, *J. Biol. Chem.* 238, 214, 1963.

94. Stark, G.R., Modification of proteins with cyanate, *Methods Enzymol.* 25, 579, 1972.

95. Shaw, D.C., Stein, W.H., and Moore, S., Inactivation of chymotrypsin by cyanate, *J. Biol. Chem.* 239, PC 671, 1964.

96. Cerami, A., and Manning, J.M., Potassium cyanate as an inhibitor of the sickling of erythrocytes *in vitro*, *Proc. Natl. Acad. Sci. USA*, 68, 1180, 1971.

97. Lee, C.K., and Manning, J.M., Kinetics of the carbamylation of the amino groups of sickle cell hemoglobin by cyanate, *J. Biol. Chem.* 248, 5861, 1973.

98. Njikam, N. et al., Carbamylation of the chains of hemoglobin S by cyanate *in vitro* and *in vivo*, *J. Biol. Chem.* 248, 8052, 1973.

99. Nigen, A.M., Bass, B.D., and Manning, J.M., Reactivity of cyanate with valine-1 (α) of hemoglobin. A probe of conformational change and anion binding, *J. Biol. Chem.* 251, 7638, 1976.

100. Plapp, B.V., Moore, S., and Stein, W.H., Activity of bovine pancreatic deoxyribonuclease A with modified amino groups, *J. Biol. Chem.* 246, 939, 1971.

101. Sekiguchi, T. et al., Chemical modification of ε-amino groups in glutamine synthetase from *Bacillus stearothermophilus* with ethyl acetimidate, *J. Biochem.* 85, 75, 1979.

102. Browne, D.J., and Kent, S.B.H., Formation of non-amidine products in the reaction of primary amines with imido esters, *Biochem. Biophys. Res. Commn.*, 67, 126, 1975.

103. Monneron, A., and d'Alayer, J., Effects of imido-esters on membrane-bound adenylate cyclase, *FEBS Lett.* 122, 241, 1980.

104. Plapp, B.V., Enhancement of the activity of horse liver alcohol dehydrogenase by modification of amino groups at the active sites, *J. Biol. Chem.* 245, 1727, 1970.

105. Dunathan, H.C., Stereochemical aspects of pyridoxal phosphate catalysis, *Adv. Enzymol.* 35, 79, 1971.

106. Shapiro, S., Enser, M., Pugh, E., and Horecker, B.L., The effect of pyridoxal phosphate on rabbit muscle aldolase, *Arch. Biochem. Biophys.* 128, 554, 1968.

107. Schnackerz, K.D., and Noltmann, E.A., Pyridoxal-5'-phosphate as a site-specific protein reagent for a catalytically critical lysine residue in rabbit muscle phosphoglucose isomerase, *Biochemistry* 10, 4837, 1971.

108. Havran, R.T., and du Vigneaud, V., The structure of acetone-lysine vasopressin as established through its synthesis from the acetone derivative of S-benzyl-l-cysteinyl-l-tyrosine, *J. Am. Chem. Soc.* 91, 2696, 1969.

109. Peach, C., and Tolbert, N.E., Active site studies of ribulose-1,5-bisphosphate carboxylase/oxygenase with pyridoxal-5'-phosphate, *J. Biol. Chem.* 253, 7864, 1978.

110. Kent, A.B., Krebs, E.G., and Fischer, E.H., Properties of crystalline phosphorylase b, *J. Biol. Chem.* 232, 549, 1958.
111. Wimmer, M.J., Mo, T., Sawyers, D.L., and Harrison, J.H., Biphasic inactivation of porcine heart mitochondrial malate dehydrogenase by pyridoxal-5′-phosphate, *J. Biol. Chem.* 250, 710, 1975.
112. Bleile, D.M., Jameson, J.L., and Harrison, J.H., Inactivation of porcine heart cytoplasmic malate dehydrogenase by pyridoxal-5′-phosphate, *J. Biol. Chem.* 251, 6304, 1976.
113. Jones, C.W., III, and Priest, D.G., Interaction of pyridoxal-5′-phosphate with aposerine hydroxymethyltransferase, *Biochim. Biophys. Acta* 526, 369, 1978.
114. Cortijo, M., Jimenez, J.S., and Lior, J., Criteria to recognize the structure and micropolarity of pyridoxal-5′-phosphate binding sites in protein, *Biochem. J.,* 171, 497, 1978.
115. Cake, M.A., DiSorbo, D.M., and Litwack, G., Effect of pyridoxal phosphate on the DNA binding site of activated hepatic glucocorticoid receptor, *J. Biol. Chem.* 253, 4886, 1978.
116. Slebe, J.C., and Martinez-Carrion, M., Selective chemical modification and 19F NMR in the assignment of a pK value to the active site lysyl residue in aspartate transaminase, *J. Biol. Chem.* 253, 2093, 1978.
117. Nishigori, H., and Toft, D., Chemical modification of the avian progesterone receptor by pyridoxal-5′-phosphate, *J. Biol. Chem.* 254, 9155, 1979.
118. Sugiyama, Y., and Mukohata, Y., Modification of one lysine by pyridoxal phosphate completely inactivates chloroplast coupling factor 1 ATPase, *FEBS Lett.* 98, 276, 1979.
119. Peters, H., Risi, S., and Dose, K., Evidence for essential primary amino groups in a bacterial coupling factor F1 ATPase, *Biochem. Biophys. Res. Commn.*, 97, 1215, 1980.
120. Gould, K.G., and Engel, P.C., Modification of mouse testicular lactate dehydrogenase by pyridoxal 5′-phosphate, *Biochem. J.* 191, 365, 1980.
121. Ogawa, H., and Fujioka, M., The reaction of pyridoxal-5′-phosphate with an essential lysine residue of saccharopine dehydrogenase (L-lysine-forming), *J. Biol. Chem.* 255, 7420, 1980.
122. Forrey, A.W. et al., Synthesis and properties of α- and ε-pyridoxyl lysines and their phosphorylated derivatives, *Biochimie* 53, 269, 1971.
123. Sober, H.A., *Handbook of Biochemistry*, 2nd ed., The Chemical Rubber Company, Cleveland, OH, 1970.
124. Moldoon, T.G., and Cidlowski, J.A., Specific modification of rat uterine estrogen receptor by pyridoxal-5′-phosphate, *J. Biol. Chem.* 2, 55, 3100, 1980.
125. Ohsawa, H., and Gualerzi, C., Structure-function relationship in *Escherichia coli* inhibition factors. Identification of a lysine residue in the ribosomal binding site of initiation factor by site-specific chemical modification with pyrodixal phosphate, *J. Biol. Chem.* 256, 4905, 1981.
126. Bürger, E., and Görisch, H., Evidence for an essential lysine at the active site of l-histidinol:NAD+ oxidoreductase, a bifunctional dehydrogenase, *Eur. J. Biochem.* 118, 125, 1981.
127. Miller, A.D., Packman, L.C., Hart, G.J., Alefounder, P.R., Abell, C., and Battersby, A.R., Evidence that pyridoxal phosphate modification of lysine residues (Lys-55 and Lys-59) causes inactivation of hydroxymethylbilane synthase (porphobilinogen deaminase), *Biochem. J.* 262, 119, 1989.
128. Basu, S., Basu, A., and Modak, M.J., Pyridoxal 5′-phosphate mediated inactivation of *Escherichia coli* DNA polymerase I: identification of lysine-635 as an essential residue for the processive mode of DNA synthesis, *Biochemistry* 27, 6710, 1988.

129. Mahrenholz, A.M., Wang, Y., and Roach, P.J., Catalytic site of rabbit glycogen synthase isozymes. Identification of an active site lysine close to the amino terminus of the subunit, *J. Biol. Chem.* 263, 10561, 1988.
130. Tamura, J.K., LaDine, J.R., and Cross, R.L., The adenine nucleotide binding site on yeast hexokinase PII. Affinity labeling of Lys-111 by pyridoxal 5'-diphospho-5'-adenosine, *J. Biol. Chem.* 263, 7907, 1988.
131. Kluger, R., Methyl acetyl phosphate. a small anionic acetylating agent, *J. Org. Chem.* 45, 2733, 1980.
132. Ueno, H., Pospischil, M.A., Manning, J.M., and Kluger, R., Site-specific modification of hemoglobin by methyl acetyl phosphate, *Arch. Biochem. Biophys.* 244, 795, 1986.
133. Ueno, H., Pospischil, M.A., and Manning, J.M., Methyl acetyl phosphate as a covalent probe for anion-binding sites in human and bovine hemoglobins, *J. Biol. Chem.* 264, 12344, 1989.
134. Means, G.E., Reductive alkylation of amino groups, *Methods Enzymol.* 47, 469, 1977.
135. Rice, R.H., Means, G.E., and Brown, W.D., Stabilization of bovine trypsin by reductive methylation. *Biochim. Biophys. Acta* 492, 316, 1977.
136. Morris, R.W., Cagan, R.H., Martenson, R.E., and Deibler, G., Methylation of the lysine residues of monellin, *Proc. Soc. Exp. Biol. Med.* 157, 194, 1978.
137. Chen, F.M.F., and Benoiton, N.L., Reductive *N,N*-dimethylation of amino acid and peptide derivatives using methanol as the carbonyl source, *Can. J. Biochem.* 56, 150, 1978.
138. Dottavio-Martin, D., and Ravel, J.M., Radiolabeling of proteins by reductive alkylation with [^{14}C] formaldehyde and sodium cyanoborohydride, *Anal. Biochem.* 87, 562, 1978.
139. Jentoft, N., and Dearborn, D.G., Labeling of proteins by reductive methylation using sodium cyanoborohydride, *J. Biol. Chem.* 254, 4359, 1979.
140. Dick, L.R., Geraldes, C.F.G.C., Sherry, A.D., Gray, C.W., and Gray, D.M., ^{13}C NMR of methylated lysines of fd gene 5 protein: evidence for a conformational change involving lysine 24 upon binding of a negatively charged lanthanide chelate, *Biochemistry* 28, 7896, 1989.
141. Brown, E.M., Pfeffer, P.E., Kumosinski, T.F., and Greenberg, R., Accessibility and mobility of lysine residues in β-lactoglobulin, *Biochemistry* 27, 5601, 1988.
142. Fretheim, K., Iwai, S., and Feeney, R.E., Extensive modification of protein amino groups by reductive addition of different sized substituents, *Int. J. Pept. Protein Res.* 14, 451, 1979.
143. Fretheim, K., Edelandsdal, B., and Harbitz, O., Effect of alkylation with different size substituents on the conformation of ovomucoid, lysozyme and ovotransferrin, *Int. J. Pept. Protein Res.* 25, 601, 1985.
144. Geoghegan, K.F., Ybarra, D.M., and Feeney, R.E., Reversible reductive alkylation of amino groups in proteins, *Biochemistry* 18, 5392, 1979.
145. Jentoft, J.E., Jentoft, N., Gerken, T.A., and Dearborn, D.G., ^{13}C NMR studies of ribonuclease A methylated with [^{13}C] formaldehyde, *J. Biol. Chem.* 254, 4366, 1979.
146. Jentoft, N., and Dearborn, D.G., Protein labeling by reductive methylation with sodium cyanoborohydride effect of cyanide and metal ions on the reaction, *Anal. Biochem.* 106, 186, 1980.
147. Geoghegan, K.F. et al., Alternative reducing agents for reductive methylation of amino groups in proteins, *Int. J. Pept. Protein Res.* 17, 345, 1981.
148. Cabacungan, J.C., Ahmed, A.J., and Feeney, R.E., Amine boranes as alternative reducing agents for reductive alkylation of proteins, *Anal. Biochem.* 124, 272, 1982.

149. Wu, H.-L., and Means, G.E., Immobilization of proteins by reductive alkylation with hydrophobic aldehydes, *Biotechnol. Bioeng.* 23, 855, 1981.
150. Acharya, A.S., and Manning, J.M., Reactivity of the amino groups of carbonmonoxy-hemoglobin S with glyceraldehyde, *J. Biol. Chem.* 255, 1406, 1980.
151. Acharya, A.S., and Manning, J.M., Amadori rearrangement of glyceraldehyde-hemoglobin Schiff base adducts. A new procedure for the determination of ketoamine adducts in proteins, *J. Biol. Chem.* 255, 7218, 1980.
152. Wilson, G., Effect of reductive lactosamination on the hepatic uptake of bovine pancreatic ribonuclease A dimer, *J. Biol. Chem.* 253, 2070, 1978.
153. Bunn, H.F., and Higgins, P.J., Reaction of monosaccharides with proteins: possible evolutionary significance, *Science* 213, 222, 1981.
154. Goldfarb, A.R., A kinetic study of the reactions of amino acids and peptides with trinitrobenzenesulfonic acid, *Biochemistry* 5, 2570, 1966.
155. Goldfarb, A.R., Heterogeneity of amino groups in proteins. I. Human serum albumin, *Biochemistry* 5, 2574, 1966.
156. Habeeb, A.F.S.A., Determination of free amino groups in proteins by trinitrobenzene-sulfonic acid, *Anal. Biochem.* 14, 328, 1966.
157. Fields, R., The rapid determination of amino groups with TNBS, *Methods Enzymol.* 25, 464, 1972.
158. Kotaki, A., and Satake, K., Acid and alkaline degradation of the TNP-amino acids and peptides, *J. Biochem.* 56, 299, 1964.
159. Cayot, P., and Tainturier, G., The quantification of protein amino groups by the trinitrobenzenesulfonic acid method: a reexamination, *Anal. Biochem.* 249, 184, 1997.
160. Coffee, C.J. et al., Identification of the sites of modification of bovine liver glutamate dehydrogenase reacted with trinitrobenzene-sulfonate, *Biochemistry* 10, 3516, 1971.
161. Goldin, B.R., and Frieden, C., Effects of trinitrophenylation of specific lysyl residues on the catalytic, regulatory and molecular properties of bovine liver glutamate dehydrogenase, *Biochemistry* 10, 3527, 1971.
162. Bates, D.J., Goldin, B.R., and Frieden, C., A new reaction of glutamate dehydrogenase: the enzyme-catalyzed formation of trinitrobenzene from TNBS in the presence of reduced coenzyme, *Biochem. Biophys. Res. Commn.* 39, 502, 1970.
163. Means, G.E., Congdon, W.I., and Bender, M.L., Reactions of 2,4,6-trinitrobenzene-sulfonate ion with amines and hydroxide ion, *Biochemistry* 11, 3564, 1972.
164. Whitaker, J.R., Granum, P.E., and Aasen, G., Reaction of ammonia with trinitrobenzene sulfonic acid, *Anal. Biochem.* 108, 72, 1980.
165. Flügge, U.I., and Heldt, H.W., Specific labelling of the active site of the phosphate translocator in spinach chloroplasts by 2,4,6-trinitrobenzene sulfonate, *Biochem. Biophys. Res. Commn.* 84, 37, 1978.
166. Flügge, U.I., and Heldt, H.W., Specific labelling of a protein involved in phosphate transport of chloroplasts by pyridoxal-5′-phosphate, *FEBS Lett.* 82, 29, 1977.
167. Parrott, C.L., and Shifrin, S., A spectrophotometric study of the reaction of borohydride with trinitrophenyl derivatives of amino acids and proteins, *Biochim. Biophys. Acta* 491, 114, 1977.
168. Salem, N., Jr., Lauter, C.J., and Trams, E.G., Selective chemical modification of plasma membrane ectoenzymes, *Biochim. Biophys. Acta* 641, 366, 1981.
169. Haniu, M. et al., Structure-function relationship of NAD(P)H:quinone reductase: characterization of NH_2-terminal blocking group and essential tyrosine and lysine residues, *Biochemistry* 27, 6877, 1988.

170. Boulton, A.E., and Hunter, W.M., The labelling of proteins to high specific radio-activities by conjugation to a ^{125}I-containing acylating agent, *Biochem. J.* 133, 529, 1973.
171. Yem, A.W. et al., Biotinylation of reactive amino groups in native recombinant human interleukin-1, *J. Biol. Chem.* 264, 17691, 1989.
172. Wilchek, M., and Bayer, E., Biotin-containing reagents, *Methods Enzymol.* 184, 123, 1990.
173. Bayer, E., and Wilchek, M., Protein biotinylation, *Methods Enzymol.* 184, 148, 1990.
174. Dixon, H.B.F., and Fields, R., Specific modification of NH_2-terminal residues by transamination, *Methods Enzymol.* 25, 409, 1972.
175. Verheij, H.M., Egmond, M.R., and de Haas, G.H., Chemical modification of the α-amino group in snake venom phospholipases A_2. A comparison of the interaction of pancreatic and venom phospholipases with lipid-water interfaces, *Biochemistry* 20, 94, 1981.

3 The Modification of Histidine Residues

Because many enzymes contain a histidine residue, which is critical for the catalytic process, the site-specific modification of this residue has been the subject of many studies. Most of these studies have been directed at the study of the catalytic mechanism of enzymes and a few at protein–protein interactions or substrate/cofactor binding. Thus, despite the importance of histidine, only a small number of reagents have been studied. Most of the current work reported in the literature is with diethylpyrocarbonate.[1-9] Several other approaches, such as photooxidation or reaction with p-bromophenacyl bromide, are discussed later in detail, as is the use of diethylpyrocarbonate.

The technique of photooxidation has not proved to be of extensive value because of problems with the specificity of modification. Histidine, methionine, and tryptophan are quite sensitive to photooxidation, whereas tyrosine, serine, and threonine are somewhat less sensitive.[10] Notwithstanding the issue of specificity, photooxidation and oxidation of histidine continue to be of considerable interest. The oxidation of proteins as an *in vivo* process is of increasing interest.[11-15] Histidine residues are also oxidized in the process of radiolytic protein footprinting.[16,17]

Photooxidation was used to identify proteins at the peptidyltransferase site of a bacterial ribosomal subunit.[18,19] Rose bengal dye was used in these experiments. Histidine is the only amino acid modified under the reaction conditions. With the exception of EF-G-GTP binding activity, the loss of biological activity is most closely related to the "fast" histidine loss. In subsequent experiments, methylene blue dye (Eastman, dye content 91%) was used.[20] Peptidyltransferase activity was lost at a more rapid rate in the presence of methylene blue than of rose bengal, but data are not presented about any differences in residues modified or whether amino acid residues other than histidine are modified in the presence of methylene blue. Other investigators have also explored the effects of photooxidation on peptidyltransferase activity in *Escherichia coli* ribosomes.[21] These experiments were performed in 0.030 M Tris, 0.020 M MgCl$_2$, and 0.220 KCl, pH 7.5 (9 mg ribosomes in 0.300 ml) with either eosin or rose bengal as the photooxidation agent. Irradiation was performed at 0 to 4°C, using a 500-W slide projector (26 cm from condenser lens to sample) for 20 min. Photooxidation has also been used to study the role of histidine residues in polypeptide chain elongation factor Tu from *E. coli*.[22] The reaction was performed in 0.05 M Tris, 0.010 M Hg (OAc)$_2$, 0.005 M 2-mercaptoethanol, and 10% glycerol, pH 7.9. Irradiation was performed at 0 to 4°C with gentle stirring, using a 375-W tungsten lamp at a distance of 15 cm. A glass plate was placed in the light beam to eliminate ultraviolet (UV) irradiation.

The rose bengal dye was removed after 5 to 30 min from the reaction by chromatography on DEAE-Sephadex A-25 or A-50 equilibrated with 0.050 M Tris, pH 7.9, 0.010 M Mg (OAc)$_2$, 0.005 M 2-mercaptoethanol, and 10% glycerol. Amino acid analysis after acid hydrolysis (6 N HCl, 22 h, 110°C, or 4 M methanesulfonic acid, 0.2% 2-aminoethylindol, 115°C, 24 h for the determination of tryptophan) demonstrated that only histidine is modified (ca. 5 of 10 residues modified; only one residue modified in the presence of guanosine diphosphate). Photooxidation with methylene blue (25 mM Tris, pH 7.9, 0.05% methylene blue; 8°C) abolished placental anticoagulant protein activity, with loss only of histidine residues (based on amino acid analysis).[23] Diol dehydrase was inactivated with first-order kinetics by photooxidation in the presence of either rose bengal or methylene blue.[24] In these experiments, substitution of a helium atmosphere markedly decreased the rate of enzyme inactivation. There was a difference in the pH dependence of the photooxidation reaction performed in the presence of rose bengal (optimum pH 6.2) as compared with the reaction in the presence of methylene blue (optimum pH >8.0). The pH dependence for the rose bengal reaction was suggested to reflect the charge status of this compound.

Despite problems with residue specificity, photooxidation continues to be of value as a method to modify histidine residues in proteins. A histidine residue was suggested to be at the active site of Ehrlich cell plasma membrane NADH-ferricyanide oxidoreductase on the basis of inactivation by either diethylpyrocarbonate or photooxidation with rose bengal.[25] Metal-catalyzed photooxidation of His-21 in human growth hormone has been reported.[26] His-21, together with His-19 and Glu-174, forms a cation-binding site in this protein. The photooxidation of histidine in the presence of tetrasulfated aluminum phthalocyanin[27] is more effective at low fluence (100 W m^{-2}) when compared with irradiation at 500 W m^{-2}. Hypericin (active ingredient of St. John's wort) has been demonstrated to act as a sensitizer in the photooxidation of α-crystallin from calf lenses.[28] Modification of histidine, methionine, and tryptophan was observed under these experimental conditions (10 mM (NH$_4$)$_2$CO$_3$, pH 7.0, >300 nm, 24 mW cm^{-2}) by mass spectrometric analysis. Photooxidation of histidine has been observed in the presence of calcein,[29] a fluorescent probe used in studies of cell viability. Xanthurenic acid acts as photosenstizer for the photooxidation of cytosolic lens protein.[30] Analysis by mass spectrometry demonstrated the site-specific modification of histidine, tryptophan, and methionine. Two derivatives of histidine, 2-imidazolone and 2-oxohistidine, were observed in alpha A-crystallin.

Histidine residues can be modified by α-halo carboxylic acids and amides (i.e., bromoacetate and bromoacetamide; Figure 3.1). The histidine residue must have either enhanced nucleophilic character[31] or have been located in a unique microenvironment such as in ribonuclease.[32–35] Other amino acids, most notably cysteine, are modified more rapidly than histidine. Liu[36] found it necessary to modify the cysteine residue at the active site of streptococcal proteinase with sodium tetrathionate before it was possible to alkylate the histidine residue. The S-sulfenylsulfonate derivative of the enzyme was inactive and resisted alkylation with iodoacetic acid, iodoacetamide, N-chloroacetyltryptophan, or N-chloroacetylglycyleucine. Alkylation of the active-site histidine was accomplished by using a positively charged

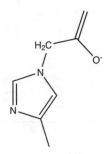

3-Carboxymethylhistidine

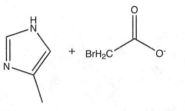

Histidine   Bromoacetic acid

+

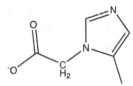

1-Carboxymethylhistidine

FIGURE 3.1 Reaction of histidine with 2-bromoacetic acid to form 1-carboxymethyl histidine and 3-carboxymethylhistidine.

reagent, α-N-bromoacetylarginine methyl ester. The chemistry of histidine alkylation with α-halo carboxylic acids and amides provides the basis for the development of peptide chloromethyl ketones for the affinity labeling of proteolytic enzymes[37–42] These are affinity labeling reagents, and the reader is directed to an excellent review by Plapp[41] for further discussion of these and related agents. Alkylation of histidine has been observed with the use of succinimidyl carbonyl poly(ethylene glycol) to modify rh-interferon α2B [100 mM sodium phosphate, pH 6.5, 22°C, 2 h, 4 Mol succinimidyl carbonyl poly(ethylene glycol)/Mol peptide].[43] Reaction with histidine was far less at a higher pH, at which alkylation of lysine is the predominant reaction.

A compound related to α-halo carboxylic acids is p-bromophenacyl bromide (Figure 3.2), which has been demonstrated in several instances to modify histidyl residues in proteins. p-Bromophenacyl bromide modifies a single histidine residue in taipoxin, with a 350-fold decrease in neurotoxicity.[44] The modification was performed in 0.1 M sodium cacodylate, pH 6.0, 0.1 M NaCl with an eightfold molar excess of p-bromophenacyl bromide at 30°C for 22 h. The reagent should be recrystallized before use. The reaction mixture was concentrated by lyophilization and

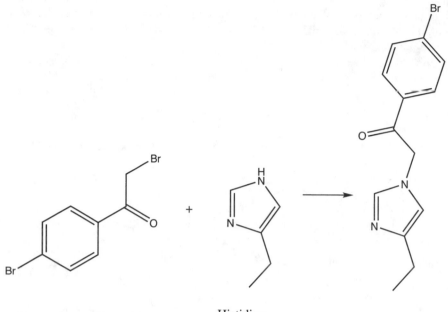

p-Bromophenacyl bromide   Histidine

FIGURE 3.2 Reaction of p-bromophenacyl bromide with histidine.

subjected to gel filtration (G-25 Sephadex in 0.1 M ammonium acetate) to remove excess reagent and buffer salts. The protein fraction was taken to dryness as a salt-free preparation and subjected to a second reaction with p-bromophenacyl bromide. The extent of modification was assessed by both amino acid analysis (loss of histidine) and spectral analysis ($\varepsilon_{271} = 17,000\ M^{-1}\ cm^{-1}$).[45] Two of seven histidine residues are modified (1 Mol/Mol in α-subunit; 1 Mol/Mol in β-subunit) under these reaction conditions.

The basic phospholipase A_2 from *Naja nigrocollis* venom has been modified with p-bromophenacyl bromide.[46] The modification was performed in 0.025 M Tris, pH 8.0, with a 10-fold molar excess of reagent at 30°C. After 40 min of reaction, the mixture was taken to pH 4.0 with glacial acetic acid and passed through a G-25 Sephadex column. Amino acid analysis after acid hydrolysis showed the loss of 1 mol of histidine per mole of enzyme with no other significant changes in composition. Subsequent analysis identified His-47 as the residue modified.

The reaction of p-bromophenacyl bromide with pancreatic phospholipase A_2 has also been studied.[48] The reaction was performed in 0.1 M sodium cacodylate, pH 6.0.[27] Only histidine residues are modified under these conditions, and it has been established that His-48 is the residue modified. Under these reaction conditions, the second-order rate constant for the reaction of p-bromophenacyl bromide with porcine pancreatic phospholipase A_2 is 125 $M^{-1}\ min^{-1}$ when compared with 79 $M^{-1}\ min^{-1}$ for phenacyl bromide and 75 $M^{-1}\ min^{-1}$ for 1-bromooctan-2-one. No reaction was observed with iodoacetamide under these reaction conditions. The same investigators[47] also explored the methylation of His-48 at the N_1-position on the imidazole ring with either methyl p-toluenesulfonate or methyl p-nitrobenzenesulfonate. Reaction with the latter reagent at pH 6.0 (0.050 M cacodylate, 40°C) is more rapid than with the former reagent.

p-Bromophenacyl bromide continues to be used to modify histidine residues in proteins.[49-54] Nakano and co-workers[49] studied the inactivation of phospholipase A_2 in low-density lipoproteins (150 mM sodium phosphate, pH 7.4, 37°C, 2 h, 100 to 400 μM reagent concentration). Battaglia and Radomiska-Pandya[51] studied the functional role of histidine in the UDP-glucuronic acid carrier by measuring the effect of chemical modification on the uptake of radiolabeled UDP-glucuronic acid in rat liver endoplasmic reticulum. Inhibition of uptake was more pronounced with either p-bromophenacyl bromide or diethylpyrocarbonate (both hydrophobic reagents) than with p-nitrobenzenesulfonic acid, methyl ester, a hydrophilic reagent. The reactions were performed in 10 mM HEPES, 0.25 M sucrose, 1 mM MgCl$_2$, pH 7.4, for 1 min. deVet and van den Bosch[53] examined the role of histidine in recombinant guinea pig alkyl-dihydroxyacetonephosphate synthase by modification with p-bromophenacyl bromide and oligonucleotide-directed mutagenesis. Modification with p-bromophenacyl bromide was performed in 10 mM Tris-HCl, 0.15% Triton X-100, pH 7.4, at room temperature. Inactivation was observed showing pseudo first-order kinetics and the enzyme was protected by substrate. A remarkable increase in the rate of inactivation was observed at pH 8.0 (50 mM Tris-HCl). Replacement of His-617 with alanine also eliminated catalytic activity. Modrow and co-workers[54] showed that the phospholipase A2-like activity of the VP1 region of parvovirus B19 is inactivated by p-bromophenacyl bromide.

Methyl p-nitrobenzenesulfonate (Figure 3.3) has been used to methylate histidine residues in ribosomal peptidyl transferase.[55] In these experiments, the ribosome preparation was modified by a 300-fold molar excess of methyl p-nitrobenzenesulfonate (from a stock solution dissolved in acetonitrile). The reaction took place

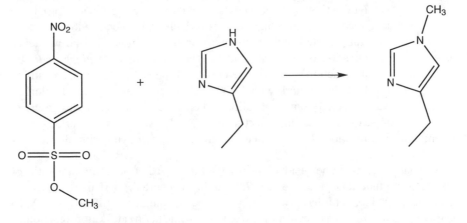

Methyl-p-nitrobenzenesulfonate N^3-methylhistidine

FIGURE 3.3 N-methylation of histidine with methyl-p-nitrobenzene sulfonate.

in 0.01 M Tris, pH 7.4, 0.008 M MgCl$_2$, 0.05 M NH$_4$Cl, and 1 μM puromycin at 24°C for 45 min. The author suggests only histidine residues are modified, but definitive evidence on this point is absent. A more recent study by Marcus and Dekker[56] examined the effect of methyl-p-nitrobenzenesulfonate on the activity of *Escherichia coli* L-threonine dehydrogenase. The reaction was performed in 100 mM potassium phosphate, pH 7.0, at 25°C. Early reactions in this study were performed in 200 mM Tris HCl, pH 8.4, but the reagent was found to be more specific for the modification of histidine at the lower pH. Examination of the effect of reagent concentration on reaction rate demonstrated saturation kinetics with a limit value for the rate of inactivation of 0.01 min^{-1}, suggesting the binding of inhibitor before the inactivation reaction. Analysis of the modified protein showed that His-90 had been methylated at the N3 position (3-methylhistidine). Subsequent studies[57] with oligonucleotide-directed mutagenesis confirmed the importance of this residue in enzyme function.

Another example of the modification of histidine by reagents that, in general, react more avidly with residues other than histidine is the reaction of D-amino acid oxidase with dansyl chloride.[58] In this study, D-amino acid oxidase was allowed to react with a fivefold molar excess of dansyl chloride (from a stock solution in acetone; final concentration of acetone in the reaction mixture did not exceed 5% final volume) in 0.05 M phosphate, pH 6.6. The reaction was terminated by the addition of benzoate, insoluble material was removed by centrifugation, and the mixture was passed through a G-25 Sephadex column equilibrated with 0.06 M phosphate and 0.010 M benzoate, pH 6.6. Reaction with dansyl chloride under these conditions resulted in virtually complete inactivation of the enzyme, with the incorporation of 1.7 mol of reagent per mole of enzyme. Substantially complete reactivation occurred with 0.5 M hydroxylamine (NH$_2$OH) at pH 6.6. This reactivation excluded reaction with primary amino functional groups such as lysine, and amino acid analysis suggested that the reaction had not occurred with an oxygen nucleophile such as tyrosine. Treatment of the enzyme with diethylpyrocarbonate also resulted in the loss of catalytic activity and reduced the amount of dansyl groups incorporated in a subsequent reaction, suggesting that dansyl chloride reacts with the same functional group that reacted with diethylpyrocarbonate. Pilone and co-workers[59] observed the inactivation of D-amino acid oxidase from *Rhodotorula gracilis* by dansyl chloride. The reaction was performed in 50 mM phosphate, pH 6.6, containing 10% glycerol at 18°C in the dark with a 300-fold molar excess of dansyl chloride. The enzyme was protected from inactivation by benzoate. Nonlinearity was observed in the time course of inactivation, reflecting the hydrolysis of dansyl chloride during the modification reaction. The modified enzyme retained activity with altered substrate specificity.

Cyanation of histidine residues in myoglobin by using an equimolar ratio of cyanogen bromide and protein at pH 7.0 has been reported.[60,61] This derivative is somewhat unstable, but it has proved useful in spectral studies (NMR, IR, UV–VIS) of this protein. There has been no further work with this modification reaction in the past decade.

Competitive labeling of the amino-terminal histidine residue in secretin with 1-fluoro-2,4-dinitrobenzene has been used to study the reactivity of this residue vs.

other nucleophiles[62] in this protein. The amino-terminal functional group has a pK_a of 8.83 and reactivity fivefold that of the model compound (histidyl-glycine), whereas the imidazolium ring has a pK_a of 8.24 and a reactivity 26-fold that of the model compound. These results were interpreted as reflecting a conformational state in which the histidine interacts with a carboxylate function.

Diethylpyrocarbonate (pyrocarbonic acid, diethyl ester, dicarbonic acid, diethyl ester, oxydiformic acid diethyl ester, ethoxyformic anhydride) was used as a pesticide and antifungal agent, but its use was banned in food and food products in 1972.[63] There has been recent interest in the use of a related compound, dimethydicarbonate (DMDC), for decontamination of grape musts.[64]

Diethylpyrocarbonate is the most extensively used reagent for the specific modification of histidine in proteins (Figure 3.4). In the pH range of 5.5 to 7.5, diethylpyrocarbonate is reasonably specific for reaction with histidyl residues. There are several studies of the reaction under more acidic conditions.[65,66] In one study, polymerization of ribonuclease was observed both in deionized water (presumably at acidic pH) and in 0.1 M Tris, pH 7.2. Maleylation of ribonuclease obviated polymerization, suggesting that the amino groups were involved in this cross-linking reaction. Miles[67] has reviewed work with diethylpyrocarbonate through 1975.

Reaction of diethylpyrocarbonate with histidine residues at a moderate excess of diethylpyrocarbonate results in substitution at one of the nitrogen positions on the imidazole ring. This reaction is associated with an increase in absorbance at 240 nm ($\varepsilon = 3200$ M^{-1} cm^{-1}). The modification is readily reversed at alkaline pH and, in particular, in the presence of nucleophiles such as hydroxylamine. Tris and other nucleophilic buffers can also reverse the modification, and their use should be avoided with diethylpyrocarbonate. Generally, treatment with neutral hydroxylamine (0.1 to 1.0 M, pH 7.0) is used to regenerate histidine. As with the deacylation of O-acetyl tyrosine by neutral hydroxylamine, the higher the concentration of hydroxylamine, the more rapid the process of decarboxyethylation. Disubstitution on the imidazole ring and carboxyethylation at both the N1 and N3 positions results in a derivative with altered spectral properties compared with those of the monosubstituted derivative. This derivative does not regenerate histidine, and treatment with neutral hydroxylamine or base results in scission of the imidazole ring. With disubstitution, a loss of histidine is detected by amino acid analysis after acid hydrolysis. Sequence analysis using Edman degradation chemistry also shows the absence of histidine, with the presence of a disubstituted derivative.[68] In these studies, a PTH derivative eluting near PTH-glycine was observed and the structure verified by mass spectrometry. The monosubstituted derivative is unstable under conditions of acid hydrolysis and yields free histidine. Mass spectrometry is of increasing value in the analysis of the chemical modification of histidine in proteins, including carboethylated histidine.[69–74] A good correlation has been shown between spectral measurements and mass spectrometry.[70]

Reaction can also occur with other nucleophiles, such as cysteine, tyrosine, and primary amino groups. Modification at sulfhydryl residues, which is not well documented with protein-bound cysteine, can be determined by a decrease in free sulfhydryl groups. Reaction of tyrosine is easily assessed by a decrease in absorbance at 275 to 280 nm, similar to that observed on O-acetylation with N-acetylimidazole.

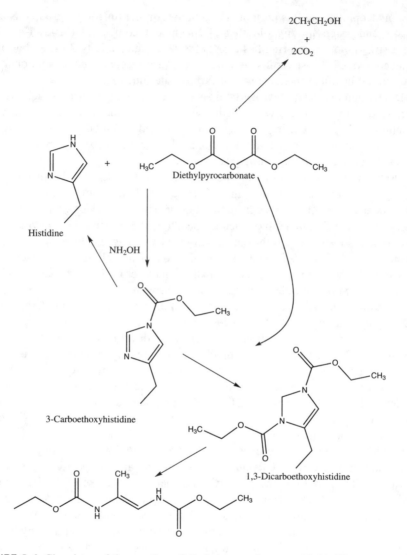

FIGURE 3.4 Chemistry of the reaction of diethylpyrocarbonate with histidine.

This modification is reversed by neutral hydroxylamine. Reaction at primary amino groups (α-amino groups; ε-amino groups of lysine) results in a derivative that is stable to hydroxylamine. An elegant study[75] has examined the reaction of diethyl-pyrocarbonate with histidyl residues in cytochrome b_5. Using NMR spectroscopy with this well-characterized protein, it has been possible to identify factors influencing histidine modification with this reagent. Three major factors are (1) the pK_a of the individual histidine residue, (2) solvent exposure of the residue, and (3) hydrogen bonding of the imidazolium ring. Furthermore, these investigators point out that tautomerization of the imidazolium ring leads to heterogeneity of modification, which in turn explains differences in the spectral properties of modified proteins.[35] Site-specific mutagenesis studies of subtilisin[38] have demonstrated the influence of neighboring charged groups on histidine ionization (and hence reactivity).

As described in some detail by Miles,[67] the reagent is very sensitive to base-catalyzed hydrolysis. At ambient temperature, the $T_{1/2}$ for the hydrolysis of diethyl-pyrocarbonate at pH 7.0 (phosphate) is less than 10 min and is markedly shorter with increasing pH. Increasing the pH not only decreases reagent stability (and thus the concentration of one component of a second-order reaction over the time period studied) but also increases the possibility of reaction at primary amine functional groups. In our laboratory, we have found it convenient to use dilute (0.025 to 0.100 M) phosphate buffer, pH 6.0, for our studies. We prepare stock solutions of diethylpyro-carbonate in anhydrous ethanol. These solutions are used within a few hours, and the actual concentration of reagent is obtained by the stoichiometry of reaction with imidazole in the pH 6.0 buffer, using the increase in absorbance at 230 nm to monitor the reaction ($\varepsilon = 3 \times 10^3\ M^{-1}\ cm^{-1}$)[65,67] both before and after a given series of experiments. Morjana and Scarborough[77] have presented data on the rate of diethyl-pyrocarbonate hydrolysis and discussed the necessity of correcting for actual reagent concentration. There is not complete agreement regarding the magnitude of the spectral change as a result of carboxyethylation. The value $\varepsilon = 3200\ M^{-1}\ cm^{-1}$ at 242 nm has been given previously. Other investigators have used the value of $\varepsilon = 3600\ M^{-1}\ cm^{-1}$ at 240 nm.[78,79] A value of $\varepsilon = 3500\ M^{-1}\ cm^{-1}$ at 242 nm has also been reported.[80] The greater the excess of diethylpyrocarbonate used, the less reliable the value for ε obtained with known stoichiometric modification of model histidine (imidazole derivative) compounds.[81] This is shown in cases in which increasing the ratio of diethylpyrocarbonate to histidine results in species with increased absorbance. It is suggested that the commonly used $\varepsilon = 3200\ M^{-1}\ cm^{-1}$ can only be used at low concentrations of diethylpyrocarbonate.

Table 3.1 presents selected examples of the use of diethylpyrocarbonate to study the function of histidyl residues in proteins. Table 3.2 presents selected examples of the combination of site-specific chemical modification with diethylpyrocarbonate and oligonucleotide-directed mutagenesis to study histidine in proteins. A detailed discussion of a limited number of studies is presented later.

A single histidine residue essential for catalysis by D-xylose isomerase has been identified by reaction with diethylpyrocarbonate.[81] Instability of N-carboethoxyhis-tidine has made unequivocal identification of the histidyl residues modified by diethylpyrocarbonate difficult. In this study, the protein was first denatured in 6.0 M guanidine hydrochloride (pH 7.0) and then digested with subtilisin in 2.0 M guani-dine hydrochloride, pH 7.0, at 30°C for 2 h. A single peptide containing the modified histidine residue was purified by HPLC, using dual-wavelength detection. In this technique, effluent is monitored by absorbance at 238 nm (the maximum in the difference spectrum between the modified and native protein) and 214 nm (peptide bond absorbance). The ratio of A_{238} to A_{214} has been used to identify peptides containing the modified histidine residues.

Horiike and co-workers[82] examined the reaction of diethylpyrocarbonate with pyridoxamine (pyridoxine)-5'-phosphate oxidase. The modification reaction was performed at pH 7.0 (0.1 M potassium phosphate containing 5% [v/v] EtOH) at 25°C generally in the presence of flavin mononucleotide (FMN). Analysis showed

TABLE 3.1

Reaction of Diethylpyrocarbonate with Histidyl Residues in Proteins

Protein	Solvent and Temperature	Reagent Excess[a]	Extent of Modification[b]	Second-Order Rate Constant[c] (1 M^{-1} min^{-1})	Other Amino Acids Modified	Ref.
Thermolysin	0.05 M CaCl$_2$, 0.1 M NaCl, 0.025 M HEPES, pH 7.2, at 25°C	—	2[d]	—	Tyr (2), Lys (8)	1
Thermolysin	0.05 M CaCl$_2$, 0.025 M HEPES, pH 5.7, at 25°C	—	1[d]	—	Tyr (1), Lys (2)	1
Bacterial luciferase	0.1 M phosphate, pH 6.1, at 0°C	600	3	146[e]		2
Crotoxin	Half-saturated NaOAc	1,000		—	—[f]	3
Prostatic acid phosphatase	0.025 M sodium barbital, 0.15 M NaCl, pH 6.9, at 25°C	5,000	12/27	7[e]		4
Pyridoxamine-5'-phosphate oxidase	0.1 M phosphate, pH 7.0, at 25°C	500	4/7	750[g]	Cys (1/6)[h]	5
Yeast enolase	0.050 M ADA, 0.001 M MgCl$_2$, 0.01 mM EDTA, pH 6.1, at 0°C	1,390	6[i]	55.0[j] 10.7[l]	—[k]	6
Fructose bisphosphatase	0.050 M acetate, pH 6.5	4,000 8,000 40,000	5.3/13 8.8/13 13/13[m]	—	—[m]	7
Ribulose bisphosphate carboxylase	0.050 or Tris, 0.001 M EDTA, 0.020 M MgCl$_2$, pH 7.0, at 30°C	—	2/2	—[n]	—[o]	8
Escherichia coli elongation factor Tu	—[o]	—		—	—	10
L-a-Hydroxy acid oxidase	0.020 M MES, pH 7.0, at 25°C	—	2[q]	690[r]	—[s]	11
Thiamin-binding protein from Saccaromyces cerevisiae	0.050 M sodium phosphate, pH 7.0, at 25°C	—	—	120.5[t]	—	12

Clostridium histolyticum collagenase — 0.050 M HEPES, 0.010 M CaCl$_2$, pH 17.5, at 22°C	—	—	—	—	13
Lactate dehydrogenase — 0.1 M sodium pyrophosphate, pH 7.2, at 10°C	—	10,380 / 522[l,u]	—	—	14
Mitochondrial nicotinamide nucleotide transhydrogenase — 0.025 M sucrose, 0.020 M MES, pH 5.9, at 23°C	—	—	—	—	15
Alcohol oxidase — 0.050 M sodium[v] phosphate, pH 7.5, at 0°C	—[w]	25.2	4	—	16
D-β-hydroxybutyrate dehydrogenase — 0.020 MES, 5 M rotenone, pH 6.0, 20°C	—	—	—	—	17
Benzodiazepine receptor — 0.01 M sodium phosphate, 0.2 M NaCl, pH 6.0, 0°C	—	—	—	—	18
Scrapie agent — 0.020 M Tris, pH 7.4, containing 1 mM EDTA and 0.2% Sarkosyl at 23°C	—[x]	—	—	—	19
Succinate dehydrogenase — 0.24 M sucrose, 0.100 M potassium phosphate, pH 6.0, at 0°C	—[y]	—	—	—	20
Ribulose bisphosphate carboxylase/oxygenase — 0.050 M Tris, 0.001 M EDTA, 0.020 M MgCl$_2$, pH 7.0, at 21°C	—	2,340	4.5[z]	—	21
Transferrin — 0.01 M potassium phosphate, 0.05 KCl, pH 6.1	—	—	54–72%[aa]	—	22
Dihydrofolate reductase — 0.05 M Tris, pH 7.5, at 10°C	—[bb]	29	6/7	50–500	23
RNA polymerase — 100 mM phosphate, pH 6.0, at ambient temp.	—	—	—	—	24
A$_1$ adenosine receptor — 20 mM potassium phosphate, pH 7.0, at ambient temp.	—	—	—	—	25
Lysyl oxidase — 200 mM potassium phosphate, pH 7.0, at ambient temp.	—	2.5 min^{-1} mM^{-1}	—	—	26

(continued)

TABLE 3.1 (continued)
Reaction of Diethylpyrocarbonate with Histidyl Residues in Proteins

Protein	Solvent and Temperature	Reagent Excess[a]	Extent of Modification[b]	Second-Order Rate Constant[c] (1 M^{-1} min^{-1})	Other Amino Acids Modified	Ref.
Malic enzyme	50 mM acetate, pH 6.0, with 1.0 mM EDTA at 25°C	—	2.8/5.0[cc] 1.8/5.0[cc]	—	—	27
Neurospora membrane ATPase	50 mM HEPES, pH 6.9, with 30% (v/v) glycerol	—	—	385–420 min^{-1} M^{-1}[dd]	—	28
Asparaginase	50 mM MES, pH 6.0, at 25°C	—	—	—	—	29
Ampicillin acylase	10 mM sodium phosphate, pH 7.0	—	—	—	—	30
Pancreatic lipase	100 mM phosphate, pH 6.0, at ambient temp.	—	—	—	—	31
Lipase fragment	100 mM phosphate, pH 6.0, at ambient temp.	—	—	—	—	32
Phenol hydroxylase	50 mM potassium phosphate, pH 6.0 at 4°C[ff]	—	—	—[ee]	—	33
Lysolecithin-lysolecithin acyltransferase	100 mM phosphate, pH 6.5	—	—	1.17 min^{-1} mM^{-1}[gg] 0.56 min^{-1} mM^{-1}	—	34
D-amino acid oxidase	50 mM potassium phosphate, pH 7.5, 30°C	—	—	254	—	35
Heparinase II (*Flavobacterium heparinum*)	Sodium phosphate	—	—	160[hh] 240[ii]	—	36
NADP-isocitrate dehydrogenase (*Cephalosporium acremonium*)	5 mM potassium phosphate, pH 7.6, with 2 mM EDTA	—	—	180	—	37
3-Deoxy-D-octulosonic acid 8-phosphate synthase	20 mM Tris-HCl, pH 7.6, 4°C	—	—	340[jj]	—	38
Horse radish peroxidase	100 mM phosphate, pH 6.5	—	—	0.66[kk]	—	39
Lacrimal gland peroxidase	100 mM potassium phosphate, pH 7.5, 25°C	—	—	216	—	40

[a] Moles per mole of protein.

[b] Residues His modified per total His in protein.

[c] For reaction with histidine.

[d] Inactivation was demonstrated to result from the modification of a single histidine residue.

[e] Assuming loss of activity is a direct indication of a single histidine modification.

[f] There was only partial recovery of activity on treatment with hydroxylamine (0.2 M, pH 7.0, 25°C). Two residues of histidine were lost as assessed by amino acid analysis after acid hydrolysis without loss of other amino acids, suggesting that disubstitution occurred on the imidazole ring of certain histidine residues.

[g] For inactivation of catalytic activity. A value of 51.6 M^{-1} sec^{-1} (3096 M^{-1} min^{-1}) was calculated for the pH-independent second-order rate constant.

[h] No direct determination of primary amino modification is reported. Activity is recovered by neutral hydroxylamine (0.09 M). Direct determination of tryptophan and tyrosine revealed no loss of these residues.

[i] Obtained from reaction at either 0 or 25°C.

[j] Data obtained at pH 6.6 [0.050 M N-(2-acetamido)iminodiacetic acid, ADA], 0°C. Two-phase reaction was observed.

[k] Spectral analysis did not indicate tyrosine modification. Possible primary amine modification was not determined. The loss in catalytic activity was reversed by 0.25 M hydroxylamine, pH 7.0.

[l] The second-order rate constant for the native enzyme is 10,920 M^{-1} min^{-1}. These values were obtained from the measurement of the rate of loss of enzyme activity.

[m] Addition of more diethylpyrocarbonate resulted in further increases in absorbance at 242 nm, suggesting disubstitution on the imidazole ring of histidyl residues. Spectral analysis did not suggest modification of tyrosine under the reaction conditions. Possible modification of primary groups was not assessed.

[n] The reaction of diethylpyrocarbonate with ribulose bisphosphate carboxylase shows saturation kinetics (k = 7.3 mM), suggesting "specific" binding of diethylpyrocarbonate to the enzyme before the reaction resulting in inactivation. Data are not presented to show a similar phenomenon with the actual reaction of histidyl residues in the enzyme.

[o] Activity was recovered by treatment with hydroxylamine (0.4 M NH$_2$OH, pH 7.0) for 48 h at 4°C increased activity from 55 to 89%; similar treatment at 25°C resulted in similar activity recovery in 1 h). The authors note that reaction of diethylpyrocarbonate with cysteine (N-acetylcysteine) also results in an increase in absorbance at 240 nm that is reversed by hydroxylamine. This reaction apparently occurs only in carboxylate buffers (e.g., acetate or succinate) and has been noted by other investigators.[9] The reaction product of diethylpyrocarbonate with cysteine is considerably less stable than N-carbethoxyimidazole derivatives.

[p] The crystalline enzyme preparation (7 nMol) [washed with 41% (NH$_4$)SO$_4$] was dissolved in 0.60 ml, 0.010 M Tris, pH 7.0, containing 5 mM MgCl$_2$, 0.100 M KCl, and 10 mM guanosine diphosphate. The pH of this solution was then adjusted to 6.0 with 1.0 M sodium cacodylate, 0.050 M MgCl$_2$.

[q] Per FMN (hence per dimer, therefore this would be four residues per tetramer).

[r] Determined from rate of loss of catalytic activity.

[s] No reaction at cysteine, tryptophan, or tyrosine is observed under these reaction conditions. Determined from rate of loss of thiamine binding activity.

[t] Thiomethylated at Cys-165 (reaction with methyl methanethiosulfonate). Enzyme remains catalytically active, but with reduced affinity for pyruvate and lactate.

(continued)

TABLE 3.1 (continued)
Reaction of Diethylpyrocarbonate with Histidyl Residues in Proteins

[u] Virtually identical values were obtained from direct measurement of histidine modification by spectroscopy and loss of catalytic activity.

[v] Diethylpyrocarbonate introduced as acetonitrile solution.

[w] Incorporation of radiolabeled diethylpyrocarbonate was closely related to extent of histidine modification as assessed by spectroscopy ($\epsilon = 3900$ M^{-1} cm^{-1} for monosubstituted derivative). Hydroxylamine treatment did not result in recovery of enzyme activity, although radiolabel was lost. The enzyme was inactivated by 10 mM hydroxylamine at neutral pH at 0°C. This is not an infrequent observation from our consideration of the literature in this area. Although very few investigators (we have not found any report) have examined the possibility of peptide bond cleavage with hydroxylamine at neutral pH, the possibility cannot be disregarded considering the cleavage of Asn-Gly bonds under more alkaline conditions.[17]

[x] Inactivation reversed by 0.100 to 0.5 M hydroxylamine. The pH of this reaction is not specified.

[y] Submitochondrial particles were used in this study. The inactivation produced by diethylpyrocarbonate is partially reversed by neutral (pH 7.0) hydroxylamine. The extent of activity recovery was dependent on hydroxylamine, with maximum activity recovery at 0.020 M hydroxylamine decreasing significantly at 0.115 M hydroxylamine.

[z] Stoichiometry determined by spectral analysis ($\epsilon = 3200$ M^{-1} cm^{-1} at 240 nm; 3.4 residues modified) is in reasonable agreement with amount of radiolabeled diethylpyrocarbonate incorporated (4.2).

[aa] Varied with species source of transferrin: human, 14:7; rabbit, 14:18; human lactotransferrin, 7:10; bovine lactotransferrin, 7:9; chicken ovotransferrin, 9:14.

[bb] There is no reaction with tyrosine under these conditions. Reaction at primary amine functions was not excluded. Only partial reactivation is obtained on treatment with hydroxylamine (ca. 50% recovery with 1.0 M hydroxylamine; no reaction at 0.1 M hydroxylamine).

[cc] Differences were noted between enzymes isolated from aged animals (old) and young animals (young). The difference in the extent of modification (5.0 residues are present in the native enzyme) is ascribed to oxidation occurring during the aging process.

[dd] Partial reactivation with hydroxylamine, but no detectable modification at tyrosyl residues.

[ee] Fast (A) and slow (B) reaction and reaction at nonhistidine residues as being involved in the loss of activity. K_A, 0.46 min^{-1}; K_B, 0.011 min^{-1}; and K_C, 0.031 min^{-1}.

[ff] Alternatively, 20 mM MES (pH 5.0 to 5.5) was used with 50 mM sodium phosphate.

[gg] Inactivation of hydrolytic reaction, 1.17 min^{-1} mM^{-1}; inactivation of transacylation, 0.56 min^{-1} mM^{-1}. It was concluded that two different histidine residues are necessary for hydrolysis and only one histidine residue for transacylation.

[hh] Heparin as substrate; corrected for reagent hydrolysis.

[ii] Heparan sulfate as substrate; corrected for reagent hydrolysis.

[jj] Rate corrected for reagent hydrolysis.

[kk] Based on the rate of activity loss which appears to be due to the disubstitution reaction.

References for Table 3.1

1. Burstein, Y., Walsh, K.A., and Neurath, H., Evidence of an essential histidine residue in thermolysin, *Biochemistry* 13, 205, 1974.
2. Cousineau, J., and Meighen, E., Chemical modification of bacterial luciferase with ethoxyformic anhydride: evidence for an essential histidyl residue, *Biochemistry* 15, 4992, 1976.
3. Jeng, T.-W., and Fraenkel-Conrat, H., Chemical modification of histidine and lysine residues of crotoxin, *FEBS Lett.* 87, 291, 1978.
4. McTigue, J.J., and van Etten, R.L., An essential active-site histidine residue in human prostatic acid phosphatase. Ethoxyformylation by diethylpyrocarbonate and phosphorylation by a substrate, *Biochim. Biophys. Acta* 523, 407, 1978.
5. Horiike, K., Tsuge, H., and McCormick, D.B., Evidence for an essential histidyl residue at the active site of pyridoxamine (pyridoxine)-5'-phosphate oxidase from rabbit liver, *J. Biol. Chem.* 254, 6638, 1979.
6. George, A.L., Jr., and Borders, C.L., Jr., Chemical modification of histidyl and lysyl residues in yeast enolase, *Biochim. Biophys. Acta* 569, 63, 1979.
7. Demaine, M.M., and Benkovic, S.J., Selective modification of rabbit liver fructose bisphosphatase, *Arch. Biochem. Biophys.* 205, 308, 1980.
8. Saluja, A.K., and McFadden, B.A., Modification of histidine at the active site of spinach ribulose bisphosphate carboxylase, *Biochem. Biophys. Res. Commn.* 94, 1091, 1980.
9. Garrison, C.K., and Himes, R.H., The reaction between diethylpyrocarbonate and sulfhydryl groups in carboxylate buffers, *Biochem. Biophys. Res. Commn.*, 67, 1251, 1975.
10. Jonák, J., and Rychlík, I., Chemical evidence for the involvement of histidyl residues in the functioning of *Escherichia coli* elongation factor Tu, *FEBS Lett.* 117, 167, 1980.
11. Meyer, S.E., and Cromartie, T.H., Role of essential histidine residues in 1-α-hydroxy acid oxidase from rat kidney, *Biochemistry* 19, 1874, 1980.
12. Nishimura, H., Sempuku, K., and Iwashima, A., Possible functional roles of carboxyl and histidine residues in a soluble thiamine-binding protein of *Saccharomyces cerevisiae*, *Biochim. Biophys. Acta* 668, 333, 1981.
13. Bond, M.D., Steinbrink, D.R., and Van Wart, H.E., Identification of essential amino acid residues in *Clostridium histolyticum* collagenase using chemical modification reactions, *Biochem. Biophys. Res. Commn.*, 102, 243, 1981.
14. Bloxham, D.P., The chemical reactivity of the histidine-195 residue in lactate dehydrogenase thiomethylated at the cysteine-165 residue, *Biochem. J.* 193, 93, 1981.
15. Phelps, D.C., and Hatefi, Y., Inhibition of the mitochondrial nicotinamide nucleotide transhydrogenase by dicyclohexylcarbodiimide and diethylpyrocarbonate, *J. Biol. Chem.* 256, 8217, 1981.
16. Cromartie, T.H., Sulfhydryl and histidinyl residues in the flavoenzyme alcohol oxidase from *Candida boidinii*, *Biochemistry*, 5416, 1981.
17. Phelps, D.C., and Hatefi, Y., Inhibition of D-β-hydroxybutyrate dehydrogenase by butanedione, phenylglyoxal and diethyl pyrocarbonate, *Biochemistry* 20, 459, 1981.
18. Burch, T.P., and Ticku, M.K., Histidine modification with diethylpyrocarbonate shows heterogeneity of benzodiazepine receptors, *Proc. Natl. Acad. Sci. USA*, 78, 3945, 1981.
19. McKinley, M.P., Masiarz, F.R., and Prusiner, S.B., Reversible chemical modification of the scrapie agent, *Science* 214, 1259, 1981.
20. Vik, S.B., and Hatefi, Y., Possible occurrence and role of an essential histidyl residue in succinate dehydrogenase, *Proc. Natl. Acad. Sci. USA* 78, 6749, 1981.
21. Saluja, A.K., and McFadden, B.A., Modification of the active site histidine in ribulose bisphosphate carboxylase/oxygenase, *Biochemistry* 21, 89, 1982.
22. Mazurier, J., Leger, D., Tordera, V., Montreuil, J., and Spik, G., Comparative study of the iron-binding properties of transferrins. Differences in the involvement of histidine residues as revealed by carbethoxylation, *Eur. J. Biochem.* 119, 537, 1981.

(continued)

TABLE 3.1 (continued)
Reaction of Diethylpyrocarbonate with Histidyl Residues in Proteins

23. Daron, H.H., and Aull, J.L., Inactivation of dihydrofolate reductase from *Lactobacillus casei* by diethylpyrocarbonate, *Biochemistry* 21, 737, 1982.

24. Abdulwajid, A.W., and Wu, F.Y.-H., Chemical modification of *Escherichia coli* RNA polymerase by diethyl pyrocarbonate: evidence of histidine requirement for enzyme activity and intrinsic zinc binding, *Biochemistry* 25, 8167, 1986.

25. Klotz, K.-N., Lohse, M.J., and Schwabe, U., Chemical modification of A1 adenosine receptors in rat brain membranes. Evidence for histidine in different domains of the ligand binding site, *J. Biol. Chem.* 263, 17522, 1988.

26. Gacheru, S.N., Trackman, P.C., and Kagan, H.M., Evidence for a functional role for histidine in lysyl oxidase catalysis, *J. Biol. Chem.* 263, 16704, 1988.

27. Gordillo, E., Ayala, A., F-Lobato, M., Bautista, J., and Machado, A., Possible involvement of histidine residues in the loss of enzymatic activity of rat liver malic enzyme during aging, *J. Biol. Chem.* 263, 8053, 1988.

28. Morjana, N.A., and Scarborough, G.A., Evidence for an essential histidine residue in the *Neurospora crassa* plasma membrane H+-ATPase, *Biochim. Biophys. Acta* 985, 19, 1989.

29. Bagert, U., and Röhm, K.-H., On the role of histidine and tyrosine residues in *E. coli* asparaginase. Chemical modification and ¹H-nuclear magnetic resonance studies, *Biochim. Biophys. Acta* 999, 36, 1989.

30. Kim, D.J., and Byun, S.M., Evidence for involvement of 2 histidine residues in the reaction of ampicillin acylase, *Biochem. Biophys. Res. Comm.* 166, 904, 1990.

31. De Caro, J.D., Guidoni, A.A., Bonicel, J.J., and Rovery, M., The histidines reacting with ethoxyformic anhydride in porcine pancreatic lipase: their relationships with enzyme activity, *Biochimie* 71, 1211, 1989.

32. De Caro, J.D., Rouimi, P., and Rovery, M., Hydrolysis of *p*-nitrophenyl acetate by the peptide chain fragment (335–449) of porcine pancreatic lipase, *Eur. J. Biochem.* 158, 601, 1986.

33. Sejlitz, T., and Neujahnr, H. Y., Chemical modification of phenol hydroxylase by ethoxyformic anhydride, *Eur. J. Biochem.* 170, 351, 1987.

34. Kinnunen, P.M., DeMichele, A., and Lange, L.G., Chemical modification of acyl-CoA:cholesterol *O*-acyltransferase. 1. Identification of acyl-CoA:cholesterol *O*-acyltransferase subtypes by differential diethyl pyrocarbonate sensitivity, *Biochemistry* 27, 7344, 1988.

35. Ranon, F., Chemical modification of histidyl residues in D-amino acid oxidase from *Rhodotorula gracilis*, *J. Biochem.* 118, 911–916, 1995.

36. Shrivers, S., Hu, Y., and Sasiesekharan, R., Heparinase II from *Flavobacterium heparinium*. Role of histidine residues in enzymatic activity as probed by chemical modification and site-directed mutagenesis, *J. Biol. Chem.* 273, 10160, 1999.

37. Olana, J. et al., Chemical modification of NADP-isocitrate dehydrogenase from *Cephalosporium acremonium*. Evidence of essential histidine and lysine groups at the active site, *Eur. J. Biochem.* 261, 640, 1999.

38. Sheflyan, G.Y. et al., Identification of essential histidine residues in 3-deoxy-D-manno-octulosonic acid 8-phosphate synthase: analysis by chemical modification and site-directed mutagenesis, *Biochemistry* 38, 14320, 1999.

39. Adak, S. et al., An essential role of active site arginine residue in iodide binding and histidine residue in electron transfer for iodide oxidation by horse radish peroxidase, *Mol. Cell. Biochem.* 218, 1, 2001.

40. Mazumdar, A. et al., Probing the role of active site histidine residue in the catalytic activity of lacrimal gland peroxidase, *Mol. Cell. Biochem.* 238, 21, 2002.

TABLE 3.2
Some Studies Using Both Oligonucleotide-Directed Mutagenesis and Site-Specific Chemical Modification to Study Histidine Function in Proteins

Protein	Ref.
Staphylococcus aureus α-toxin	1
4-Chlorobenzoyl-coenzyme A dehalogenase	2
Aminoglycosidase-3′-phosphotransferase Type III	3
Pyridine nucleotide transhydrogenase (*Escherichia coli*)	4
Heparinase II (*Flavobacterium heparinium*)	5
Galactose mutarotase	6
Maize branching enzymes	7
Murine bifunctional ATP sulfurylase/adenosine 5′-phosphosulfate kinase	8
Human liver arginase	9
3-Deoxy-D-manno-octulosonic acid 8-phosphate synthase	10
Methane monoxygenase component	11
Human ecto-nucleoside triphosphate diphosphohydrolase-3	12
Vinculin	13
Human pituitary glutaminyl cyclase	14
Apoptic nuclease C	15
ArsA ATPase	16
Human liver arginase	17
3-Hydroxyacyl ACP:CoA transacylase (*Pseudomonas putida*)	18
Rat kidney acylase	19
Plant vacuolar (II+) pyrophosphatase	20

References for Table 3.2

1. Menzies, B.E., and Kernalle, D.S., Site-directed mutagenesis of the alpha-toxin gene of *Staphylococcus aureus*. role of histidines in the toxin activity *in vitro* in murine model, *Infect. Immun.* 62, 1843, 1994.
2. Yang, G. et al., Identification of active-site residues essential to 4-chlorobenzoyl-coenzyme A dehalogenase catalysis by chemical modification and site directed mutagenesis, *Biochemistry* 38, 10879, 1996.
3. Thompson, P.R., Hughes, D.W., and Wright, G.D., Mechanism of aminoglycosidase 3′ phosphotransferase type IIIa: His188 is not a phosphate-accepting residue, *Chem. Biol.* 3, 747, 1996.
4. Bragg, P.D., and Hou, C., The role conserved histidine residues in the pyridine nucleotide transferase of *Escherichia coli*, *Eur. J. Biochem.* 241, 611, 1996.
5. Shriver, Z., Hu, Y., and Sasisekharan, R., Heparinase II from *Flavobacterium heparinium*. Role of histidine residues in enzymatic activity as probed by chemical modification and site-directed mutagenesis, *J. Biol. Chem.* 273, 10160, 1998.
6. Beebe, J.A., and Frey, P.A., Galactose mutarotase: purification, characterization, and investigation of two important histidine residues, *Biochemistry* 37, 14989, 1998.
7. Funane, K. et al., Analysis of essential histidine residues of maize branching enzymes by chemical modification and site-directed mutagenesis, *J. Protein Chem.* 17, 579, 1998.

TABLE 3.2 (continued)
Some Studies Using Both Oligonucleotide-Directed Mutagenesis and Site-Specific Chemical Modification to Study Histidine Function in Proteins

8. Deyrup, A.T. et al., Chemical modification and site-directed mutagenesis of conserved HxxH and PP-loop motif arginines ad histidines in the murine bifunctional ATP sulfurylase/adenosine-5′-phosphosufate kinase, *J. Biol. Chem.* 274, 28929, 1999.

9. Cervagjel, N. et al., Chemical modification and site-directed mutagenesis of human liver arginase: evidence that the imidazole group of histidine-141 is not involved in substrate binding, *Arch. Biochem. Biophys.* 371, 202, 1999.

10. Sheflyan, G.Y. et al., Identification of essential histidine residues in 3-deoxy-D-manno-octulosonic acid 8-phosphate synthase, *Biochemistry* 38, 14320, 1999.

11. Waller, B.J., and Lipscomb, J.D., Methane monooxygenase component alters the kinetics of steps throughout the catalytic cycle, *Biochemistry* 40, 2220, 2001.

12. Hicks-Berger, C.A., et al., The importance of histidine residues in human ecto-nucleoside triphosphate diphosphohydrolase-3 as determined by site-directed mutagenesis, *Biochim. Biophys. Acta* 1547, 72, 2001.

13. Miller, G.J., and Ball, E.H., Conformational change in the vinculin C-terminal depends on a critical histidine residue (His-906), *J. Biol. Chem.* 276, 28829, 2001.

14. Baleman, R.D., Jr., et al., Evidence for essential histidines in human pituitary glutaminyl cyclase, *Biochemistry* 40, 11246, 2001.

15. Meiss, G., Identification of functionally relevant histidine residues in apoptotic nuclease CAD, *Nucleic Acids Res.* 29, 3901, 2001.

16. Bhattacharjee, H., and Rosen, P.P., Structure-function analysis of the ArsA ATPase: contribution of histidine residues, *J. Bioenerg. Biomemb.* 33, 459, 2001.

17. Orellana, M.S. et al., Insights into the interaction of human liver arginase with tightly and weakly bound manganese ions by chemical modification and site-directed mutagenesis studies, *Arch. Biochem. Biophys.* 403, 155, 2002.

18. Hoffman, N. et al., Biochemical characterization of the *Pseudomonas putide* 3-hydroxyacyl ACP: CoA transacylase, which directs intermediates of fatty acid *de novo* biosynthesis, *J. Biol. Chem.* 277, 43926, 2002.

19. Durand, A. et al., Rat kidney acylase I: further characterization and mutation studies on the involvement of Glu147 in the catalytic process, *Biochimie* 85, 953, 2003.

20. Hsiao, Y.Y. et al., Roles of histidine residues in plant vacuolar H(+)-pyrophosphatase, *Biochim. Biophys. Acta* 1608, 199, 2004.

that the reaction was second order, with a rate constant of 12.5 M^{-1} sec^{-1} (750 M^{-1} min^{-1}). The loss of catalytic activity appeared to be correlated with the modification of one of the four histidyl residues in this protein.

Saluja and McFadden[83] explored the reaction of diethylpyrocarbonate with spinach ribulose bisphosphate carboxylase. One interesting observation is that the plot of half-inactivation time vs. the reciprocal of diethylpyrocarbonate concentration suggested that saturation kinetics existed consistent with the "affinity" binding of reagent before protein modification. The study by Bloxham[84] on the reactivity of the active-site histidine in lactate dehydrogenase is of interest. The rate of reaction of the histidine residue in the native enzyme was compared to the thiomethyl derivative (prepared by reaction with methyl methanethiosulfonate). There was a substantial decrease in the nucleophilic character of the active-site histidine (His-195). Cromartie[85] examined the modification of alcohol oxidase with diethylpyrocarbonate in 0.050 M sodium phosphate, pH 7.5, 0°C. The UV difference spectrum of the enzyme before (a) and after (b) with the addition of diethylpyrocarbonate did not provide evidence for tyrosine modification under these reaction conditions. Treatment with neutral hydroxylamine (0.010 to 0.100 M) did not result in the recovery of catalytic activity, although most of the radiolabeled reagent was removed ([1-^{14}C]-diethylpyrocarbonate). It was observed that 0.010 M hydroxylamine caused an 80% loss of alcohol oxidase activity at 0°C. As mentioned previously, the reaction of diethylpyrocarbonate at pH values above 7.0 increases the possibility of amino group modification. This is demonstrated by the observation of Bond and co-workers[86] on the reaction of diethylpyrocarbonate with a bacterial collagenase. Reaction of the enzyme with diethylpyrocarbonate resulted in a loss of catalytic activity. Reaction with hydroxylamine did not markedly restore enzymatic activity. Hydroxylamine by itself did not have a deleterious effect on catalytic activity, as treatment with this reagent did restore activity lost on the modification of tyrosyl residues with N-acetyl-imidazole. These investigators consider modification of the ε-amino groups of lysine with diethylpyrocarbonate to be a more likely cause for the irreversible loss of enzymic activity. Daron and Aull[87] studied the reaction of diethylpyrocarbonate with dihydrofolate reductase (Lactobacillus casei). Catalytic activity was lost on reaction with diethylpyrocarbonate, but partially recovered on reaction with 1.0 M hydroxyl-amine (pH 7.5) but not with 0.1 M hydroxylamine (pH 7.5.). Clearly, the reaction of proteins with hydroxylamine after modification with diethylpyrocarbonate must be carefully studied to obtain meaningful results.

Lin and co-workers[88] studied the inactivation of an aminopeptidase isolated from Pronase E with diethylpyrocarbonate. Reaction of the aminopeptidase with diethyl-pyrocarbonate in the absence of calcium ions (50 mM sodium phosphate, pH 6.0) results in the modification of two histidine residues. When the assay (L-leucine-p-nitroanilide) is performed in the absence of calcium ions, activity is lost with the modification of one histidine residue; assay in the presence of calcium ions (5 mM) demonstrates that the loss of activity is associated with the modification of two histidine residues. The reaction of the aminopeptidase with diethylpyrocarbonate (50 mM sodium phosphate, pH 8.0, 25°C) was associated with the modification of two histidine residues; the presence of 5 mM calcium ions reduced the extent of modification to one histidine residue. It is suggested that the loss of activity observed on reaction with diethylpyrocarbonate is due to the modification of two histidine residues (His(1), pK_a 6.9; His(2), pK_a 7.7). Reaction of the modified enzyme with hydroxylamine (1 M in 50 mM sodium phosphate, pH 7.0, 4 h) resulted in the partial recovery of enzyme activity. As a result of differential recovery of enzyme activity depending on whether the assay is performed in the presence or absence of calcium ions, it was suggested that hydroxylamine caused the decarboethoxylation of His(2) and not His(1). Acebal and co-workers[89] studied the modification of histidine residues in a D-amino acid oxidase from *Rhodotorula gracilis*. Complete reversal of inactivation was seen with 0.05 M NaOH, whereas less recovery (80%) was seen with 0.5 M hydroxylamine in 50 mM sodium phosphate, pH 7.5 (20 min at 30°C). An excellent difference spectrum for the reaction of diethylpyrocarbonate with the D-amino acid oxidase is presented. When beef liver carnitine octanolyl transferase was reacted with diethylpyrocarbonate (20 mM potassium phosphate, pH 6.0, 4°C), the loss of activity was biphasic.[90] The residue modified in the first phase is not a histidine residue, but more likely a serine residue with a lysine modified in the second, slower phase of inactivation. Histidine residues (2) are modified during the reaction, but both are modified at a slower rate (k_{obs} = 0.07 min^{-1}) than that for the first phase of inactivation (k_{obs} = 3.25 min^{-1}). Diethylpyrocarbonate has been used to study the role of histidyl residues in heparin binding by rat selanoprotein P.[91] Modification was performed in 0.2 M sodium phosphate, pH 7.0, for 2 h. The presence of heparin blocked the modification reaction, as measured by change in absorbance at 240 nm. The reaction was dependent on pH, with the rate of histidine modification increasing with increasing pH. Modified histidine residues were determined by mass spectrometry.

In addition to the direct modifications of lysine, cysteine, and tyrosine by carboethoxylation, which are occasionally observed as side reactions in the use of diethylpyrocarbonate for the modification of histidine residues in proteins, an additional side reaction involving lysyl residues has been observed. Sams and Matthews[92] reported isopeptide bond formation between the ε-amino group of lysine and an adjacent carboxylic acid. This reaction has been previously observed with ribonuclease.[66] A reaction mechanism has not been described, but it likely involves transient diethylpyrocarbonate modification of the carboxyl group.[93]

An increasing number of studies have used diethylpyrocarbonate to study more complex systems.[94–99] Given the studies discussed previously, it would be wise to be extremely cautious in the interpretation of these results without strong analytical data to support modification of histidine in these studies.

4-Hydroxy-2-nonenal and 4-oxo-2-nonenal are aldehydes derived from the oxidation of lipids.[100] These aldehydes react with proteins to form a variety of products.[101–103] Mass spectrometry can be used to identify the various products.[104] In this study, reduction with borohydrides was evaluated as an approach to stabilizing the alkylated products in the protein. In the absence of reducing agent, 4-hydroxy-2-nonenal modification of histidine in the oxidized B chain of insulin proceeded via a Michael addition as the predominant reaction. If the reducing agent, sodium cyanoborohydride, was added at the start of the reaction, the mechanism shifted to Schiff base formation at the amino-terminal residue. Monoclonal antibodies to the 4-hydroxy-2-nonenal adduct with histidine have been developed.[105,106] As mentioned previously, the reaction of 4-hydroxy-2-nonenal with histidine proceeds via a Michael-type addition reaction,[107,108] yielding a cyclic hemiacetal.[109,110] Doorn and Peterson[111] compared the reaction of 4-hydroxy-2-nonenal and 4-oxo-2-nonenal with model peptides containing different nucleophilic amino acids. They observed that cysteine was more reactive with either aldehyde than was histidine or lysine. Histidine was more reactive than lysine, whereas arginine was not modified by 4-hydroxy-2-nonenal but did react with 4-oxo-2-nonenal.

Several groups have reported the modification of histidine with Woodward's reagent K (*N*-ethyl-5-phenylisoxaxolium-3-sulfonate).[112–114] This reagent is usually considered specific for carboxyl groups (see Chapter 5). One group[114] observed saturation kinetics in the inactivation of an acylphosphatase with Woodward's reagent K, suggesting the formation of a reversible complex before the inactivation reaction.

REFERENCES

1. Turk, T., and Macek, P., The role of lysine, histidine and carboxyl residues in biological activity of equinatoxin II, a pore forming polypeptide from the sea anemone *Actinia equina* L., *Biochim. Biophys. Acta* 1119, 5, 1992.
2. Mattsson, P., Pohjalainen, T., and Korpela, T., Chemical modification of cyclomaltodextrin glucanotransferase from *Bacillus circulans* var. *alkalophilus*, *Biochim. Biophys. Acta* 1122, 33, 1992.
3. Park, I.-S., and Hausinger, R.P., Diethylpyrocarbonate reactivity of *Klebsiella aerogenes* urease: effect of pH and active site ligands on the rate of inactivation, *J. Protein Chem.* 12, 51, 1993.
4. Bhattacharyya, D.K., Bandyopadhyay, U., and Bannerjee, R.K., Chemical and kinetic evidence for an essential histidine residue in the electron transfer from aromatic donor to horseradish peroxidase compound I, *J. Biol. Chem.* 268, 22292, 1993.
5. Basuki, W., Furuichi, K., Iizuka, M., Ito, K., and Minamiura, N., Differences between isozymes of glucose isomerase of *Streptomyces phaeochromogenes* by chemical modification, *Biosci. Biotech. Biochem.* 57, 1341, 1993.
6. Britten, C.J., and Bird, M.T., Chemical modification of an alpha 3-fusosyltransferase: definition of amino acid residues essential for enzyme activity, *Biochim. Biophys. Acta* 1334, 57, 1997.
7. Olano, J. et al., Chemical modification of NADP-isocitrate dehydrogenase from *Cephalosporium acremanonium*: evidence of essential histidine and lysine groups at the active site, *Eur. J. Biochem.* 261, 640, 1999.
8. Adak, S. et al., An essential role of active site arginine residue in iodide binding and histidine residue in electron transfer for iodide oxidation by horseradish peroxidase, *Mol. Cell. Biochem.* 218, 1, 2001.
9. Mardanyan, S.S. et al., Interaction of adenosine deaminase with inhibitors: chemical modification by diethyl pyrocarbonate, *Biochemistry (Moscow)* 67, 770, 2002.
10. Bond, J.S., Francis, S.H., and Park, J.H., An essential histidine in the catalytic activities of 3-phosphoglyceraldehyde dehydrogenase, *J. Biol. Chem.* 245, 1041, 1970.
11. Levine, K.L. et al., Carbonyl assay for determination of oxidatively modified proteins, *Methods Enzymol.* 233, 346, 1994.
12. Resnick, A.Z., and Packer, L., Oxidative damage to proteins: spectrophotometric method for carbonyl assay, *Methods Enzymol.* 233, 357, 1994.
13. Requena, J.R. et al., Copper-catalyzed oxidation of the recombinant SHa (29-231) prion protein, *Proc. Natl. Acad. Sci. USA* 98, 7170, 2001.
14. Schöneich, C., and Williams, T.D., Cu(II)-catalyzed oxidation of β-amyloid peptide targets His[13] and His[14] over His[6]: detection of 2-oxo-histidine by HPLC-MS/MS, *Chem. Res. Toxicol.* 15, 717, 2002.

15. Ghezzi, P., and Bonetto, V., Redox proteomics: identification of oxidatively modified proteins, *Proteomics* 3, 1145, 2003.
16. Rashidzaden, H. et al., Solution structure and interdomain interactions of the *Saccharomyces cerevesiae* "TATA binding protein" (TBP) proved by radiolytic protein footprinting, *Biochemistry* 42, 3655, 2003.
17. Xu, G., Takamoto, K., and Chance, M.R., Radiolytic modification of basic amino acid residues in peptides: probes for examining protein-protein interactions, *Anal. Chem.* 75, 6995, 2003.
18. Fahnestock, S.R., Evidence of the involvement of a 50S ribosomal protein in several active sites, *Biochemistry* 14, 5321, 1975.
19. Auron, P.E., Erdelsky, K.J., and Fahnestock, S.R., Chemical modification studies of a protein at the peptidyltransferase site of the *Bacillus stearothermophilus* ribosome. The 50S ribosomal subunit is a highly integrated functional unit, *J. Biol. Chem.* 253, 6893, 1978.
20. Dohme, F., and Fahnestock, S.R., Identification of proteins involved in the peptidyl transferase activity of ribosomes by chemical modification, *J. Mol. Biol.* 129, 63, 1979.
21. Cerna, J., and Rychlik, I., Photoinactivation of peptidyl transferase binding sites, *FEBS Lett.* 102, 277, 1979.
22. Nakamura, S., and Kaziro, Y., Selective photooxidation of histidine residues in polypeptide chain elongation factor Tu from *E. coli*, *J. Biochem. (Tokyo)* 90, 1117, 1981.
23. Funakoshi, T., Abe, M., Sakata, M., Shoji, S., and Kubota, Y., The functional site of placental anticoagulant protein: essential histidine residue of placental anticoagulant protein, *Biochem. Biophys. Res. Commn.* 168, 125, 1990.
24. Kuno, S., Fukui, S., and Toraya, T., Essential histidine residues in coenzyme B12-dependent diol dehydrase: dye-sensitized photooxidation and ethoxycarbonylation, *Arch. Biochem. Biophys.* 277, 211, 1990.
25. Medina, M.M. et al., Involvement of essential histidine residue(s) in the activity of Ehrlich cell plasma membrane NADH-ferricyanide oxidoreductase, *Biochim. Biophys. Acta* 1190, 20, 1994.
26. Chung, S.H. et al., Metal-catalyzed photooxidation of histidine in human growth hormone, *Anal. Biochem.* 244, 221, 1997.
27. Moor, A.C. et al., *In vitro* fluence rate effects in photodynamic reactions with AlPcS4 as sensitizer, *Photochem. Photobiol.* 66, 860, 1997.
28. Schey, K.L. et al., Photooxidation of lens alpha-crystallin by hypericin (active ingredient of St. John's wort), *Photochem. Photobiol.* 72, 200, 2000.
29. Beghetto, C. et al., Implications of the generation of reactive oxygen species by photoactivated calcein for mitochondrial studies, *Eur. J. Biochem.* 267, 5585, 2000.
30. Roberts, J.E. et al., Photooxidation of lens proteins with xanthurenic acid: a putative chromaphore for cataractogenesis, *Photochem. Photobiol.* 74, 740, 2001.
31. Inagami, T., and Hatano, H., Effect of alkylguanidines on the inactivation of trypsin by alkylation and phosphorylation, *J. Biol. Chem.* 244, 1176, 1969.
32. Stark, G.R., Stein, W.H., and Moore, S., Relationships between the conformation of ribonuclease and its reactivity toward iodoacetate, *J. Biol. Chem.* 236, 436, 1961.
33. Heinrikson, R.L., Stein, W.H., Crestfield, A.M., and Moore, S., The reactivities of the histidine residues at the active site of ribonuclease toward halo acids of different structures, *J. Biol. Chem.* 240, 2921, 1965.
34. Fruchter, R.G., and Crestfield, A.M., The specific alkylation by iodoacetamide of histidine 12 in the active site of ribonuclease, *J. Biol. Chem.* 242, 5807, 1967.

35. Lin, M.C., Stein, W.H., and Moore, S., Further studies on the alkylation of the histidine residues of pancreatic ribonuclease, *J. Biol. Chem.* 243, 6167, 1968.

36. Liu, T.Y., Demonstration of the presence of a histidine residue at the active site of streptococcal proteinase, *J. Biol. Chem.* 242, 4029, 1967.

37. Kettner, C., and Shaw, E., Inactivation of trypsin-like enzymes with peptides of arginine chloromethyl ketone, *Methods Enzymol.* 80, 826, 1981.

38. Bock, P.E., Active-site-selective labeling of blood coagulation proteinases with fluorescence probes by the use of thioester peptide chloromethyl ketones. II. Properties of thrombin derivatives as reporters of prothrombin fragment 2 binding and specificity of the labeling approach for other proteinases, *J. Biol. Chem.* 267, 14974, 1992.

39. Bock, P.E., Active site selective labeling of serine proteases with spectroscopic probes using thioester peptide chloromethyl ketones: demonstration of thrombin labeling using N^α [T-[(acetylthio)acetyl]-D-Phe-Pro-Arg-Ch$_2$Cl, *Biochemistry* 27, 6633, 1988.

40. Williams, E.B., Krishnaswamy, S., and Mann, K.G., Zymogen/enzyme discrimination using peptide chloromethyl ketones, *J. Biol. Chem.* 264, 7536, 1989.

41. Plapp, B.V., Application of affinity labeling for studying structure and function of enzymes, *Methods Enzymol.* 87, 469, 1992.

42. Pechenov, A. et al., Potential transition state analogue inhibitors for the penicillin-binding proteins, *Biochemistry* 42, 579, 2003.

43. Wylie, D.C. et al., Carboxyalkylated histidine is a pH dependent product of pegylation with SC-PEG, *Pharm. Res.* 18, 1354, 2001.

44. Fohlman, J., Eaker, D., Dowdall, M.J., Lüllmann-Rauch, R., Sjödin, T., and Leander, S., Chemical modification of taipoxin and the consequences for phospholipase activity, pathophysiology, and inhibition of high-affinity choline uptake, *Eur. J. Biochem.* 94, 531, 1979.

45. Halpert, J., Eaker, D., and Karlsson, E., The role of phospholipase activity in the action of a presynaptic neurotoxin of *Notechis scutatus scutatus* (Australian Tiger Snake), *FEBS Lett.* 61, 72, 1976.

46. Yang, C.C., and King, K., Chemical modification of the histidine residue in basic phospholipase A2 from the venom of *Naja nigricollis*, *Biochim. Biophys. Acta* 614, 373, 1980.

47. Volwerk, J.J., Pieterson, W.A., and de Haas, G.H., Histidine at the active site of phospholipase A2, *Biochemistry* 13, 1446, 1974.

48. Verheij, H.M., Volwerk, J.J., Jansen, E.H.J.M., Puyk, W.C., Dijkstra, B.W., Drenth, J., and de Haas, G.H., Methylation of histidine-48 in pancreatic phospholipase A2. Role of histidine and calcium ion in the catalytic mechanism, *Biochemistry* 19, 743, 1980.

49. Sakori, T. et al., Phospholipase A(2) activity in non-glycated and glycated low density lipoproteins, *Biochim. Biophys. Acta* 1301, 85, 1996.

50. Diaz-Oreino, C., and Gutierrez, J.M., Chemical modification of histidine and lysine residues of myotoxic phospholipasesA2 isolated from *Bothrops asper* and *Bothrops godmani* snake venoms: effects on enzymatic and pharmacological properties, *Toxion* 35, 241, 1997.

51. Battaglia, E., and Radominska-Pandya, A., A functional role for histidyl residues of the UDP-glucuronic acid carrier in rate liver endoplasmic reticulum membranes, *Biochemistry* 37, 258, 1998.

52. Toyama, M.M. et al., Amino acid sequence of piratoxin-I, a myotoxin from *Bothrops piragai* snake venom, and its biological activity after alkylation with *p*-bromophenacyl bromide, *J. Protein Chem.* 17, 713, 1998.

53. de Vet, E.C., and van den Bosch, H., Characterization of recombinant guinea pig alkyl-dihydroxyacetonephosphate synthase expressed in *Escherichia coli*: kinetics, chemical modification, and mutagenesis, *Biochim. Biophys. Acta* 1436, 299, 1999.

54. Dorsch, S. et al., The VP1 unique region of parvovirus B19 and its constituent phospholipase A2-like activity, *J. Virol.* 76, 2014, 2002.

55. Glick, B.R., The chemical modification of *Escherichia coli* ribosomes with methyl *p*-nitrobenzenesulfonate. Evidence for the involvement of a histidine residue in the functioning of the ribosomal peptidyl transferase, *Can. J. Biochem.* 58, 1345, 1980.

56. Marcus, J.P., and Dekker, E.E., Identification of a second active site residue in *Escherichia coli* L-threonine dehydrogenase: methylation of histidine-90 with methyl-*p*-nitrobenzenesulfonate, *Arch. Biochem. Biophys.* 316, 413, 1995.

57. Johnson, A.R., and Dekker, E.E., Site-directed mutagenesis of histidine-90 in *Escherichia coli* L-threonine dehydrogenase alters its substrate specificity, *Arch. Biochem. Biophys.* 351, 8, 1998.

58. Nishino, T., Massey, V., and Williams, C.H., Jr., Chemical modifications of D-amino acid oxidase. Evidence for active site histidine, tyrosine, and arginine residues, *J. Biol. Chem.* 255, 3610, 1980.

59. Gadda, G., Beretta, G.L., and Pilone, M.S., Reactivity of histidyl residues in D-amino acid oxidase from *Rhodotorula gracilis*, *FEBS Lett.* 363, 307, 1995.

60. Morishima, I., Shiro, Y., Adachi, S., Yano, Y., and Orii, Y., Effect of the distal histidine modification (cyanation) of myoglobin on the ligand binding kinetics and the heme environmental structures, *Biochemistry* 28, 7582, 1989.

61. Shiro, Y., and Morishima, I., Modification of the heme distal side chain in myoglobin by cyanogen bromide. Heme environmental structures and ligand binding properties of the modified myoglobin, *Biochemistry* 23, 4879, 1984.

62. Hefford, M.A., and Kaplan, H., Chemical properties of the histidine residue of secretin: evidence for a specific intramolecular interaction, *Biochim. Biophys. Acta* 998, 267, 1989.

63. 21 CFR 189.150, Diethyl pyrocarbonate (DEPC), August 2, 1972.

64. Delfini, C. et al., Fermentability of grape must after inhibition with dimethydicarbonate, *J. Agric. Food Chem.* 50, 5601, 2002.

65. Melchior, W.B., Jr., and Fahrney, D., Ethoxyformylation of proteins. Reaction of ethoxyformic anhydride with α-chymotrypsin, pepsin and pancreatic ribonuclease at pH 4, *Biochemistry* 9, 251, 1970.

66. Wolf, B., Lesnaw, J.A., and Reichmann, M.E., A mechanism of the irreversible inactivation of bovine pancreatic ribonuclease by diethylpyrocarbonate. A general reaction of diethylpyrocarbonate with proteins, *Eur. J. Biochem.* 13, 519, 1970.

67. Miles, E.W., Modification of histidyl residues in proteins by diethylpyrocarbonate, *Methods Enzymol.* 47, 431, 1977.

68. Welsch, D.J., and Nelsestuen, G.L., Irreversible degradation of histidine-96 of prothrombin fragment 1 during protein acetylation: another unusually reactive site in the kringle, *Biochemistry* 27, 7513, 1988.

69. Glocker, M.O. et al., Selective biochemical modification of function residues in recombinant human macrophage colony-stimulating factor beta (rhM-CSFbeta): identification by mass spectrometry, *Biochemistry* 35, 14625, 1996.

70. Dage, J.L., Sun, H., and Halsall, H.B., Determination of diethyl pyrocarbonate-modified amino acid residues in alpha-1-acid glycoprotein by high-performance liquid chromatography electrospray ionization mass spectrometry and matrix-assisted laser desorption/ionization time-of-flight mass spectrometry, *Anal. Biochem.* 257, 176, 1998.

71. Kalkum, M., Prxybylski, M., and Glocker, M.O., Structural characterization of functional histidine resiues and carbethoxylated derivates in peptides and proteins by mass spectrometry, *Bioconjug. Chem.* 9, 226, 1998.

72. Krell, T. et al., Chemical modification monitored by electrospray mass spectrometry: a rapid and simple method for identifying and studying functional residues in enzymes, *J. Pept. Res.* 51, 201, 1998.

73. Qin, K. et al., Mapping Cu(II) binding sites in prion protein by diethyl pyrocarbonate modification of matrix-assisted laser desorption time-of-flight (MALDI-TOF) mass spectrometric footprinting, *J. Biol. Chem.* 277, 1981, 2002.

74. Willard, B.B., and Kintes, M., Effects of internal histidine residues on the collision-induced fragmentation of triply protonated tryptic peptides, *J. Am. Soc. Mass Spectrom.* 12, 1262, 2001.

75. Altman, J., Lipka, J.J., Kuntz, I., and Waskell, L., Identification by proton nuclear magnetic resonance of the histidines in cytochrome b5 modified by diethyl pyrocarbonate, *Biochemistry* 28, 7516, 1989.

76. Bycroft, M., and Fersht, A.R., Assignment of histidine resonances in the ^{1}H NMR (500 MHz) spectrum of subtilisin BPN' using site-directed mutagenesis, *Biochemistry* 27, 7390, 1988.

77. Morjana, N.A., and Scarborough, G.A., Evidence for an essential histidine residue in the *Neurospora crassa* plasma membrane H+-ATPase, *Biochim. Biophys. Acta* 985, 19, 1989.

78. Holbrook, J.J., and Ingram, V.A., Ionic properties of an essential histidine residue in pig heart lactate dehydrogenase, *Biochem. J.* 131, 729, 1973.

79. Cousineau, J., and Meighen, E., Chemical modification of bacterial luciferase with ethoxyformic anhydride: evidence for an essential histidyl residue, *Biochemistry* 15, 4992, 1976.

80. Roosemont, J.L., Reaction of histidine residues in proteins with diethylpyrocarbonate: differential molar absorptivities and reactivities, *Anal. Biochem.* 88, 314, 1978.

81. Vangrysperre, W., Ampe, C., Kersters-Hilderson, H., and Tempst, P., Single active-site histidine in D-xylose isomerase from *Streptomyces violaceoruber*. Identification by chemical derivatization and peptide mapping, *Biochem. J.* 263, 195, 1989.

82. Horiike, K., Tsuge, H., and McCormick, D.B., Evidence for an essential histidyl residue at the active site of pyridoxamine (pyridoxine)-5'-phosphate oxidase from rabbit liver, *J. Biol. Chem.* 254, 6638, 1979.

83. Saluja, A.K., and McFadden, B.A., Modification of histidine at the active site of spinach ribulose bisphosphate carboxylase, *Biochem. Biophys. Res. Commn.* 94, 1091, 1980.

84. Bloxham, D.P., The chemical reactivity of the histidine-195 residue in lactate dehydrogenase thiomethylated at the cysteine-165 residue, *Biochem. J.* 193, 93, 1981.

85. Cromartie, T.H., Sulfhydryl and histidinyl residues in the flavoenzyme alcohol oxidase from *Candida boidinii*, *Biochemistry* 20, 5416, 1981.

86. Bond, M.D., Steinbrink, D.R., and Van Wart, H.E., Identification of essential amino acid residues in *Clostridium histolyticum* collagenase using chemical modification reactions, *Biochem. Biophys. Res. Commn.* 102, 243, 1981.

87. Daron, H.H., and Aull, J.L., Inactivation of dihydrofolate reductase from *Lactobacillus casei* by diethyl pyrocarbonate, *Biochemistry* 21, 737, 1982.

88. Yong, S.-H., Wu, C.-H., and Lin, W.-Y., Chemical modification of amino peptidase isolated from Pronase. *Biochem J.* 302, 595, 1994.

89. Ranon, F. et al., Chemical modification of histidyl residues in D-amino acid oxidase from *Rhodotorula gracilis*, *J. Biochem.* 118, 911, 1995.

90. a'Bháird, N.N., Yankovskaya, V., and Ramsay, R.R., Active site residues of beef liver carnitine octanoyl transferase (COT) and carnitine palmitoyl transferase (CPT-II), *Biochem. J.* 330, 1029, 1998.
91. Hondal, K.J., et al., Heparin binding histidine and lysine residues of rat selenoprotein P. *J. Biol. Chem.* 276, 15823, 2001.
92. Sams, C.F., and Matthews, K.S., Diethyl pyrocarbonate reaction with the lactose repressor protein affects both inducer and DNA binding, *Biochemistry* 27, 2277, 1988.
93. Wold, F., Bifunctional reagents, *Methods Enzymol.* 25, 423, 1972.
94. Meves, H, Slowing of ERG current deactivation in NG108-15 cells by the histidine-specific reagent diethyl pyrocarbonate, *Neuropharmacology* 41, 220-228, 2001.
95. Monkelow, T.J., and Henderson, L.M., Inhibition of the neutrophils NADPH oxidase and associate H+ channel by diethyl pyrocarbonate, a histidine-modifying agent: evidence for at least two target sites, *Biochem. J.* 358, 315, 2001.
96. Seebungkert, B., and Lynch, J.W., A common inhibitory binding site for zinc and oderants at the voltage-gated K(+) channel of rat olfactory receptor neurons, *Eur. J. Neurosci.* 14, 353, 2001.
97. Rondel, M., Ortega, J.M., and Losada, M., Factors determining the special redox properties of photosynthetic cytochrome b559. *Eur. J. Biochem.* 268, 4961, 2001.
98. Verri, A. et al., Molecular characteristics of small intestinal and renal brush border thiamin transporters in rats, *Biochim. Biophys. Acta* 1558, 187, 2002.
99. Hasse, H., and Beyersmann, D., Intracellular zinc distribution and transport in C_6 rat glioma cells, *Biochem. Biophys. Res. Commn.* 296, 923, 2002.
100. Uchida, K., 4-Hydroxy-2-nonenal: a product and mediator of oxidative stress, *Prog. Lipid Res.* 42, 318, 2003.
101. Hidalgo, F.J., Alaiz, M., and Zamero, R., A spectrophotometric method for the determination of proteins damaged by oxidized lipids, *Anal. Biochem.* 262, 129, 1998.
102. Refsgaard, H.H.F., Tsai, L., and Stadtman, E.R., Modification of proteins by poly-unsaturated fatty acid peroxidation products, *Proc. Natl. Acad. Sci. USA* 97, 611, 2000.
103. Oe, T. et al., A novel lipid peroxide-derived cyclic covalent modification to histone H4, *J. Biol. Chem.* 278, 42098, 2003.
104. Fenaiile, F., Tabet, J.C., and Guy, P.A., Identification of 4-hydroxy-2-nonenal-modi-fied peptides within unfractionated digests using matrix-assisted laser desorption/ion-ization time-of-flight mass spectrometry, *Anal. Chem.* 76, 867, 2004.
105. Toyokuni, S. et al., The monoclonal antibody specific for the 4-hydroxy-2-nonenal histidine adduct, *FEBS Lett.* 359, 189, 1995.
106. Waeg, G., Dimsity, G., and Esterhaven, H., Monoclonal antibodies for detection of 4-hydoxy-nonenal modified proteins, *Free Radic. Res.* 25, 149, 1996.
107. Bruennen, B.A., Jones, H.V., and German, J.B., Maximum entropy deconvolution of heterogeneity in protein modification: protein adducts of 4-hydroxy-2-nonenal, *Rapid Commn. Mass Spectrom.* 8, 509, 1994.
108. Uchida, K. et al., Michael addition-type 4-hydroxy-2-nonenal adducts in modified low-density lipoproteins: markers for atherosclerosis, *Biochemistry* 33, 12487, 1994.
109. Tsai, L., and Sokoloski, F.A., The reaction of 4-hydroxy-2-nonenal with N-α-acetyl-L-histidine, *Free Radic. Biol. Med.* 19, 39, 1995.
110. Hashimoto, M. et al., Structural basis of protein-bound endogenous aldehydes. Chem-ical and immunochemical characterization of configurational isomers of a 4-hydroxy-2-nonenal-histidine adduct, *J. Biol. Chem.* 278, 5044, 2003.

111. Doorn, J.A., and Peterson, D.R., Covalent modification of amino acid nucleophiles by the lipid peroxidation products 4-hydroxy-2-nonenal and 4-oxo-2-nonenal, *Chem. Res. Toxicol.* 15, 1445, 2002.

112. Johnson, A.R., and Dekker, E.E., Woodward's reagent K inactivation of *Escherichia coli* L-threonine dehydrogenase: increased absorbance at 340-350 nm is due to modification of cysteine and histidine residues, not asparate or glutamate carboxyl groups, *Protein Sci.* 5, 382, 1996.

113. Bustos, P. et al., Woodward's reagent K reacts with histidine and cysteine residues in *Escherichia coli* and *Saccharomyces cerevisiae* phosphoenolpyruvate carboxykinase, *J. Protein Chem.* 15, 467, 1996.

114. Paoli, P. et al., Mechanism of acylphosphatase inactivation by Woodward's reagent K, *Biochem J.* 328, 855, 1997.

4 The Modification of Arginine

The group-specific modification of arginine is relatively easy to achieve; it is somewhat more difficult to obtain the site-specific modification of arginine. Arginine residues are usually involved in binding sites rather than catalytic sites. Thus, modification of arginine rarely results in the total loss of activity. As a result, it is critical to take residual activity into consideration in determining the kinetics of inactivation. The author has found the approach developed by Levy and co-workers[1] quite useful. This approach has been used in a number of studies on the site-specific modification of proteins.[2-8] The reader is also referred to the excellent review by Plapp.[9]

Present approaches to the site-specific modification of arginyl residues in proteins use three reagents: phenylglyoxal (and derivatives such as p-hydroxyphenyl-gloxal),[10] 2,3-butanedione,[11] and 1,2-cyclohexanedione.[12] A review of the literature from 1994 to 2004 suggests that phenylglyoxal is the most extensively used reagent for the site-specific chemical modification of arginine in proteins. Other reagents for the site-specific modification of arginine have been described, but their current use is negligible. The reader is referred to an earlier work for a more complete listing of reagents.[13]

The extent of arginine modification is generally determined by amino acid analysis after acid hydrolysis, but conditions generally need to be modified to prevent loss of the arginine derivative.[12] The Sakaguchi reaction[14,15] has been used, after acid hydrolysis, to determine the extent of arginine modification by 2,3-butanedione and 1,2-cyclohexanedione in rat liver microsomal stearoyl-coenzyme A desaturase.[16] Reference to the use of the Sakaguchi reaction for the analysis of proteins has not been found for 1994–2004. The Sakaguchi reaction has been used in histochemical studies[17] and for the assay guanidine compounds (guanidinoacetic acid) in screening for hepatic guanidinoacetate-methyl transferase activity.[18,19]

A fluorometric method for the determination of arginine by using 9,10-phenanthrenequinone[20] has been described. This method is ca. 1000-fold more sensitive than the Sakaguchi reaction, but some concern remains on the absolute accuracy of the reagent for the determination of arginine in peptide linkage. This reagent has also been used to study the modification of arginine in proteins by methylglyoxal,[21,22] glycation,[22,23] and oxidation.[22] As with other site-specific chemical modifications of proteins, there has been increasing use of mass spectrometry to characterize the chemical modification of arginine in proteins.[24-29]

The use of phenylglyoxal (Figure 4.1) was developed by Takahashi[10] in 1968 and has since been applied to the study of the role of arginyl residues in proteins, as shown in Table 4.1.

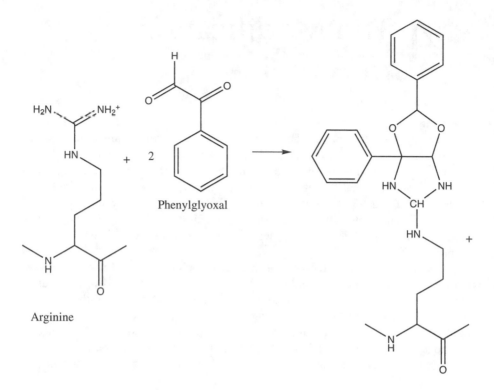

FIGURE 4.1 Reaction of phenylglyoxal with arginine in proteins.

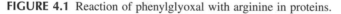

As with other modifying agents, there has been increased use of phenylglyoxal to study complex systems in the past decade.[30-44] It is extremely difficult to consider these studies to be useful for protein chemistry, because there is usually insufficient analytical information to support the conclusions. However, with cautious interpretation, these studies can be useful. The use of multiple arginine reagents might prove instructive. Many of these studies fail to recognize that phenylglyoxal, like glyoxal, reacts with ε-amino groups at a significant rate.[10] Polymerization was noted in a sample incubated for 21 h. The amino-terminal lysine residue was rapidly modified under these conditions. The possible effect of light on the reaction of phenylglyoxal with arginine, as has been reported for 2,3-butanedione,[1] has not been studied.[45,46]

As established by Takahashi, the stoichiometry of the reaction involves the reaction of 2 Mol of phenylglyoxal with 1 Mol of arginine. The [^{14}C]-labeled reagent can be easily prepared.[10,47] A facile modification of the original Riley and Gray[47] method, which omits the vacuum distillation step, has been reported by Schloss and co-workers.[48] Radiolabeled acetophenone was added to an equal amount (on the basis of weight) of selenium dioxide in dioxane–water (30:1). The mixture was refluxed for 3 h, after which solvent was removed under a stream of nitrogen. The residue was taken up in boiling water, and activated charcoal was added. The hot slurry was filtered through celite. The phenylglyoxal crystallized spontaneously from the filtrate on cooling. The synthesis of phenyl [2-^{3}H]-glyoxal[49] has been reported.

TABLE 4.1
Reaction of Phenylglyoxal with Arginyl Residues in Peptides and Proteins

Protein	Solvent	Reagent Excess[a]	Extent of Modification	Ref.
Pancreatic RNase	0.1 M N-ethylmorpholine acetate, pH 8.0	—	2–3/4[b,c]	1
Porcine carboxypeptidase B	0.3 M borate, pH 7.9	200[d]	1[e]	2
Aspartate transcarbamylase	0.125 M potassium bicarbonate, pH 8.3 or 0.1 M N-ethylmorpholine, pH 8.3	—	2.2[f,g]	3
Pyruvate kinase	0.1 M triethanolamine, pH. 7.0	—	3/28.33[h]	4
Horse liver alcohol dehydrogenase	—	—	—	5, 6
Mitochondrial ATPase	0.097 M sodium borate, 0.097 M EDTA, pH 8.0	—	4[i]	7
Adenylate kinase	0.1 M triethanolamine · HCl, pH 7.0	—	—[j]	8, 9
Rhodospirillum rubrum chromatophores	0.05 M borate, pH 8.0	—	—[k]	10
Glutamic acid decarboxylase	0.05 M sodium borate[l]	—	—	11
Ribulose bisphosphate carboxylase	0.066 M sodium[m] bicarbonate, 0.050 M Bicine, 0.1 M EDTA, pH 8.0	—	2–3/35[n]	12
Yeast hexokinase	0.035 M Veronal, pH 7.5	—	1/18[o]	13
Propionyl CoA carboxylase	0.050 M borate, pH 8.0	—	—	14
β-Methylcrotonyl CoA carboxylase	0.050 M borate, pH 8.0	—	—	14
Superoxide dismutase	0.125 M sodium bicarbonate, pH 8.0	—	1/4[p]	15
Myosin (subfragment 1)	0.1 M potassium bicarbonate, pH 8.0	—	1.7/35[q]	16
Thymidylate synthetase	0.125 M bicarbonate, pH 8.0[r]	—	3.6/12	17
Glutamate apodecarboxylase	0.125 M sodium[s] bicarbonate, pH 7.5	—	1/23[t]	18
Adenylate kinase (yeast)[u]	0.025 M HEPES, pH 7.5	—	—	19
Cardiac myosin S-1	0.1 M N-ethylmorpholine acetate, pH 7.6	—	2.8/42[v]	20
Cystathionase	0.125 M bicarbonate, pH 7.9	—	18/45	21
Fatty acid synthetase	0.1 M sodium phosphate, 0.0005 M dithioerythritol, 0.001 M EDTA, pH 7.6	—	4/106	22
Yeast inorganic pyrophosphatase	0.08 M N-ethylmorpholine acetate, pH 7.0	—	1/6	23
Porcine phospholipase A	0.125 M potassium bicarbonate, pH 8.5	—	1.4/4[w]	24

(continued)

TABLE 4.1 (continued)
Reaction of Phenylglyoxal with Arginyl Residues in Peptides and Proteins

Protein	Solvent	Reagent Excess[a]	Extent of Modification	Ref.
Superoxide dismutase[x]	0.100 M sodium bicarbonate, pH 8.3	50–100	0.88/4.0[y]	25, 26[z]
p-Hydroxybenzoate hydroxylase	0.050 M potassium phosphate, pH 8.0	250	2–3/24[aa]	27
Thymidylate synthetase	0.200 M N-ethylmorpholine, pH 7.4[cc]	65	2/12[bb]	28
Acetylcholine esterase	0.025 M borate, 0.005 phosphate, 0.050 M NaCl, pH 7.0	—	3/31[dd]	29
γ-Aminobutyrate aminotransferase	0.05 M Tris, pH 8.5	—	—	30
D-β-Hydroxybutyrate dehydrogenase	0.05 M HEPES, pH 7.5	—	—[ee]	31
Ornithine transcarboxylase	0.05 M Bicine, 0.1 M KCl, 0.0001 M EDTA, pH 8.05	—	—[ff]	32
Coenzyme B_{12}-dependent diol dehydrase	0.05 M borate, pH 8.0	—	—	33
Transketolase	0.125 M sodium bicarbonate, pH 7.6	—	4/34[gg]	34
Aldehyde reductase	20 mM sodium phosphate, pH 7.0			35
ATP citrate lyase	0.050 M HEPES,[hh] pH 8.0	—	8.5/40	36
Malic enzyme	0.037 M borate,[ii] pH 7.5	—	—	37
Pyridoxamine-5′-phosphate oxidase	0.1 M potassium phosphate, pH 8.0, containing 5% ETOH	—[jj]	6/40	38
Ornithine transcarboxylase	0.125 M potassium bicarbonate, pH 8.3	—[kk]	1.5/[ll]	39
Acetate kinase	0.050 M triethanolamine, pH 7.6	—[mm]	—[nn]	40
Pancreatic phospholipase A_2	0.2 M N-ethylmorpholine, pH 8.0	30	1.0–1.2[oo]	41
Phosphatidylcholine transfer protein	0.1 M sodium bicarbonate, pH 8.0	—	4/10[pp]	42
Aldehyde reductase	0.020 M phosphate,[qq] pH 7.0	—	0.6/16[rr]	43
Choline acetyltransferase	0.050 M HEPES, pH 7.8	—	—[ss]	44
ADP-glucose synthetase	0.05 M potassium phosphate, 0.00025 M EDTA, pH 7.5	110	1[tt]	45
Pyruvate oxidase	0.1 M sodium phosphate, 0.010 M magnesium chloride, pH 7.8	—	2.5/5[uu]	46
Calcineurin	50 mM Tris, pH 7.5, with 0.1 M EDTA, 0.1 mM NiCl$_2$ and 0.3 mM CaCl$_2$	10,000[vv]	—	47
Carbon monoxide	20 mM sodium phosphate, pH 8.2, with 4 mM dithiothreitol	—	—	48
Epithelial sodium channel	pH 8.1[ww]	—	—	49

TABLE 4.1 (continued)
Reaction of Phenylglyoxal with Arginyl Residues in Peptides and Proteins

Protein	Solvent	Reagent Excess[a]	Extent of Modification	Ref.
Calcium "pump"	100 mM N-ethylmorpholine, 40 mM KCl, 6 mM HEPES, 0.7 mM MgCl$_2$, 0.2 mM EGTA, pH 7.7	—	—	50
Calmodulin dependent	40 mM HEPES, pH 7.5 with 7% glycerol, 0.1 M EDTA, and 0.3 mM CaCl$_2$	333[xx]	—	51
Inositol 1,4,5-triphosphate 3-kinase A				52
Rat microsomal acyl transferase				53
ADP-glucose pyrophosphatase	50 mM HEPES, pH 8.0, 25°C			54
HlyC, protein acyl transferase				55
Urease				56
Alfalfa phytase	50 mM bicarbonate, pH 8.0, 37°C			57
Peptidylglycine α-amidating monooxygenase	0.4 M NaHCO$_3$, pH 8.5, 25°C, dark, 60 min			58
N-Acetylneuraminate synthase (*Streptoococcus agalactiae*)	120 mM NaHCO$_3$, pH 8.0, room temp.			59

[a] Reagent/protein.
[b] After 3 h at 25°C.
[c] Had modification of α-amino group and lysine residues.
[d] Reagent/arginine.
[e] After 1 h at 37°C.
[f] After 3 h at 25°C.
[g] 1.3/8 in regulatory chain.
[h] 20 min at 37°C with 23.8 mM phenylglyoxal, protein 1 mg/ml.
[i] 30 min of reaction at 30°C; the presence of efrapeptin, a low-molecular-weight antibiotic that is a potent inhibitor of oxidative phosphorylation, prevented the modification of one fast-reacting arginyl residue.
[j] A single arginine residue is modified (Arg-97).
[k] A single site appeared to be modified with a second-order rate constant of 1.6 M^{-1} min^{-1}.
[l] pH not given; reaction at 23°C, kinetic evidence for stoichiometric inactivation.
[m] Solvent made metal-free using BioRad Chelex; reaction performed with and without MgCl$_2$.
[n] Analysis of sulfhydryl groups after phenylglyoxal modification showed no loss of cysteine. These investigations noted that modification with phenylglyoxal is apparently more specific than with 2,3-butanedione.
[o] The authors claim 1:1 stoichiometry of phenylglyoxal with the arginyl residue from analysis of dependence of pseudo first-order rate constant vs. reciprocal of reagent (phenylglyoxal concentration). Partial reactivation of modified enzyme was observed, reflecting lability of modified arginine residues. Reaction also shows saturation kinetics, reflecting specific affinity of reagent for enzyme possibly from hydrophobic interaction. These authors suggest that this phenomenon is observed with the reaction of other hydrophobic reagents with this enzyme. A similar phenomenon has been observed with trinitrobenzenesulfonic acid (see Chapter 10).

(continued)

TABLE 4.1 (continued)
Reaction of Phenylglyoxal with Arginyl Residues in Peptides and Proteins

[p] Reaction was performed at 25°C for 1 h.

[q] Reaction was performed at 25°C with 3 mM phenylglyoxal for 3 min.

[r] Rates of enzyme inactivation were dependent upon buffer; at 5.9 mM phenylglyoxal, the following data were obtained: bicarbonate ($T_{1/2}$ = 6.0 min), MOPS ($T_{1/2}$ = 11.5 min), borate ($T_{1/2}$ = 34.0), and phosphate ($T_{1/2}$ = 48.0 min) at 25°C.

[s] These investigators noted a significant buffer effect on the reaction; the following second-order rate constants were obtained with the following reagent/solvent conditions (reactions performed at 23°C): 0.69 M^{-1} min^{-1} with 2,3-butanedione 0.050 M borate, pH 8.0; 33.78 M^{-1} min^{-1} with glyoxal/0.125 M sodium bicarbonate, pH 8.0; 31.00 M^{-1} min^{-1} with methylglyoxal/0.125 M sodium bicarbonate, pH 8.0; and 107.68 M^{-1} min^{-1} with phenylglyoxal 0.125 M sodium bicarbonate, pH 8.0.

[t] 300-fold excess of reagent, 0.083 M sodium bicarbonate, pH 8.1, 7 min, 23°C.

[u] 2,3-Butanedione or 1,2-cyclohexanedione appeared to be more effective than phenylglyoxal in this system.

[v] 6 min, 22°C, 50% loss of activity.

[w] Determined at 99% inactivation (25°C) of phospholipase activity (release of fatty acid from egg yolk in water with 3 mM CaCl$_2$ and 1.4 mM sodium deoxycholate). These investigators (see Ref. 24, Table 4.1) examined the possibility of amino-terminal alanine modification; no loss of alanine was observed with 75% inactivation (0.9 Mol Arg modified per mole of protein), whereas enzyme samples with a greater extent of inactivation had some loss of amino-terminal alanine (quantity not given). These investigators examined the pH dependence of enzyme inactivation by phenylglyoxal (presumably, a direct measure of the rate of arginine modification) and reported the following second-order rate constants (M^{-1} min^{-1}): pH 6.5, 0.3; pH 7.5, 1.5; pH 8.5, 3.3; and pH 9.5, 3.9. These investigators also showed that phenylglyoxal ($T_{1/2}$ = 1 min) was more effective than 2,3-butanedione ($T_{1/2}$ = 20 min) and 1,2-cyclohexanedione ($T_{1/2}$ = 120 min).

[x] Cu, Zn superoxide dismutase from *Saccharomyces cerevisiae*.

[y] Determined at 80% loss of enzymatic activity, using reaction of the modified enzyme with 9,10-phenanthrenequinone. This value corresponded to that determined by the incorporation of radiolabeled phenylglyoxal assuming a 2:1 adduct. Amino acid analysis with samples prepared by using normal hydrolytic conditions (6 N HCl, 110°C, 20 h) suggested only ca. 50% of this extent of arginine modification. When thioglycolic acid was included during the hydrolysis, values for the extent of arginine modification approached those determined by the fluorescence technique and radiolabel incorporation.

[z] The study uses reaction with 4-hydroxy-3-nitrophenylglyoxal, a chromophoric derivative of phenylglyoxal, to identify the specific arginine residue modified. The rate of reaction with this derivative is ca. sixfold less than that with the parent phenylglyoxal.

[aa] Reaction was performed at 25°C for 60 to 120 min. Loss of lysine residues was not observed under these reaction conditions. Amino acid analysis (hydrolysis in 6 N HCl, 110°C, 24 h) correlated well with radiolabeled phenylglyoxal incorporation assuming 2:1 stoichiometry (i.e., amino acid analysis gave 3.6 Mol arginine lost per mole of enzyme, whereas 7.54 Mol radiolabel was incorporated).

[bb] The presence of substrate, 2′-deoxyuridylate, prevents the modification of 1 Mol of arginine per mole of enzyme. Note that these results differ from those reported in Ref. 17, Table 4.1. There were differences in solvent conditions. It is not clear why this would account for the differences observed in these two studies. Note that the investigators in Ref. 17 obtained similar stoichiometry with 2,3-butanedione.

[cc] These investigators (see Ref. 28, Table 4.1) examined the reaction at pH 7.4 (rate of inactivation of 32 M^{-1} min^{-1}). There was a ca. 100-fold increase in the rate of inactivation.

[dd] The modification with phenylglyoxal is associated with a ca. 15% loss of enzyme activity. Treatment with 2,3-butanedione under similar reaction conditions results in the modification of ca. one more mole of arginine per mole enzyme with a ca. 75% loss of catalytic activity.

TABLE 4.1 (continued)
Reaction of Phenylglyoxal with Arginyl Residues in Peptides and Proteins

[ee] Stoichiometry was not established, but data are consistent with the loss of activity resulting from the modification of a single arginine residue. Submitochondrial vesicles were used as the source of enzyme in these studies. A second-order rate constant of 1.03 M^{-1} min^{-1} was obtained from the reaction with phenylglyoxal. A value of 0.8 M^{-1} min^{-1} was obtained for reaction with 1,2-cyclohexanedione (0.050 M borate, pH 7.5) and a value of 4.6 M^{-1} min^{-1} was obtained for 2,3-butanedione in the borate buffer system.

[ff] See a more complete discussion of this study in Table 4.2. For inactivation by phenylglyoxal, a second-order rate constant of 56 M^{-1} min^{-1} was obtained at pH 8.04. The reactions were performed in the dark.

[gg] Analysis of Tsou plots[235,236] indicates at least two classes of residues react at different rates.

[hh] Most studies were performed in this solvent at 30°C, with a second-order rate constant of 0.33 M^{-1} sec^{-1}. The rate was reduced in potassium phosphate ($k = 0.25$ M^{-1} sec^{-1}) and borate ($k = 0.078$ M^{-1} sec^{-1}).

[ii] Under these conditions at 24°C, a second-order rate constant of $k = 7.08$ M^{-1} min^{-1}, assuming that the rate of inactivation is directly related to the modification of arginine. with 2,3-butanedione in 0.48 M borate, a second-order rate constant of $k = 5.4$ M^{-1} min^{-1} is compared to 1.69 M^{-1} min^{-1} with methylglyoxal and 0.032 M^{-1} min^{-1} with 2,4-pentancdione.

[jj] The rate of inactivation at 25°C for the apoenzyme was determined to be 3.7 and 11.1 M^{-1} min^{-1} for the holoenzyme.

[kk] A second-order rate constant of $k = 4.6$ M^{-1} min^{-1} at 25°C was obtained under these conditions.

[ll] Based on incorporation of radiolabeled phenylglyoxal, 1.5 arginine residues are modified per 35,000 chain after 3 h of reaction. There are likely different classes of reactive arginyl residues, the more reactive groups being directly associated with catalytic activity.

[mm] Saturation kinetics are observed with phenylglyoxal, suggesting the formation of an enzyme inhibitor complex before reaction with an arginine residues.

[nn] With 95% loss of catalytic activity, there is 94% modification of arginine.

[oo] This study shows that this level of arginine modification is associated with 80% loss of amino-terminal alanine. It was necessary to protect the α-amino group of the amino-terminal alanine with a t-butyloxycarbonyl group to avoid modification under these reaction conditions. The use of radiolabeled cyclohexandedione established Arg-6 as the primary site of modification.

[pp] Reaction was performed for 30 min at 25°C. Extent of modification based on radiolabel incorporation and amino acid analysis.

[qq] For reaction at 30°C, a second-order rate constant of $k = 2.6$ M^{-1} min^{-1}, assuming that the loss of activity seen with phenylglyoxal directly reflects the loss of an arginine residues.

[rr] Determined from both amino acid analysis and radiolabel incorporation.

[ss] Phenylglyoxal was much more effective than 2,3-butanedione or camphorquinone-10-sulfonic acid.

[tt] Assuming 2:1 stoichiometry of phenylglyoxal to arginine; reaction at 25°C. Phenylglyoxal is much more effective than 1,2-cyclohexanedione (twofold molar excess of 1,2-cyclohexanedione had $T_{1/2} = 24$ min).

[uu] From radiolabel incorporation assuming 2:1 stoichiometry. There are clearly at least two classes of reactive arginine residues. When the reaction is performed at pH 6.0, inactivation with phenylglyoxal can be partially reversed on dilution in pH 6.0 buffer.

[vv] Inactivation rate constant of 1.5 M^{-1} min^{-1} at pH 7.5, 30°C.

[ww] Reaction was performed with an undefined quantity of Tris buffer. The inactivation reaction was markedly increased by the presence of sodium ions.

[xx] Inactivation rate constant of 132 M^{-1} min^{-1} at pH 7.5.

(continued)

TABLE 4.1 (continued)
Reaction of Phenylglyoxal with Arginyl Residues in Peptides and Proteins

References for Table 4.1

1. Takahashi, K., The reaction of phenylglyoxal with arginine residues in proteins, *J. Biol. Chem.* 243, 6171, 1968.
2. Werber, M.M., and Sokolovsky, M., Chemical evidence for a functional arginine residue in carboxypeptidase B, *Biochem. Biophys. Res. Commn.* 48, 384, 1972.
3. Kantrowitz, E.R., and Lipscomb, W.N., Functionally important arginine residues of aspartate transcarbamylase, *J. Biol. Chem.* 252, 2873, 1977.
4. Berghäuser, J., Modifizierung von argininresten in pyruvat-kinase, *Hoppe-Seyler's Physiol. Chem.* 358, 1565, 1977.
5. Lange, L.G., III, Riordan, J.F., and Vallee, B.L., Functional argininyl residues as NADH binding sites of alcohol dehydrogenases, *Biochemistry* 13, 4361, 1974.
6. Jörnvall, H., Lange, L.G., III, Riordan, J.F., and Vallee, B.L., Identification of a reactive arginyl residue in horse liver alcohol dehydrogenase, *Biochem. Biophys. Res. Commn.* 77, 73, 1977.
7. Kohlbrenner, W.E., and Cross, R.L., Efrapeptin prevents modification by phenylglyoxal of an essential arginyl residue in mitochondrial adenosine triphosphatase, *J. Biol. Chem.* 253, 7609, 1978.
8. Berghäuser, J., A reactive arginine in adenylate kinase, *Biochim. Biophys. Acta* 397, 370, 1975.
9. Berghäuser, J., and Schirmer, R.H., Properties of adenylate kinase after modification of Arg-97 by phenylglyoxal, *Biochim. Biophys. Acta* 537, 428, 1978.
10. Vallejos, R.H., Lescano, W.I.M., and Lucero, H.A., Involvement of an essential arginyl residue in the coupling activity of *Rhodospirillum rubrum* chromatophores, *Arch. Biochem. Biophys.* 190, 578, 1978.
11. Tunnicliff, G., and Ngo, T.T., Functional role of arginine residues in glutamic acid decarboxylase from brain and bacteria, *Experientia* 34, 989, 1978.
12. Schloss, J.V., Norton, I.L., Stringer, C.D., and Hartman, F.C., Inactivation of ribulosebisphosphate carboxylase by modification of arginyl residues with phenylgloxal, *Biochemistry* 17, 5626, 1978.
13. Philips, M., Pho, D.B., and Pradel, L.-A., An essential arginyl residue in yeast hexokinase, *Biochim. Biophys. Acta* 566, 296, 1979.
14. Wolf, B., Kalousek, F., and Rosenberg, L.E., Essential arginine residues in the active sites of propionyl CoA carboxylase and beta-methylcrotonyl CoA carboxylase, *Enzyme* 24, 302, 1979.
15. Malinowski, D.P., and Fridovich, I., Chemical modification of arginine at the active site of the bovine erythrocyte superoxide dismutase, *Biochemistry* 18, 5909, 1979.
16. Mornet, D., Pantel, P., Audemard, E., and Kassab, R., Involvement of an arginyl residue in the catalytic activity of myosin heads, *Eur. J. Biochem.* 100, 421, 1979.
17. Cipollo, K.L., and Dunlap, R.B., Essential arginyl residues in thymidylate synthetase from amethopterin-resistant *Lactobacillus casei*, *Biochemistry* 18, 5537, 1979.
18. Cheung, S.-T., and Fonda, M.L., Kinetics of the inactivation of *Escherichia coli* glutamate apodecarboxylase by phenylglyoxal, *Arch. Biochem. Biophys.* 198, 541, 1979.
19. Varimo, K., and Londesborough, J., Evidence for essential arginine in yeast adenylate cyclase, *FEBS Lett.* 106, 153, 1979.
20. Morkin, E., Flink, I.L., and Banerjee, S.K., Phenylglyoxal modification of cardiac myosin S-1. Evidence for essential arginine residues at the active site, *J. Biol. Chem.* 254, 12647, 1979.
21. Portemer, C., Pierre, Y., Loriette, C., and Chatagner, F., Number of arginine residues in the substrate binding sites of rat liver cystathionase, *FEBS Lett.* 108, 419, 1979.
22. Poulose, A.J., and Kolattukudy, P.E., Presence of one essential arginine that specifically binds the 2′-phosphate of NADPH on each of the ketoacetyl reductase and enoyl reductase active sites of fatty acid synthetase, *Arch. Biochem. Biophys.* 199, 457, 1980.
23. Bond, M.W., Chiu, N.Y., and Cooperman, B.S., Identification of an arginine residue important for enzymatic activity within the covalent structure of yeast inorganic pyrophosphatase, *Biochemistry* 19, 94, 1980.

TABLE 4.1 (continued)
Reaction of Phenylglyoxal with Arginyl Residues in Peptides and Proteins

24. Vensel, L.A., and Kantrowitz, E.R., An essential arginine residue in porcine phospholipase A2, *J. Biol. Chem.* 255, 7306, 1980.

25. Borders, C.L., Jr., and Johansen, J.T., Essential arginyl residues in Cu, Zn superoxide dismutase from *Saccharomyces cerevisiae, Carlsberg Res. Commn.* 45, 185, 1980.

26. Borders, C.L., Jr., and Johansen, J.T., Identification of Arg-143 as the essential arginyl residue in yeast Cu, Zn superoxide dismutase by the use of a chromophoric arginine reagent, *Biochem. Biophys. Res. Commn.* 96, 1071, 1980.

27. Shoun, H., Beppu, T., and Arima, K., An essential arginine residue at the substrate-binding site of *p*-hydroxybenzoate hydroxylase, *J. Biol. Chem.* 255, 9319, 1980.

28. Belfort, M., Maley, G.F., and Maley, F., A single functional arginyl residue involved in the catalysis promoted by *Lactobacillus casei* thymidylate synthetase, *Arch. Biochem. Biophys.* 204, 340, 1980.

29. Müllner, H., and Sund, H., Essential arginine residue in acetylcholinesterase from *Torpedo californica, FEBS Lett.* 119, 283, 1980.

30. Tunnicliff, G., Essential arginine residues at the pyridoxal phosphate binding site of brain α-aminobutyrate aminotransferase, *Biochem. Biophys. Res. Commn.* 97, 160, 1980.

31. El Kebbaj, M.S., Latruffe, N., and Gaudemer, Y., Presence of an essential arginine residue in D-β-hydroxybutyrate dehydrogenase from mitochondrial inner membrane, *Biochem. Biophys. Res. Commn.* 96, 1569, 1980.

32. Marshall, M., and Cohen, P.P., Evidence for an exceptionally reactive arginyl residue at the binding site for carbamyl phosphate in bovine ornithine transcarbamylase, *J. Biol. Chem.* 255, 7301, 1980.

33. Kuno, S., Toraya, T., and Fukui, S., Coenzyme B12-dependent diol dehydrase: chemical modification with 2,3-butanedione and phenylglyoxal, *Arch. Biochem. Biophys.* 205, 240, 1980.

34. Kremer, A.B., Egan, R.M., and Sable, H.Z., The active site of transketolase. Two arginine residues are essential for activity, *J. Biol. Chem.* 255, 2405, 1980.

35. Branlant, G., Tritsch, D., and Biellmann, J.-F., Evidence for the presence of anion-recognition sites in pig-liver aldehyde reductase. Modification by phenyl glyoxal and *p*-carboxyphenyl glyoxal of an arginine located close to the substrate binding site, *Eur. J. Biochem.* 116, 505, 1981.

36. Ramakrishna, S., and Benjamin, W.B., Evidence for an essential arginine residue at the active site of ATP citrate lyase from rat liver, *Biochem. J.* 95, 735, 1981.

37. Chang, G.-G., and Huang, T.-M., Modification of essential arginine residues of pigeon liver malic enzyme, *Biochim. Biophys. Acta* 660, 341, 1981.

38. Choi, J.-D., and McCormick, D.B., Roles of arginyl residues in pyridoxamine-5'-phosphate oxidase from rabbit liver, *Biochemistry* 20, 5722, 1981.

39. Fortin, A.F., Hauber, J.M., and Kantrowitz, E.R., Comparison of the essential arginine residue in *Escherichia coli* ornithine and aspartate transcarbamylases, *Biochim. Biophys. Acta* 662, 8, 1981.

40. Wong, S.S., and Wong, L.-J., Evidence for an essential arginine residue at the active site of *Escherichia coli* acetate kinase, *Biochim. Biophys. Acta* 660, 142, 1981.

41. Fleer, E.A.M., Puijk, W.C., Slotboom, A.J., and DeHaas, G.H., Modification of arginine residues in porcine pancreatic phospholipase A2, *Eur. J. Biochem.* 116, 277, 1981.

42. Akeroyd, R., Lange, L.G., Westerman, J., and Wirtz, K.W.A., Modification of the phosphatidylcholine-transfer protein from bovine liver with butanedione and phenylglyoxal. Evidence for one essential arginine residue, *Eur. J. Biochem.* 121, 77, 1981.

43. Branlant, G., Tritsch, D., and Biellmann, J.-F., Evidence for the presence of anion-recognition sites in pig-liver aldehyde reductase. Modification by phenylglyoxal and *p*-carboxyphenyl glyoxal of an arginyl residue located close to the substrate-binding site, *Eur. J. Biochem.* 116, 505, 1981.

44. Mautner, H.G., Pakyla, A.A., and Merrill, R.E., Evidence for presence of an arginine residue in the coenzyme A binding site of choline acetyltransferase, *Proc. Natl. Acad. Sci. USA* 78, 7449, 1981.

45. Carlson, C.A., and Preiss, J., Involvement of arginine residues in the allosteric activation of *Escherichia coli* ADP-glucose synthetase, *Biochemistry* 21, 1929, 1982.

(continued)

TABLE 4.1 (continued)
Reaction of Phenylglyoxal with Arginyl Residues in Peptides and Proteins

46. Koland, J.G., O'Brien, T.A., and Gennis, R.B., Role of arginine in the binding of thiamin pyrophosphate to *Escherichia coli* pyruvate oxidase, *Biochemistry* 21, 2656, 1982.
47. King, M.M., and Heiny, L.P., Chemical modification of the calmodulin-stimulated phosphatase, calcineurin, by phenylglyoxal, *J. Biol. Chem.* 262, 10658, 1987.
48. Shanmugasundaram, T., Kumar, G.K., Shenoy, B.C., and Wood, H.G., Chemical modification of the functional arginine residues of carbon monoxide dehydrogenase from *Clostridium thermoaceticum, Biochemistry* 28, 7112, 1989.
49. Garty, H., Yeger, O., and Asher, C., Sodium-dependent inhibition of the epithelial sodium channel by an arginyl-specific reagent, *J. Biol. Chem.* 263, 5550, 1988.
50. Missiaen, L., Raeymaekers, L., Droogmans, G., Wuytack, F., and Casteels, R., Role of arginine residues in the stimulation of the smooth-muscle plasma-membrane Ca^{2+} pump by negatively charged phospholipids, *Biochem. J.* 264, 609, 1989.
51. King, M.M., Conformation-sensitive modification of the type II calmodulin-dependent protein kinase by phenylglyoxal, *J. Biol. Chem.* 263, 4754, 1988.
52. Communi, D. et al., Active site labeling of inositol 1,4,5-triphosphate 3-kinase by phenylglyoxal, *Biochem. J.* 310, 109, 1995.
53. Weinander, R. et al., Structural and functional aspects of rat microsomal glutathione transferase. The roles of cysteine 49, arginine 107, lysine 67, histidine and tyrosine residues, *J. Biol. Chem.* 272, 8871, 1997.
54. Sheng, J., and Preiss, J., Arginine[294] is essential for the inhibition of *Anaaena* PCC 7120 ADP-glucose pyrophosphatase by phosphate, *Biochemistry* 36, 13077, 1997.
55. Trent, M.S., Worsham, L.M., and Ernst-Fonberg, M.L., HlyC, the internal protein acyltransferase that activates hemolysin toxin: the role of conserved tyrosine and arginine residues in enzymatic activity as probed by chemical modification and site-directed mutagenesis, *Biochemistry* 38, 8831, 1999.
56. Pearson, M.A. et al., Kinetic and structural characterization of urease active site variants, *Biochemistry* 39, 8575, 2000.
57. Ullah, A.H.J. et al., Cloned and expressed fungal *phyA* gene in alfalfa produces a stable phytase, *Biochem. Biophys. Res. Commn.* 290, 1343, 2002.
58. Santini, M. et al., Expression and characterization of human bifunctional peptidylglycine α-amidating monooxygenase, *Protein Exp. Purif.* 28, 293, 2003.
59. Suryanti, V., Nelson, A., and Berry, A., Cloning, over-expression, purification, and characterization of *N*-acetylneuraminate synthase from *Streptococcus agalactiae. Protein Expresss. Purif.* 27, 346, 2003.

Borders and co-workers[50] reported the synthesis of a chromophoric derivative, 4-hydroxy-3-nitrophenylglyoxal. The adduct between 4-hydroxy-3-nitrophenylgyloxal and arginine absorbs light at 316 nm ($\varepsilon = 1.09 \times 10^4 M^{-1}$ cm^{-1}). The derivative is unstable to acid hydrolysis (6 N HCl, 110°C, 24 h) but can be stabilized by the inclusion of thioglycolic acid. Fresh solutions of the reagent should be used to avoid problems associated with the decomposition of the reagent. The same group subsequently used this reagent to identify the reactive arginine in yeast Cu,Zn superoxide dismutase.[51] The reaction of 4-hydroxy-3-nitrophenylglyoxal (50 mM Bicine, 100 mM NaHCO$_3$, pH 8.3) with yeast Cu,Zn superoxide dismutase is slower (0.57 M^{-1}min^{-1}) than that observed with phenylglyoxal (2.8 M^{-1} min^{-1}). A similar difference in the rate of reaction with the two reagents was observed with creatinine kinase.[50,52] 4-Hydroxy-3-nitrophenylglyoxal has been used to measure creatinine.[53] This reagent is used infrequently but is useful.[54,55]

TABLE 4.2
Use of *p*-Hydroxyphenylglyoxal to Modify Arginine Residues in Proteins

Protein	Conditions	Ref.
Bacterial peptidyl dipeptiase-4	0.2 *M* *N*-ethylmorpholine, pH 8.0, 25°C	1
Recombinant soluble CD4[a]	25 m*M* NaHCO$_3$, 15 h, 25°C, dark	2
Amadoriase II from *Aspergillus* sp.	10 m*M* HEPES, pH 8.0, room temperature, dark	3

[a] Analysis of modification by mass spectrometry.

References for Table 4.2

1. Lanzillo, J.J., Dascrathy, Y., and Fanburg, B.L., Detection of essential arginine in bacterial peptidyl dipeptidase-4: arginine is not the anion binding site, *Biochem. Biophys. Res. Commn.* 160, 243, 1989.
2. Hager-Braun, C., and Tomer, K.B., Characterization of the tertiary structure of soluble CD4 bound to glycosylated full-length HIVgp120 by chemical modification of arginine residues and mass spectrometric analysis, *Biochemistry* 41, 1759, 2002.
3. Wu, X. et al., Alternation of substrate specificity through mutation of two arginine residues in the binding site of amadoriase II from *Aspergillus* sp., *Biochemistry* 41, 5548, 2002.

p-Hydroxylphenylglyoxal[56] has also been used as a spectrophotometric reagent to study this reaction. Feeney and colleagues[56] developed this reagent to detect available arginine residues in proteins. As with phenylglyoxal, it reacts with arginine under mild conditions (pH 7 to 9, 25°C, 30 to 60 min). The concentration of the resulting adduct (2:1 stoichiometry) with arginine can be determined at 340 nm (ε = 1.83 × 10^4 M^{-1}cm^{-1}). The modification is slowly reversed under basic conditions. The general characteristics of the reaction of *p*-hydroxyphenylglyoxal are similar to those described for phenylglyoxal. Most studies on the modification of arginine in proteins have used multiple reagents, but a few have used *p*-hydroxyphenylglyoxal. These are presented in Table 4.2.

Several studies have compared *p*-hydroxyphenylglyoxal and phenylglyoxal. It can be argued that *p*-hydroxyphenylglyoxal is more hydrophilic than phenylglyoxal. Béliveau and co-workers[57] showed that the rate of inactivation of phosphate transport by rat kidney brush border membranes was more rapid (k_{obs} = 0.052 s^{-1}) with phenylglyoxal than with *p*-hydroxyphenylglyoxal (k_{obs} = 0.012 s^{-1}). Mukovyama and colleagues[58] compared the modification of an L-phenylalanine oxidase from *Pseudomonas* sp. P.501 by phenylglyoxal and *p*-hydroxyphenylglyoxal. The rate for the phenylglyoxal modification of a single essential arginine residue was 10.6 M^{-1}min^{-1}, whereas the rate observed for the modification by 4-hydroxy-3-nitrophenylglyoxal was 15.1 M^{-1}min^{-1}. The most remarkable observation on differences between *p*-hydroxyphenylglyoxal and phenylglyoxal comes from studies by Ericksson and co-workers.[59] These investigators observed that treatment of mitochondria with phenylglyoxal (10 m*M* HEPES, 250 m*M* sucrose, 10 m*M* succinate, 100 μ*M* EGTA, 3 μ*M* rotenone, pH 8.0) results in the closing of the permeability pore, whereas reaction with *p*-hydroxyphenylglyoxal results in pore opening.

The reaction of arginine with phenylglyoxal is greatly accelerated in bicarbonate–carbonate buffer systems.[60] The reaction of methylglyoxal with arginine is also enhanced by bicarbonate, but a similar effect is not seen with either glyoxal or 2,3-butanedione. The molecular basis for this specific buffer effect is not clear at this time, and it is also not known whether reaction with α-amino functional groups occurs at a different rate than with other solvent systems used for this modification of arginine with phenylglyoxal. Yamasaki and co-workers[61] reported that p-nitrophenylglyoxal reacts with arginine in 0.17 M sodium pyrophosphate and 0.15 M sodium ascorbate, pH 9.0, to yield a derivative that absorbs at 475 nm. There was also reaction with histidine. (The imidazole ring is critical for this reaction in that the L-methyl derivative yielded a derivative that absorbed at 475 nm, whereas the 3-methyl derivative did not.) Free sulfhydryl groups also yielded a product with absorbance at 475 nm, but its absorbance was only 3% that of the arginine. Branlant and co-workers[62] used p-carboxyphenyl glyoxal in bicarbonate buffer at pH 8.0 to modify aldehyde reductase. A second-order rate constant of 26 M^{-1} min^{-1} was observed in 80 mM bicarbonate and 2.9 M^{-1} min^{-1} in 20 mM sodium phosphate, pH 7.0. Saturation kinetics was observed with this reagent under certain conditions. Phenylglyoxal also inactivated this enzyme with a second-order rate constant of 2.6 M^{-1} min^{-1} in 20 mM sodium phosphate, pH 7.0. Inactivation was reversible on dialysis. Further work with p-carboxyphenylglyoxal has not been reported.

Eun[63] examined the effect of borate on the reaction of arginine with phenylglyoxal and p-hydroxyphenylglyoxal. The base buffer used in these studies was 0.1 M sodium pyrophosphate, pH 9.0. Spectroscopy was used to follow the rate of arginine modification. The rate of modification of either free arginine or N-acetyl-L-arginine with phenylglyoxal was 10 to 15 times faster than that of p-hydroxyphenylglyoxal in the base buffer system. The inclusion of sodium borate (10 to 50 mM) markedly increased the rate of the reaction (ca. 20-fold) of p-hydroxyphenylglyoxal with either arginine or N-acetyl-L-arginine, whereas there was only a slight enhancement of the phenylglyoxal reaction. In a related study,[64] the effect of phenylglyoxal on sodium-channel gating in frog myelinated nerve was compared with that of p-hydroxyphenylglyoxal or p-nitrophenylglyoxal. Both p-hydroxyphenylglyoxal and p-nitrophenylglyoxal had a lesser effect than phenylglyoxal in reduced sodium current. The results are discussed in terms of the differences in hydrophobicity of the reagents, but it is clear that the intrinsic difference in reagent effectiveness described by Eun might be responsible, in part, for the observed differences. Phenylglyoxal has been used to modify arginine residues in *Klebsiella aerogenes* urease.[65] Previous studies have shown Arg-336 to be present at the enzyme active site. The R336Q variant shows greatly reduced enzyme activity (decreased k_{cat}, normal K_m). Modification of this variant with phenylglyoxal resulted in a further decrease of activity, suggesting that the modification of nonactive site residues can eliminate activity.

2,3-Butanedione (Figure 4.2) is the second well-characterized reagent for the selective modification of arginyl residues in proteins. Yankeelov and co-workers introduced the use of this reagent.[10,66] There were problems with the specificity of the reaction (c.f. Ref. 67) and the time required for modification until the observation of Riordan[67] that borate had a significant effect on the nature of the reaction of 2,3-butanedione with arginyl residues in proteins.

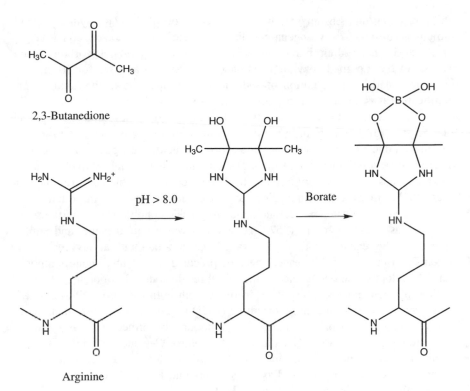

2,3-Butanedione

Arginine

FIGURE 4.2 Reaction of 2,3-butanedione with arginine in proteins.

The ability of 2,3-butanedione to act as a photosensitizing agent for the destruction of amino acids and proteins in the presence of oxygen was emphasized in work by Fliss and Viswanatha.[45] As would be expected from consideration of early photo-oxidation work, tryptophan and histidine are lost most rapidly with methionine; cystine and tyrosine are lost at a much slower rate. Loss is not seen on irradiation in the absence of 2,3-butanedione (open symbols). Azide (10 mM), a singlet oxygen scavenger, greatly reduces the rate of loss of amino acids. The absence of oxygen also greatly reduces the rate of loss of sensitive amino acids. These observations have been confirmed and extended by other laboratories.[46,68]

Several recent studies on the use of 2,3-butanedione to modify proteins are worthy of comment. Janssen and co-workers[69] modified penicillin acylase (*Escherichia coli*) with 2,3-butanedione (50 mM borate, pH 8.0, 25°C). The modified enzyme had decreased specificity (k_{cat}/K_m) for 2-nitro-5[(phenacetyl)amino]benzoic acid. Oligonucleotide-directed mutagenesis of two arginine residues yielded similar results, although there was dependence on the replacement amino acid. The kinetic parameters of R263K were similar to those of the wild type and were inactivated by 2,3-butanedione. The R263L variant showed a greater than 1000-fold decrease in k_{cat}/K_m. It is of interest that whereas phenylglyoxal also inhibited the enzyme, analysis was complicated by observation that phenylglyoxal was also a competitive inhibitor. Clark and Ensign[70] studied the inactivation of 2-[(R)-2-hydroxypropyl-thio]ethanesulfonate dehydrogenase with 2,3-butanedione. The reaction was performed in 50 mM sodium borate, pH 9.0, at 25°C. In experiments for amino acid analysis of the modified enzyme, reaction was performed at 30°C for 30 min and then terminated by the addition of sodium borohydride to trap the adducts. The modification was specific for arginine, with a second-order rate constant of 0.031 $M^{-1}sec^{-1}$ for the loss of enzyme activity. Leitner and Linder[29] developed an approach for the general labeling of guanidino groups in proteins via reaction with 2,3-butanedione in the presence of an arylboronic acid (e.g., phenylboronic acid) under alkaline conditions (pH 8 to 10). The sample is then subjected to electrospray ionization mass spectrometry without further processing. Table 4.3 examines the recent studies that use 2,3-butanedione to modify arginyl residues in proteins.

The use of 1,2-cyclohexanedione under very basic conditions to modify arginyl residues was demonstrated in 1967.[71] However, it was not until Patthy and Smith[12] reported on the reaction of 1,2-cyclohexanedione in borate with arginyl residues in proteins that the use of this reagent became practical. These investigators reported that 1,2-cyclohexanedione reacted with arginyl residues in 0.2 M borate, pH 9.0. At alkaline pH, reaction of 1,2-cyclohexanedione with arginine forms N^5-(4-oxo-1,3-diazaspiro[4,4]non-2-yliodene)-L-ornithine (CHD-arginine), a reaction that cannot be reversed. Between pH 7.0 and 9.0, a compound is formed from arginine and 1,2-cyclohexanedione, N^7-N^8-(1,2-dihydroxycyclohex-1,2-ylene)-L-arginine (DHCH-arginine). This compound is stabilized by the presence of borate and is unstable in the presence of buffers such as Tris. This compound is readily converted back to free arginine in 0.5 M hydroxylamine, pH 7.0.

Patthy and Smith subsequently used this reagent to identify functional residues in bovine pancreatic ribonuclease A and egg-white lysozyme.[72] The extent of modification of arginine residues in protein by 1,2-cyclohexanedione is generally assessed by amino acid analysis after acid hydrolysis. Under the conditions normally used for acid hydrolysis (6 N HCl, 110°C, 24 h), the borate-stabilized reaction product between arginine and 1,2-cyclohexanedione is unstable, and there is partial regeneration of arginine and formation of unknown degradation products.[12] Acid hydrolysis in the presence of an excess of mercaptoacetic acid (22 μl/ml of hydrolysate) prevents the destruction of DHCD-arginine.[11]

Ullah and Sethumadhaven[73] demonstrated differences in the susceptibility of two phytases from *Aspergillus ficus* to modification of arginine. Phytase A was rapidly inactivated by either 1,2-cyclohexanedione (borate, pH 9.0) or phenylglyoxal (NaHCO$_3$, pH 7.5). Phytase B was resistant to inactivation by 1,2-cyclohexanedione and less susceptible that Phytase A to inactivation with phenylglyoxal. Calvete and colleagues[74] used a novel approach to modify arginine residues in bovine seminal plasma protein PDC-109. The protein was bound to a heparin-agarose column and the 1,2-cyclohexanedione (in 16 mM Tris, 50 mM NaCl, 1.6 mM EDTA, 0.025% NaN$_3$, pH 7.4) circulated through the column overnight at room temperature. The modified protein was eluted with 1.0 M NaCl. Residues shielded from modification were presumed to be at the heparin-binding site. Table 4.4 lists some of the enzymes in which structure–function relationships have been studied by reaction with 1,2-cyclohexanedione.

TABLE 4.3
Use of 2,3-Butanedione to Modify Arginyl Residues in Proteins

Protein	Solvent	Reagent Excess[a]	Stoichiometry	Ref.
Carboxypeptidase A	0.05 M borate, 1.0 M NaCl, pH 7.5	—[b]	2/10	1
Chymotrypsin	0.1 M phosphate, pH 6.0	100[c]	1/3[d]	2
Thymidylate synthetase	0.050 M borate, pH 8.0	—	—[e]	3
Prostatic acid phosphatase	0.050 M borate, pH 8.0	—	—[f]	4
Purine nucleoside phosphorylase	0.0165 M borate, pH 8.0	—	—[g]	5
Yeast hexokinase PII	0.050 M borate, pH 8.3	—	4.2/18[h]	6
Isocitrate dehydrogenase	0.05 M MES, pH 6.2, 20% glycerol, 0.0021 M MnSO$_4$	—	1.6/13.4[i]	7
Stearylcoenzyme A desaturase	0.050 M sodium borate, pH 8.1	2,500	2[j]	8
Superoxide dismutase	0.050 M borate, pH 9.0	—[k]	1.3/4[l]	9
Energy-independent transhydrogenase	0.050 M sodium[m] borate, pH 7.8	—	—	10
Enolase	0.050 M borate, pH 8.3, 0.001 M Mg (OAc)$_2$, 0.01 mM EDTA	260	3/16[n]	11
NADPH-dependent aldehyde reductase	0.050 M borate, pH 7.0	—[o]	1/18[p]	12
Aryl sulfatase A	0.050 M[q] NaHCO$_3$, pH 8.0	—	—	13
Na$^+$, K$^+$-ATPase	0.04 M TES, 0.02 M borate, pH 7.4	—	—	14
Carbamate kinase	0.005 M triethanolamine, 0.050 M borate, pH 7.5	2,000	1.2/3.0[r]	15
Thymidylate synthetase	0.050 M borate, 0.001 M EDTA, pH 8.0	1,201	2.1/12[s]	16
(K$^+$ + H$^+$)-ATPase	0.125 M sodium borate, pH 7.0	—	—[t]	17
Cu, Zn superoxide dismutase	0.050 M borate, pH 8.3	—[u]	—	18
Fatty acid synthetase	0.020 M borate, 0.200 M KCl, 0.001 M dithiothreitol, 0.001 mM EDTA, pH 7.6	—	—[v]	19
Acetylcholinesterase	0.005 M phosphate, 0.025 M borate, 0.050 M NaCl, pH 7.0	—	4/31[w]	20
Coenzyme B$_{12}$-dependent diol dehydrase	0.050 M borate, pH 8.5	—	—[x]	21
Ornithine transcarbamylase	0.05 M Bicine,[y] 0.1 mM EDTA, 0.1 M KCl, pH 7.67	—	0.88/11[z]	22
Glycogen phosphorylase	0.020 M sodium tetraborate, 1 mM EDTA, pH 7.5	—	—	23

(continued)

TABLE 4.3 (continued)
Use of 2,3-Butanedione to Modify Arginyl Residues in Proteins

Protein	Solvent	Reagent Excess[a]	Stoichiometry	Ref.
Cytochrome c	0.05 M sodium bicarbonate, pH 7.5	9,900[dd]	2/2[hh]	24
Bacteriorhodopsin	0.100 M borate, pH 8.2	66,700	4/79[cc]	25
α-Ketoglutarate dehydrogenase	0.050 M sodium borate, pH 8.0	—[dd]	—[ee]	26
Acetate kinase	0.050 M borate, pH 8.6	—	—	27
Malic enzyme	0.045 M borate,[ff] pH 7.5	—	—	28
Glucose phosphate isomerase	0.05 M sodium borate, pH 8.7	—	7.8/30[gg]	29
Saccharopine dehydrogenase	0.08 M HEPES, 0.2 M KCl, 0.01 M borate, pH 8.0	—[hh]	8/38[ii]	30
Testicular hyaluronidase	0.050 M borate, pH 8.3	—[ii]	3.6/28	31
Glutathione reductase	0.050 M sodium borate, pH 8.3, 1 mM EDTA	20,000	5.3[kk]	32,33
Escherichia coli fibrillar adhesins	100 mM sodium borate with 0.9% NaCl, pH 7.5	—	—	34
Inter-α-trypsin inhibitor	10 mM HEPES, 175 mM NaCl, 100 mM sodium borate, pH 7.4	—	—	35
D-Glyceraldehyde-3-phosphate dehydrogenase	100 mM medinal[ll] buffer, pH 8.3 with 5 mM EDTA and 2 mM dithiothreitol	—	—	36
Stonustoxin (Syananceja horrida)	0.05 M borate, pH 8.1, 60 min, room temp.			37
Barley root oxalate oxidase	50 mM sodium borate, pH 9.0, 37°C			38
ADP-glucose pyrophosphorylase (Rhodobacter sphaerodies 2.4.1)	50 mM HEPES, 20 mM potassium borate, 1 mM DTE, pH 7.5, dark			39
Helix stablilizing nucleod protein HSNP-C, (Sulfolbolus acidocalderius)	20 mM triethanolamine, pH 8.0, 37°C, 30 min			40
2-[(R)-2-hydroxypropyl thio]ethanesulfonate dehydrogenase	50 mM sodium borate, pH 9.0, 25°C			41

[a] Mole reagent per mole protein unless otherwise indicated.

[b] This study demonstrated that, in the presence of borate, there is essentially no difference in the reaction of 2,3-butanedione monomer and butadione trimer. Note that commercially available 2,3-butanedione should be distilled immediately before use.

[c] This study used 2,3-butanedione trimer prepared by allowing 2,3-butanedione (40 ml) to stand with 80 g untreated Permutit under dry air (after shaking to obtain an even dispersion of 2,3-butanedione in Permutit) for 4 to 6 weeks at ambient temperature. The mixture was extracted with anhydrous ether. The ether extract was taken to an oil with dry air. The oil was allowed to stand for 5 to 7 days to permit crystallization of the timer.

TABLE 4.3 (continued)
Use of 2,3-Butanedione to Modify Arginyl Residues in Proteins

[d] In the absence of light, there is also some loss of lysine; there is no loss of catalytic activity. In the presence of sunlight, there was rapid inactivation of the enzyme with loss of lysine, arginine (less than in the dark), and tyrosine. With the exception of tyrosine modification, the changes in amino acid composition in the reaction exposed to light were less than those for the dark reaction despite the more significant loss of activity. Study of the wavelength dependence demonstrates that light at 300 nm is most effective. 2,3-Butanedione monomer was not effective in this photoinactivation process.

[e] Stoichiometry of reaction was not established. Inactivation was reversed by gel filtration in 0.05 M Tris, 0.010 M 2-mercaptoethanol, pH 8.0.

[f] 30°C.

[g] Ambient temperature. Calf spleen enzyme had 26 arginine residues modified at 98% loss of activity. Reaction with arginyl residues (as judged by loss of catalytic activity) was 50% as rapid with 2,3-butanedione in borate ($T_{1/2} = 40.3$ min) as with phenylglyoxal in Tris buffer ($T_{1/2} = 19.2$ min).

[h] Reaction was performed at 25°C. Determined by amino acid analysis after acid hydrolysis (6 N HCl, 110°C, 18 h). MgATP (5 mM) did not protect against either modification or loss of enzymatic activity, but MgATP and glucose reduced extent of modification from 3.3 arginine residues per subunit (65% inactivation) to 2.1 residues per subunit (20% inactivation). Inactivation was also observed with phenylglyoxal in 0.050 M Bicine, pH 8.3. Stoichiometry with this modification was not established.

[i] Determined by amino acid analysis. As indicated, the maximum value obtained is 1.6 residues modified out of an average of 13.4 arginyl residues per subunit.

[j] The modification was performed at 25°C. The presence of stearyl-CoA greatly decreased the rate and extent of inactivation by 2,3-butanedione. When the modified enzyme is taken into 0.020 Tris (acetate), 0.100 M NaCl, pH 8.1, by gel filtration, there is the rapid recovery of activity and the concomitant decrease in the extent of arginine modification. A similar extent of modification and loss of catalytic activity was seen with 1,2-cyclohexanedione in 0.1 M sodium borate, pH 8.1.

[k] Inactivation occurred at a rate of 10.9 M^{-1} min^{-1} under these conditions (compared to 4.0 M^{-1} min^{-1} with phenylglyoxal in bicarbonate/carbonate and 6.6 M^{-1} min^{-1} with 1,2-cyclohexanedione in 0.050 M borate, pH 9.0). Inactivation with 2,3-butanedione is not observed in 0.05 M bicarbonate/carbonate, pH 9.0, at 25°C; however, there is reduced modification of arginine (0.4 residues per subunit as compared to 1.3 residues per subunit with 77% inactivation).

[l] The majority of arginine modification could be reversed by the removal of reagent and borate solvent by dialysis vs. 0.05 M potassium phosphate, pH 7.8. Enzymatic activity was also recovered as a result of the dialysis procedure. These investigators were able to obtain evidence supporting the selective modification of Arg-141 by 2,3-butanedione, 1,2-cyclohexandione, or phenylglyoxal.

[m] The modification was performed at 22°C. These studies were performed with bacterial membrane preparations. Stoichiometry was not established. Analysis of the rates of inactivation suggested that inactivation was due to modification of a single arginine residue. NADH, which stimulates the transhydrogenation of 3-acetylpyridine-NAD by NADPH, protects the enzyme from inactivation.

[n] The modification was performed at 25°C. The extent of modification was determined by amino acid analysis after acid hydrolysis. The extent of modification reported was obtained after 75 min of reaction concomitant with 85% loss of activity. The presence of substrate, α-phosphoglycerate, reduced the extent of modification to 2 Mol arginine per subunit with only 5% loss of catalytic activity.

(continued)

TABLE 4.3 (continued)
Use of 2,3-Butanedione to Modify Arginyl Residues in Proteins

[o] A second-order rate constant of 0.0635 M^{-1} min^{-1} was obtained for the loss of enzymic activity on reaction with 2,3-butanedione in 0.050 M borate, pH 7.0 at 25°C. This presumably reflects the modification of a single arginine residue (see Footnote p). The inactivation of the enzyme by 1,2-cyclohexanedione, methylglyoxal, and phenylglyoxal is compared with that by 2,3-butanedione (all at 10 mM in 0.05 M borate, pH 7.0). Butanedione is clearly most effective, followed by phenylglyoxal, methylglyoxal, and 1,2-cyclohexanedione. The authors note that the enzyme under study, aldehyde reductase, can utilize methylglyoxal and phenylglyoxal as substrates, precluding their rigorous evaluation in this study.

[p] Obtained by amino acid analysis after acid hydrolysis (6 N HCl, 110°C, 24 h). The control preparation yielded 17.8 ± 1 Arg, while the modified enzyme yielded 16.7 ± 1 Arg. The presence of cofactor yielded a preparation with 17.5 ± 1 Arg.

[q] The reactions are reported at 25°C. Borate buffers could not be used, because borate is a competitive inhibitor of the enzyme and prevents inactivation in bicarbonate buffer. Reaction with phenylglyoxal in the same solvent.

[r] Reaction was performed at 25°C. Stoichiometry was established by amino acid analysis after acid hydrolysis (6 N HCl, 100°C, 20 h). Arginine is the only amino acid modified under these reaction conditions. These values were obtained at 80% inactivation. The presence of ADP reduced activity loss to 55%, with the extent of arginine modification reduced to 0.4 to 0.5 residues.

[s] Reaction was performed at 25°C for 90 min. Stoichiometry was determined by amino acid analysis after acid hydrolysis (6 N HCl, 110°C, 24 h).

[t] The use of isolated membrane fraction prevented the establishment of stoichiometry in these studies. Analysis of the dependence of reaction rate on concentration of 2,3-butanedione is consistent with the modification of a single arginine residue. As expected, the stability of modification is dependent on the presence of borate. Gel filtration into HEPES (0.125 M, pH 7.0) and subsequent inactivation at 37°C resulted in the recovery of a substantial amount of catalytic activity. Similar results were obtained with imidazole and Tris buffers under similar reaction conditions. This reactivation does not occur when the incubation following gel filtration is performed at 0°C instead of 37°C.

[u] A reaction rate with a second-order constant of k = 5.2 M^{-1} min^{-1} is obtained at 25°C. Inactivation is dependent on the presence of borate, as inactivation is not observed with use of Bicine buffer. Dialysis with 0.025 M phosphate, pH 7.0, for 21 h at 4°C results in an increase in activity of 14 to 85%, whereas complete recovery of activity is achieved after 21 h of dialysis.

[v] Stoichiometry was not established for the reaction with 2,3-butanedione. As shown in Table 4.1, reaction with phenylglyoxal modifies ca. 4 of the 106 arginyl residues in each subunit of fatty acid synthetase. Loss of the biological activity as determined by fatty acid synthetase activity, ketoreductase activity, or enoylreductase activity was considerably more rapid with phenylglyoxal than with 2,3-butanedione. These reactions are performed in borate buffer for the studies with 2,3-butanedione and in phosphate buffer for the studies with phenylglyoxal (both buffers at pH 7.6 with the reactions performed at 30°C).

[w] Reactions were performed at 25°C. The modification of arginyl residues is associated with a ca. 70% loss of enzymatic activity. The presence of N-phenylpyridinium-2-aldoxine iodide reduces the extent of arginine modification by ca. 1 Mol/Mol of enzyme, with concomitant protection of enzymatic activity. Note that modification of this enzyme with phenylglyoxal results in the modification of 3 Mol of arginine per mole of enzyme with 17% loss of enzymatic activity (see Table 4.1). It is not clear when modification of a particular arginyl residue with the two reagents is a mutually exclusive event.

[x] Reactions were performed at 25°C. Rigorous evaluation of the stoichiometry of the reaction is not available. Analysis of the dependence of first-order rate constants on reagent concentration (double-logarithmic relationships) is consistent with the modification of a single arginyl residue. The inactivation was reversed by 100-fold dilution into 0.05 M potassium phosphate, pH 8.5, at 25°C.

TABLE 4.3 (continued)
Use of 2,3-Butanedione to Modify Arginyl Residues in Proteins

[y] The inactivation of ornithine transcarbamylase is readily reversible in this solvent; the presence of borate precludes reactivation observed on dilution of modified enzyme in solvent. A value of 179 M^{-1} min^{-1} for the second-order rate constant for reaction of 2,3-butanedione with ornithine transcarbamylase under these conditions was recorded.

[z] Obtained at 88% inactivation.

[aa] Reaction was performed at 22°C.

[bb] Determined by amino acid analysis. The reaction is readily reversible, even in the presence of borate.

[cc] Determined by amino acid analysis. Constructed Scatchard plot shows that two residues were not available for modification with 2,3-butanedione.

[dd] Second-order rate constant $k = 2.95$ M^{-1} min^{-1} in this solvent, assuming that loss in catalytic activity is a measure of reaction with arginine.

[ee] Stoichiometry was not established. Kinetic analysis suggests that inactivation of catalytic activity results from the modification of a single arginine residue.

[ff] Modification reaction was performed at 24°C. Very little inactivation is observed if the reaction is performed in Tris buffer at the same pH. Reactivation of enzyme modified in borate buffer is observed when the inactivated enzyme is diluted in borate buffer.

[gg] The reaction was performed at 25°C for 4 h. The presence of the competitive inhibitor, 6-phosphogluconate, protected 1 Mol of arginine per mole of enzyme from modification, suggesting that there is a single arginine residue critical for catalytic activity. A 20-fold increase in inhibitor concentration resulted in the modification of more than 95% of the total arginine residues.

[hh] Second-order rate constant k = 7.5 M^{-1} min^{-1} at 25°C was obtained from the analysis of reaction rate data. pH dependence study showed optimal rate of inactivation at pH 8.2.

[ii] Determined by amino acid analysis on 95+% inactivated enzyme. Plotting loss of activity vs. arginine residues modified suggests that inactivation is due to the modification of a single arginine residue. Inactivation occurs with loss of sulfhydryl content.

[jj] Second order rate constant of $k = 13.57$ M^{-1} min^{-1} was obtained at 20°C. Inactivation was much less rapid in 0.050 M HEPES, pH 8.3 ($T_{1/2}$ = 30 min in borate; 11.5 min in HEPES).

[kk] Reactions were performed at 30°C. Modification was associated with 80 to 90% inactivation. Reaction with phenylglyoxal (0.050 M sodium phosphate, 1 mM EDTA, pH 7.6) at 2000-fold molar excess led to the modification of two arginyl residues at a level of 90% inactivation. The extent of arginine was determined by spectrophotometric analysis (increase in absorbance at 250 nm, $\varepsilon = 11,000$ M^{-1} cm^{-1}.

[ll] Reaction readily reversible, reflecting the absence of borate in the buffer.

References for Table 4.3

1. Riordan, J.F., Functional arginyl residues in carboxypeptidase A. Modification with butanedione, *Biochemistry* 12, 3915, 1973.
2. Fliss, H., Tozer, N.M., and Viswanatha, T., The reaction of chymotrypsin with 2,3-butanedione trimer, *Can. J. Biochem.* 53, 275, 1975.
3. Cipollo, K.L., and Dunlap, R.B., Essential arginyl residues in thymidylate synthetase, *Biochem. Biophys. Res. Commn.* 81, 1139, 1978.
4. McTigue, J.J., and Van Etten, R.L., An essential arginine residue in human prostatic acid phosphatase, *Biochim. Biophys. Acta* 523, 422, 1978.
5. Jordan, F., and Wu, A., Inactivation of purine nucleoside phosphorylase by modification of arginine residues, *Arch. Biochem. Biophys.* 190, 699, 1978.

(continued)

TABLE 4.3 (continued)
Use of 2,3-Butanedione to Modify Arginyl Residues in Proteins

6. Borders, C.L., Jr., Cipollo, K.L., and Jordasky, J.F., Role of arginyl residues in yeast hexokinase PII, *Biochemistry* 17, 2654, 1978.
7. Hayman, S., and Colman, R.F., Effect of arginine modification on the catalytic activity and allosteric activation by adenosine diphosphate of the diphosphopyridine nucleotide specific isocitrate dehydrogenase of pig heart, *Biochemistry* 17, 4161, 1978.
8. Enoch, H.G., and Strittmatter, P., Role of tyrosyl and arginyl residues in rat liver microsomal stearyl-coenzyme A desaturase, *Biochemistry* 17, 4927, 1978.
9. Malinowski, D.P., and Fridovich, I., Chemical modification of arginine at the active site of the bovine erythrocyte superoxide dismutase, *Biochemistry* 18, 5909, 1979.
10. Homyk, M., and Bragg, P.D., Steady-state kinetics and the inactivation by 2,3-butanedione of the energy-independent transhydrogenase of *Escherichia coli* cell membranes, *Biochim. Biophys. Acta* 571, 201, 1979.
11. Borders, C.L., Jr., and Zurcher, J.A., Rabbit muscle enolase also has essential argininyl residues, *FEBS Lett.* 108, 415, 1979.
12. Davidson, W.S., and Flynn, T.G., A functional arginine residue in NADPH-dependent aldehyde reductase from pig kidney, *J. Biol. Chem.* 254, 3724, 1979.
13. James, G.T., Essential arginine residues in human liver arylsulfatase A, *Arch. Biochem. Biophys.* 197, 57, 1979.
14. Grisham, C.M., Characterization of essential arginyl residues in sheep kidney (Na++ K+)-ATPase, *Biochem. Biophys. Res. Commn.*, 88, 229, 1979.
15. Pillai, R.P., Marshall, M., and Villafranca, J.J., Modification of an essential arginine of carbamate kinase, *Arch. Biochem. Biophys.* 199, 16, 1980.
16. Cipollo, K.L., and Dunlap, R.B., Essential arginyl residues in thymidylate synthetase from amethopterin-resistant *Lactobacillus casei*, *Biochemistry* 18, 5537, 1979.
17. Schrijen, J.J., Luyben, W.A.H.M., DePont, J.J.H.M., and Bonting, S.L., Studies on (K+ + H+)-ATPase. I. Essential arginine residue in its substrate binding center, *Biochim. Biophys. Acta* 597, 331, 1980.
18. Belfort, M., Maley, G.F., and Maley, F., A single functional arginyl residue involved in the catalysis promoted by *Lactobacillus casei* thymidylate synthetase, *Arch. Biochem. Biophys.* 204, 340, 1980.
19. Poulose, A.J., and Kolattukudy, P.E., Presence of one essential arginine that specifically binds the 2′-phosphate of NADPH on each of the ketoacyl reductase and enoyl reductase active sites of fatty acid synthetase, *Arch. Biochem. Biophys.* 199, 457, 1980.
20. Müllner, H., and Sund, H., Essential arginine residue in acetylcholinesterase from *Torpedo californica*, *FEBS Lett.* 119, 283, 1980.
21. Kuno, S., Toraya, T., and Fukui, S., Coenzyme B12-dependent diol dehydrase: chemical modification with 2,3-butanedione and phenylglyoxal, *Arch. Biochem. Biophys.* 205, 240, 1980.
22. Marshall, M., and Cohen, P.P., Evidence for an exceptionally reactive arginyl residue at the binding site for carbamyl phosphate in bovine ornithine transcarbamylase, *J. Biol. Chem.* 255, 7301, 1980.
23. Dreyfus, M., Vandenbunder, B., and Buc, H., Mechanism of allosteric activation of glycogen phosphorylase probed by the reactivity of essential arginyl residues. Physicochemical and kinetic studies, *Biochemistry* 19, 3634, 1980.
24. Pande, J., and Myer, Y.P., The arginines of cytochrome c. The reduction-binding site for 2,3-butanedione and ascorbate, *J. Biol. Chem.* 255, 11094, 1980.
25. Tristram-Nagle, S., and Packer, L., Effects of arginine modification on the photocycle and proton pumping of bacteriorhodopsin, *Biochem. Int.* 3, 621, 1981.
26. Gomazkova, V.S., Stafeeva, O.A., and Severin, S.E., The role of arginine residues in the functioning of α-ketoglutarate dehydrogenase from pigeon breast muscle, *Biochem. Int.* 2, 51, 1981.
27. Wong, S.S., and Wong, L.-J., Evidence for an essential arginine residue at the active site of *Escherichia coli* acetate kinase, *Biochim. Biophys. Acta* 660, 142, 1981.
28. Chang, G.-G., and Huang, T.-M., Modification of essential arginine residues of pigeon liver malic enzyme, *Biochim. Biophys. Acta* 600, 341, 1981.

TABLE 4.3 (continued)
Use of 2,3-Butanedione to Modify Arginyl Residues in Proteins

29. Lu, H.S., Talent, J.M., and Gracy, R.W., Chemical modification of critical catalytic residues of lysine, arginine and tryptophan in human glucose phosphate isomerase, *J. Biol. Chem.* 256, 785, 1981.
30. Fujioka, M., and Takata, Y., Role of arginine residue in saccharopine dehydrogenase (L-lysine forming) from baker's yeast, *Biochemistry* 20, 468, 1981.
31. Gacesa, P., Savitsky, M.J., Dodgson, K.S., and Olavesen, A.H., Modification of functional arginine residues in purified bovine testicular hyaluronidase with butane-2,3-dione, *Biochim. Biophys. Acta* 661, 205, 1981.
32. Boggaram, V., and Mannervik, B., Essential arginine residues in the pyridine nucleotide binding sites of glutathione reductase, *Biochim. Biophys. Acta* 701, 119, 1982.
33. Takahashi, K., Further studies on the reactions of phenylgloxal and related reagents with proteins, *J. Biochem.* 81, 403, 1977.
34. Jacobs, A.A.C., van Mechelen, J.R., and De Graaf, F.K., Effect of chemical modification on the K99 and K88ab fibrillar adhesins of *Escherichia coli*, *Biochim. Biophys. Acta* 832, 148, 1985.
35. Swaim, M.W., and Pizzo, S.V., Modification of the tandem reactive centres of human inter-trypsin inhibitor with butanedione and *cis*-dichlorodiamineplatinum (II), *Biochem. J.* 254, 171, 1988.
36. Asryants, R.A., Kuzminskaya, E.V., Tishkov, V.I., Douzhenkova, I.V., and Nagradova, N.K., An examination of the role of arginine residues in the functioning of D-glyceraldehyde-3-phosphate dehydrogenase, *Biochim. Biophys. Acta* 997, 159, 1989.
37. Chen, D., Haemolytic activity of stonustoxin from stonefish (*Synanceja horrida*) venom: pore formation and the role of cationic amino acid residues, *Biochem. J.* 375, 685, 1997.
38. Katsira, V.Ph., and Clonis, Y.D., Chemical modification of barley root oxalate oxidase shows the presence of a lysine, a carboxylate, and disulfides essential for enzyme activity, *Arch. Biochem. Biophys.* 356, 117, 1998.
39. Meyer, C.R., Characterization of ADP-glucose pyrophosphorylase from *Rhodobacter sphaeroides* 2.4.1.: evidence for the involvement of arginine in allosteric regulation. *Arch. Biochem. Biophys.* 372, 178, 1999.
40. Celestina, F., and Suryanarayana, T., Biochemical characterization and helix stabilizing properties of HSNP-C′ from the thermoacidophilic archeon *Sulfolobus acidocaldarius*, *Biochem. Biophys. Res. Commn.* 267, 614, 2000.
41. Clark, D.D., and Ensign, S.A., Characteriztion of the 2-[(R)-2-hydroxypropyl]ethanesulfonate dehydrogenase from *Xanthobacter* strain PY2: product inhibition, pH dependence of kinetic parameters, site-directed mutagenesis, rapid equilibrium inhibition, and chemical modification, *Biochemistry* 41, 2727, 2002.

Arginine is susceptible to oxidation and reaction with products of the oxidation or lipids and carbohydrates to form a variety of products.[75–78] An excellent review of the methods for the analysis of the various oxidation products is available.[76] Methylglyoxal is formed during aerobic glycolysis and reacts with arginine to form imidazolium adducts.[79–81] Ascorbic acid has been shown to react with N^{α}-acetyl-L-arginine to form N^{α}-acetyl-N^{δ}-[4-(1,2-dihydroxy-3-propyliden)-3-imidazolin-5-on-2-yl]-L-ornithine.[82] Arginine is also subject to methylation.[83,84]

Takahashi[85] developed ninhydrin as a reagent for the modification of arginine residues in proteins. The modification reaction was performed in 0.1 M morpholine acetate, pH 8.0, at 25°C in the dark. Modification of both arginine and lysine occurs under these reaction conditions. Specific modification of arginine residues can be accomplished by first modifying the lysine residues with citraconic anhydride. Loforenza and co-workers[86] observed that thiamin transport in renal brush border

TABLE 4.4
Reaction of Arginyl Residues in Proteins with 1,2-Cyclohexanedione

Protein	Solvent	Reagent Excess	Extent of Modification	Ref.
Ribonuclease A	0.2 M sodium borate, pH 9.0	50,000	3/4	1
Lysozyme	0.2 M sodium borate, pH 9.0	50,000	11/11	1
Kunitz bovine trypsin inhibitor	0.2 M sodium borate, pH 9.0	—	5.5/6	2
Threonine dehydrogenase	25 µM triethanolamine, 25 µM sodium borate with pH 2.5 µM 2-mercaptoethanol, pH 7.4	—	—[a]	3
Phosphoenolpyruvate carboxykinase	65 mM Tris-Cl, pH 7.4	—	—[b]	4

[a] Rate of inactivation with 1,2-cyclohexanedione is less than that observed with corresponding molar excesses of either phenylglyoxal or 2,3-butanedione.
[b] Rate constant for inactivation of 0.313 M^{-1} min^{-1}, pH 7.4, at 22°C.

References for Table 4.4

1. Patthy, L., and Smith, E.L., Identification of functional arginine residues in ribonuclease A and lysozyme, *J. Biol. Chem.* 250, 565, 1975.
2. Menegatti, E., Ferroni, R., Benassi, C.A., and Rocchi, R., Arginine modification in Kunitz bovine trypsin inhibitor through 1,2-cyclohexanedione, *Int. J. Pept. Protein Res.* 10, 146, 1977.
3. Epperly, B.R., and Dekker, E.E., Inactivation of *Escherichia coli* L-threonine dehydrogenase by 2,3-butanedione. Evidence for a catalytically essential arginine residue, *J. Biol. Chem.* 264, 18296, 1989.
4. Cheng, K.-C., and Nowak, T., Arginine residues at the active site of avian liver phosphoenolpyruvate carboxykinase, *J. Biol. Chem.* 264, 3317, 1989.

membranes was decreased by treatment with phenylglyoxal but not with ninhydrin. Earlier studies[87] on an enterokinase from kidney beans (*Phaseolus vulgaris*) showed inactivation with either 1,2-cyclohexanedione or ninhydrin. Use of ninhydrin under basic conditions (0.4 M NaOH, alkaline ninhydrin) provides a specific method to detect arginine peptides.[87,88]

REFERENCES

1. Levy, H.M., Leber, P.D., and Ryan, E.M., Inactivation of myosin by 2,4-dinitrophenol and protection by adenosine triphosphate and other phosphate compounds, *J. Biol. Chem.* 238, 3654, 1963.
2. Zanetti, G., Gozzer, C., Sacchi, G., and Curti, B., Modification of arginyl residues in ferrodoxin-NADP reductase from spinach leaves, *Biochim. Biophys. Acta* 569, 127, 1979.

3. Carrillo, N., Arana, J.L., and Vallejos, R.H., An essential carboxyl group at the nucleotide binding site of ferredoxin-NADP+ oxidoreductase, *J. Biol. Chem.* 256, 6823, 1981.

4. Bhagwat, A.S., and Ramakrishna, J., Essential histidine residues of ribulosebisphosphate carboxylase indicated by reaction with diethylpyrocarbonate and rose bengal, *Biochim. Biophys. Acta* 662, 181, 1981.

5. Chang, G.-G., and Huang, T.-M., Modification of essential arginine residues of pigeon liver malic enzyme, *Biochim. Biophys. Acta* 660, 341, 1981.

6. Dunham, K.R., and Selman, B.R., Regulation of spinach chloroplast coupling factor 1 ATPase activity, *J. Biol. Chem.* 256, 212, 1981.

7. Gaccsa, P., Savitzky, M.J., Dodgson, K.S., and Olavesen, A.H., Modification of functional arginine residues in purified bovine testicular hyaluronidase with butane-2,3-dione, *Biochim. Biophys. Acta* 661, 205, 1981.

8. Lundblad, R.L., Noyes, C.M., Featherstone, G.L., Harrison, J.H., and Jenzano, J.W., The reaction of bovine alpha-thrombin with tetranitromethane, *J. Biol. Chem.* 263, 3729, 1988.

9. Plapp, B.V., Site-directed mutagenesis: a tool for the study of enzyme catalysis, *Methods Enzymol.* 249, 91, 1995.

10. Takahashi, K., The reaction of phenylglyoxal with arginine residues in proteins, *J. Biol. Chem.* 243, 6171, 1968.

11. Yankeelov, J.A., Jr., Mitchell, C.D., and Crawford, T.H., A simple trimerization of 2,3-butanedione yielding a selective reagent for the modification of arginine in proteins, *J. Am. Chem. Soc.* 90, 1664, 1968.

12. Patthy, L., and Smith, E.L., Reversible modification of arginine residues. Application to sequence studies by restriction of tryptic hydrolysis to lysine residues, *J. Biol. Chem.* 250, 557, 1975.

13. Lundblad, R.L., *Chemical Reagents for Protein Modification*, 2nd ed., CRC Press, Boca Raton, FL, 1991, p. 173.

14. Sakaguchi, S., and Weber, C.J., A modification of Sakaguchi's reaction for the quantitative determination of arginine, *J. Biol. Chem.* 86, 217, 1930.

15. Izumi, Y., New Sakaguchi reaction, *Anal. Biochem.* 10, 218, 1965.

16. Enoch, H.G., and Strittmatter, P., Role of tyrosyl and arginyl residues in rat liver microsomal stearyl-coenzyme A desaturase, *Biochemistry* 17, 4927, 1978.

17. Mahmoud, L.H., and el-Alfy, N.M., Electron microscopy and histochemical studies on four Egyptian helminthes eggs of medical importance, *J. Egypt. Soc. Paristol.* 33, 229, 2003.

18. Schulze, A. et al., Sakaguchi reaction: a useful method for screening guanidinoacetate:methyl transferase deficiency, *J. Inhert. Metab. Dis.* 19, 706, 1996.

19. Schulze, A. et al., Creatinine deficiency syndrome caused by guanidinoacetate methyl transferase deficiency: diagnostic tools for a new inborn error of metabolism, *J. Pediatr.* 131, 626, 1997.

20. Smith, R.E., and MacQuarrie, R., A sensitive fluorometric method for the determination of arginine using 9,10-phenanthrenequinone, *Anal. Biochem.* 90, 246, 1978.

21. Fan, X. et al., Methylglyoxal-bovine serum albumin stimulates tumor necrosis factor alpha secretion in RAW 264.7 cells though activation of mitogen-activating protein kinase, nuclear factor B and intracellular reactive oxygen species formation, *Arch. Biochem. Biophys.* 409, 274, 2002.

22. Knott, H.H. et al., Glycation and glycoxidation of low-density lipoproteins by glucose and low-moleculare mass aldehydes. Formation of modified and oxidized proteins, *Eur. J. Biochem.* 270, 3572, 2003.

23. Miele, C. et al., Human glycated albumin affects glucose metabolism in L6 skeletal muscle cells by impairing insulin-induced insulin receptor substrate (IRS) signaling through a protein kinase C alpha-mediated mechanism, *J. Biol. Chem.* 278, 47376, 2003.

24. Schepens, I. et al., The role of active site arginines of sorghum NADP-malate dehydrogenase in thioredoxin-dependent activation and activity, *J. Biol. Chem.* 275, 35792, 2000.

25. Linder, M.D. et al., Ligand-selective modulation of the permeability transition pore by arginine modification. Opposing effects of *p*-hydroxyphenylgloxal and phenylglyoxal, *J. Biol. Chem.* 277, 937, 2002.

26. Hagen Braun, C., and Tomer, K.B., Characterization of the tertiary structure of soluble CD4 bound to glycosylated full-length HIVgp120 by chemical modification of arginine residues and mass spectrometric analysis, *Biochemistry* 41, 1759, 2002.

27. Wu, X. et al., Alteration of substrate specificity through mutation of two arginine residues in the binding site of amadoriase II from *Aspergillus* sp., *Biochemistry* 41, 4453, 2002.

28. Xu, G. Takamoto, K., and Chance, M.R., Radiolytic modification of basic amino acid residues in peptides: probes for examining protein-protein interactions, *Anal. Chem.* 75, 6995, 2003.

29. Leitner, A., and Linder, W., Probing of arginine residues in peptides and proteins using selective tagging and electrospray ionization mass spectrometry. *J. Mass. Spectrom.* 38, 891, 2003.

30. Ericksson, O. et al., Inhibition of the mitochondrial cyclosporine A-sensitive permeability pores by the arginine reagent phenylglyoxal, *FEBS Lett.* 409, 361, 1997.

31. Ericksson, O., Fontaine, E., and Bernardi, P., Chemical modification of arginines by 2,3-butanedione and phenylglyoxal causes closure of the mitochondrial permeability transition pore, *J. Biol. Chem.* 273, 12664, 1998.

32. Skysgaard, J.M., Modification of Cl⁻ transport in skeletal muscle of *Rana temporaria* with the arginine-binding reagent phenylglyoxal, *J. Physiol.* 510, 591, 1998.

33. Cook, L.J. et al., Effects of methylglyoxal on rat pancreatic beta-cells, *Biochem. Pharmacol.* 55, 1361, 1999.

34. Emmons, C., Transport characterization of the apical anion exchanger of rabbit cortical collecting duct beta-cells, *Am. J. Physiol.* 276, F635, 1999.

35. Tamari, I. et al., Anion antiport mechanism is involved in the transport of lactic acid across intestinal epithelial brush-border membrane, *Biochim. Biophys. Acta* 1468, 285, 2000.

36. Raw, P.E., and Gray, J.C., The effect of amino acid-modifying reagents on chloroplast protein import and the formation of early import intermediates, *J. Exp. Bot.* 52, 57, 2001.

37. Crenastri, D. et al., Inhibitors of the Cl⁻/HCO₃⁻ exchanger activate an anion channel with similar features in the epithelial cells of rabbit gall bladder: analysis in the intact epithelium, *Pfuegers Arch.* 441, 456, 2001.

38. Verri, A. et al., Molecular characteristics of small intestinal and renal brush border thiamin transporters in rats, *Biochim. Biophys. Acta* 1558, 187, 2002.

39. Forster, I.E. et al., Modulation of renal type IIa Na⁺/P_i cotransporter kinetics by the arginine modifier phenylglyoxal, *J. Membr. Biol.* 187, 85, 2002.

40. Hermann, A. et al., Involvement of amino acid side chains of membrane proteins in the binding of glutathione to pig cerebral cortical membranes, *Neurochem. Res.* 27, 389, 2002.

41. Castagna, M. et al., Inhibitors of the lepidopteran amino acid co-transporter KAAT1 by phenylglyoxal. Role of Arg76, *Insect Mol. Biol.* 11, 389, 2002.

42. Kucera, I., Inhibition by phenylglyoxal of nitrate transport in *Peroacoccus denitrificans*: a comparison with the effect of a protonophorous uncoupler, *Arch. Biochem. Biophys.* 409, 327, 2003.

43. Sacchi, V.F. et al., Glutamate 59 is critical for transport function of the amino acid cotransporter KAAT1, *Am. J. Physiol.* 285, C623, 2003.

44. Winters, C.J., and Adreoli, T.E., Chloride channels is basolateral TAL membranes XVIII. Phenylglyoxal induces functional mcCIC-Ka activity in basolateral MTAL membranes, *J. Membr. Biol.* 195, 63, 2003.

45. Fliss, H., and Viswanatha, T., 2,3-Butanedione as a photosensitizing agent: application to α-amino acids and α-chymotrypsin, *Can. J. Biochem.* 57, 1267, 1979.

46. Mäkinen, K.K., Mäkinen, P.-L., Wilkes, S.H., Bayliss, M.E., and Prescott, J.M., Photochemical inactivation of *Aeromonas* aminopeptidase by 2,3-butanedione, *J. Biol. Chem.* 257, 1765, 1982.

47. Riley, H.A., and Gray, A.R., Phenylglyoxal. In *Organic Syntheses, Collective Vol. 2* (Blatt, A.H., Ed.). John Wiley & Sons, New York, 1943, p. 509.

48. Schloss, J.V., Norton, I.L., Stringer, C.D., and Hartman, F.C., Inactivation of ribulosebisphosphate carboxylase by modification of arginyl residues with phenylglyoxal, *Biochemistry* 17, 5626, 1978.

49. Augustus, B.W., and Hutchinson, D.W., The synthesis of phenyl[2-3H]glyoxal, *Biochem. J.*, 177, 377, 1979.

50. Borders, C.L., Jr., Pearson, L.J., McLaughlin, A.E., Gustafson, M.E., Vasiloff, J., An, F.Y., and Morgan, D.J., 4-Hydroxy-3-nitrophenylglyoxal. A chromophoric reagent for arginyl residues in proteins, *Biochim. Biophys. Acta* 568, 491, 1979.

51. Borders, C.L., Jr., and Johansen, J.T., Identification of Arg-143 as the essential arginine residue in yeast Cu, Zn superoxide dismutase by the use of a chromophoric arginine reagent, *Biochem. Biophys. Res. Commn.* 96, 1071, 1980.

52. Borders, C.L., Jr., and Riordan, J.F., An essential arginyl residue at the nucleotide binding site of creatinine kinase, *Biochemistry* 14, 4699, 1975.

53. Aminlari, M., and Vaseghi, T., A new colorimetric method for determination of creatinine phosphokinase, *Anal. Biochem.* 164, 397, 1987.

54. Wijnands, R.A., Müller, F., and Visser, A.J.W., Chemical modification of arginine residues in *p*-hydroxybenzoate hydroxylase from *Pseudomonas fluorescens*: a kinetic and fluorescence study, *Eur. J. Biochem.* 163, 535, 1987.

55. Julian, T., and Zaki, L., Studies on the inactivation of anion transport in human blood red cells by reversibly and irreversibly acting arginine-specific reagents, *J. Membr. Biol.* 102, 217, 1988.

56 Yamasaki, R.B., Vega, A., and Feeney, R.E., Modification of available arginine residues in proteins by *p*-hydroxyphenylglyoxal, *Anal. Biochem.* 109, 32, 1980.

57. Strevey, J. et al., Characterization of essential arginine residues implicated in the renal transport of phosphate and glucose, *Biochim. Biophys. Acta* 1106, 110, 1992.

58. Mukouyama, E.B., Hirose, T., and Suzuki, H., Chemical modification of L-phenylalanine oxidase from *Pseudomonas* sp. 5012 by phenylglyoxal. Identification of a single arginine residue, *J. Biochem.* 123, 1097, 1998.

59. Linder, M.D. et al., Ligand-modulation of the permeability transition pore by arginine modification. Opposing effects of *p*-hydroxyphenylglyoxal and phenylglyoxal, *J. Biol. Chem.* 277, 937, 2002.

60. Cheung, S.-T., and Fonda, M.L., Reaction of phenylglyoxal with arginine. The effect of buffers and pH, *Biochem. Biophys. Res. Commn.* 90, 940, 1979.

61. Yamasaki, R.B., Shimer, D.A., and Feeney, R.E., Colorimetric determination of arginine residues in proteins by *p*-nitrophenylglyoxal, *Anal. Biochem.* 111, 220, 1981.

62. Branlant, G., Tritsch, D., and Biellmann, J.-F., Evidence for the presence of anion-recognition sites in pig-liver aldehyde reductase. Modification by phenylglyoxal and p-carboxyphenyl glyoxal of an arginyl residue located close to the substrate-binding site, *Eur. J. Biochem.* 116, 505, 1981.

63. Eun, H.-M., Arginine modification by phenylglyoxal and (p-hydroxyphenyl)glyoxal: reaction rates and intermediates, *Biochem. Int.* 17, 719, 1988.

64. Meves, H., Rubly, N., and Stämpfli, R., The action of arginine specific reagents on ionic and gating currents in frog myelinated nerve, *Biochim. Biophys. Acta* 943, 1, 1988.

65. Pearson, M.A. et al., Kinetic and structural characterization of urease active site variants, *Biochemistry* 39, 8575, 2000.

66. Yankeelov, J.A., Jr., Modification of arginine in proteins by oligomers of 2,3-butane-dione, *Biochemistry* 9, 2433, 1970.

67. Riordan, J.F., Functional arginyl residues in carboxypeptidase A. Modification with butanedione, *Biochemistry* 12, 3915, 1973.

68. Gripon, J.-C., and Hofmann, T., Inactivation of aspartyl proteinases by butane-2,3-dione. Modification of tryptophan and tyrosine residues and evidence against reaction of arginine residues, *Biochem. J.* 193, 55, 1981.

69. Alkena, W.B.L. et al., Role of αArg^{145} and Arg^{263} in the active site of penicillin acylase of *Escherichia coli*, *Biochem. J.* 365, 303, 2003.

70. Clark, D.D., and Ensign, S.A., Characteriztion of the 2-[(R)-2-hydroxypropyl]ethane-sulfonate dehydrogenase from *Xanthobacter* strain PY2: product inhibition, pH dependence of kinetic parameters, site-directed mutagenesis, rapid equilibrium inhi-bition, and chemical modification, *Biochemistry* 41, 2727, 2002.

71. Toi, K., Bynum, E., Norris, E., and Itano, H. A., Studies on the chemical modification of arginine. I. The reaction of 1,2-cyclohexanedione with arginine and arginyl residues of proteins, *J. Biol. Chem.* 242, 1036, 1967.

72. Patthy, L., and Smith, E.L., Identification of functional arginine residues in ribonu-clease A and lysozyme, *J. Biol. Chem.* 250, 565, 1975.

73. Ullah, A.H.J., and Sethumadhaven, K., Differences in the active site environment of *Aspergillus ficus* phytases, *Biochem. Biophys. Res. Commn.* 243, 458, 1998.

74. Calvete, J.J. et al., Characterization of the conformation and quaternary structure-dependent heparin-binding region of bovine seminal plasma protein PDC-109, *FEBS Lett.* 444, 260, 1999.

75. Baynes, J.W., and Thorpe, S.R., Role of oxidative stress in diabetic complications: a new perspective on an old paradigm, *Diabetes* 48, 1, 1995.

76. Reubsaet, J.L.E. et al., Analytical techniques used to study the degradation of proteins and peptides: chemical instability, *J. Pharm. Biomed. Anal.* 17, 955, 1998.

77. Thornalley, P.J., Glutathione-dependent detoxification of alpha-oxoaldehyde by the glyoxalse system: involvement in disease mechanisms and antiproliferative activity of glyoxalase I inhibitors, *Chem. Biol. Interact.* 111/112, 137, 1998.

78. Deyl, Z., and Miksik, I., Post-translational non-enzymatic modification of proteins. I. Chromatography of marker adducts with special emphasis on glycation reactions, *J. Chromatogr. B.* 699, 287, 1997.

79. Henle, T. et al., Detection and identification of a protein-bound imidalone resulting from the reaction of arginine and methylglyoxal, *Z. Lebensm. Uners. Forsch.* 199, 55, 1994.

80. Westwood, M.E. et al., Methylglyoxal-modified arginine residues: a signal for recep-tor-mediated endocytosis and degradation of proteins by monocytic THP-1 cells, *Biochim. Biophys. Acta* 1316, 84, 1997.

81. Degenhardt, T.P., Thorpe, S.R., and Baynes, J.W., Chemical modification of proteins by methylglyoxal, *Cell. Mol. Biol.* 44, 1139, 1998.
82. Pischetsrieder, M., Reaction of L-ascorbic acid with L-arginine derivatives, *J. Agric. Food Chem.* 44, 2081, 1996.
83. Kim, S., Park, G.H., and Paik, W.K., Recent advances in protein methylation: enzymatic methylation of nucleic acid binding proteins, *Amino Acids* 15, 291, 1998.
84. Urnov, F.D., Methylation and the genome: the power of a small amendment, *J. Nutr.* 132 (Suppl. 8), 2450S, 2002.
85. Takahashi, K., Specific modification of arginine residues in proteins with ninhydrin, *J. Biochem.* 80, 1173, 1976.
86. Verri, A. et al., Molecular characteristics of small intestinal and renal brush border thiamin transporters in rats, *Biochim. Biophys. Acta* 1558, 187, 2002.
87. Rhodes, G.R., and Boppana, V.K., High-performance liquid chromatographic analysis of arginine-containing peptides in biological fluids by means of a selective post-column reaction with fluorescence detection, *J. Chromatogr.* 444, 123, 1988.
88. Boppana, V.K., and Rhodes, G.R., High-performance liquid chromatographic determination of an arginine-containing octapeptide antagonist of vasopressin in human plasma by means of a selective post-column reaction with fluorescence detection, *J. Chromatogr.* 507, 79, 1990.

5 The Modification of Carboxyl Groups

The site-specific modification of carboxyl groups in proteins is somewhat difficult to achieve, as is differentiation between aspartyl residues and glutamyl residues. Most of the current work uses water-soluble carbodiimides, as described later. The reader is recommended to several recent articles addressing the modification of carboxyl groups in proteins.[1-4]

Parsons and co-workers used triethyloxonium fluoroborate to modify the β-carboxyl groups of an aspartic residue essential for the enzymatic activity of lysozyme.[5,6] Paterson and Knowles[7] used trimethyloxonium fluoroborate to determine the number of carboxyl groups in pepsin essential for catalytic activity. This chapter discusses in some depth the rigorous precautions necessary to prepare this reagent. This reagent is highly reactive, and considerable care is required for its introduction into the reaction mixture containing protein. The reaction is performed at pH 5.0 (0.020 M sodium citrate, pH maintained at 5.0 with 2.5 M NaOH). These investigators also report the preparation of the ^{14}C-labeled reagent from sodium methoxide and [^{14}C]-methyl iodide. Llewellyn and Moczydlowski[8] modified saxiphilin with trimethyloxonium tetrafluoroborate. As trimethyloxonium tetrafluoroborate is unstable in water,[9] the reaction was performed under stringent conditions. Trimethyloxonium tetrafluoroborate was placed in a tube, which was flushed with nitrogen and then sealed. A buffered solution (100 mM Tris-HCl, 100 mM NaCl, pH 8.6) of saxiphilin and bovine serum albumin was introduced and the reaction allowed to proceed for 10 min. Cherbavez[10] observed that modification of batrochotoxin-activated Na channels with trimethyloxonium tetrafluoroborate altered function. Eschada and co-workers[11] used triethyloxonium tetrafluoroborate to modify carboxyl groups at the active site of a -xylosidase from *Trichoderma reese* QM9414. The reagent in dichloromethane was added to the protein in 20 mM MES, pH 5.1.

Woodward and co-workers[12] developed N-ethyl-5-phenylisoxazolium-3'-sulfonate (Woodward's reagent K; Figure 5.1) and various other N-alkyl-5-phenylisoxazolium fluoroborates as reagents for the activation of carboxyl groups for synthetic purposes. Anfinsen and co-workers[13] studied the kinetics of the aqueous hydrolysis of this reagent and reaction with staphylococcal nuclease. This study demonstrated that Woodward's reagent K is very unstable in aqueous solution above pH 3.0. Studies on the rate of enzyme inactivation by this reagent should be corrected for reagent hydrolysis to obtain accurate second-order rate constants. Bodlaender and co-workers[14] used N-ethyl-5-phenylisoxazolium-3-sulfonate, the N-methyl and N-ethyl derivatives of 5-phenylisoxazolium fluoroborate or N-methylbenzisoxazolium fluoroborate, to activate carboxyl groups on trypsin for subsequent modification with

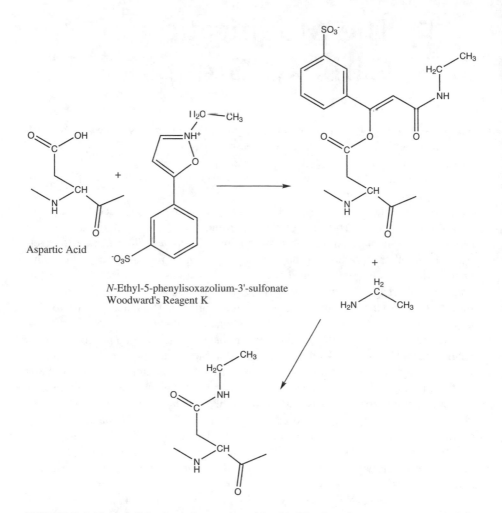

FIGURE 5.1 The modification of aspartic acid with Woodward's reagent K (*N*-ethyl-5-phenylisoxazolium-3′-sulfonate). The formation of the ketoketimine intermediate is shown with the subsequent reaction with a nucleophile (ethyl amine) to form a stable modified derivative.

methylamine or ethylamine. The extent of modification obtained ranged from ca. 3 residues modified with *N*-methyl-5-phenylisoxazolium fluoroborate or *N*-ethyl-5-phenylisoxazolium fluoroborate, pH 3.8, 20°C, 80 min, to ca. 11 residues modified with *N*-methyl-5-phenylisoxazolium fluoroborate, pH 6.0, 20°C, 10 min. Reagent decomposition occurs quite rapidly, even at ice-bath temperature (2°C). The modification appears fairly selective for carboxyl groups, although some modification of lysine was observed under conditions in which extensive modification was obtained (250-fold molar excess of *N*-methyl-5-phenylisoxazolium fluoroborate, pH 4.75, 72 min, 20°C, methylamine as the attacking nucleophile).

Saini and Van Etten[15] reported the reaction of N-ethyl-5-phenylisoxazolium-3'-sulfonate with human prostatic acid phosphatase. The modification was performed with a 4000- to 10,000-fold molar excess of reagent in 0.020 M pyridinesulfonic acid, pH 3.6, at 25°C. Ethylamine was used as the attacking nucleophile to determine the extent of modification. A substantial number of carboxyl groups in the protein were modified under these experimental conditions. Arana and Vallejos[16] compared the reaction of chloroplast coupling factor with Woodward's reagent K and dicyclohexylcarbodiimide. Reaction with Woodward's reagent K was accomplished at 25°C in 0.040 M Tricine, pH 7.9, and reaction with dicyclohexylcarbodiimide was accomplished at 30°C in 0.040 M MOPS, pH 7.4. ATP and derivatives such as ADP and inorganic phosphate protected against the loss of activity occurring on reaction with Woodward's reagent K, but they had no effect on inactivation by dicyclohexylcarbodiimide. The reverse was seen with divalent cations such as Ca^{2+}. The modification of an essential carboxyl group in pancreatic phospholipase A_2 by Woodward's reagent K has been reported.[17] The reaction was performed in 0.01 M sodium phosphate, pH 4.75 (pH stat), at 25°C. A second-order rate constant of $k_2 = 25.5$ M^{-1} min^{-1} was obtained for the loss of catalytic activity. This rate inactivation increased more than twofold in the presence of 30 mM CaCl$_2$ (69.3 M^{-1} min^{-1}). Quantitative information on the extent of modification is obtained with [^{14}C]-glycine ethyl ester. Treatment with a water-soluble carbodiimide, 1-(3'-dimethylaminopropyl)-3-ethylcarbodiimide, results in the loss of catalytic activity in a reaction with characteristics different from those seen with Woodward's reagent K. Kooistra and Sluyterman[18] modified guanidated mercuripapain with N-ethylbenziosoxazolium tetrafluoroborate at pH 4.2, 0°C, to yield a protein modified with N-ethylsalicylamide esters. When the active ester groups on the protein were allowed to undergo amminolysis (2.0 M ammonium acetate, pH 9.2) and the esters were converted to the corresponding amides (an isosteric modification), conversion to the amides resulted in the loss of a bulky substituent group, with an increase in activity. Johnson and Dekker[19] reported the modification of histidine and cysteine residues in L-threonine dehydrogenase. The lack of specificity of Woodward's reagent K can be used to possible advantage. Puri and Colman[20] took advantage of Woodward's reagent K reacting with carboxyl groups to yield enol esters, which cannot be reduced with sodium borohydride, whereas reaction with other functional groups such as amino groups and histidine forms unsaturated ketones, which can be reduced. The use of tritiated sodium borohydride (sodium borotritiide) enables the incorporation of radiolabel into a modified protein. Woodward's reagent K has been used by several groups[21,22] to modify Glu-681 in human erythrocyte band 3 protein.

The use of carbodiimide-mediated modification of carboxyl functional groups in proteins[23,24] (Figure 5.2) is likely the most extensively used method for the study of such functional groups. Carbodiimides most likely react with protonated carboxyl groups to yield an activated intermediate, most likely an acylisourea, which then reacts with a nucleophile such as an amine.[25]

Carbodiimides are also used for zero-length cross-linking of proteins between proximate lysine residues and carboxyl groups.[26,27] Water-insoluble carbodiimides such as N,N'-dicyclohexylcarbodiimide continue to be useful for the site-specific

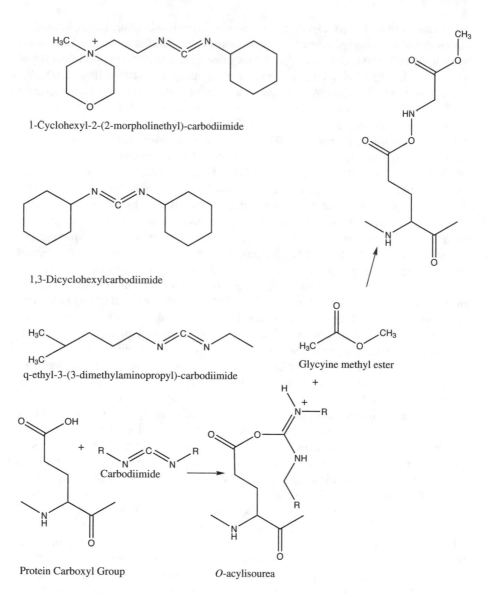

1-Cyclohexyl-2-(2-morpholinethyl)-carbodiimide

1,3-Dicyclohexylcarbodiimide

q-ethyl-3-(3-dimethylaminopropyl)-carbodiimide

Glycyine methyl ester

Carbodiimide

Protein Carboxyl Group   *O*-acylisourea

FIGURE 5.2 Structures of some commonly used carbodiimides and a scheme for the reaction of carbodiimides with carboxyl groups in proteins.

modification of carboxyl groups,[2,28,29] but most current work uses water-soluble carbodiimides such as 1-ethyl-3-(3-dimethylaminopropylcarbodiimide (*N*-ethyl-*N'*-(dimethylaminopropyl) carbodiimide; EDC). Several studies comparing the application of water-insoluble (hydrophobic) carbodiimides and water-soluble carbodiimides (hydrophilic) are discussed later. Sheehan and Hlavka[30,31] developed water-soluble carbodiimides. The first application to proteins was the zero-length cross-linking of collagen.[31] This study used 1-ethyl-3-(2-morpholinyl-(4)-ethyl)carbodiimide metho-p-toluenesulfonate in unbuffered aqueous solution. Riehm and Scheraga[10] advanced the use water-soluble carbodiimides for proteins in a study on the modification of ribonuclease with 1-cyclohexyl-3-(2-morpholinylethyl)carbodiimide at pH 4.5 with a pH-stat. A number of different products were obtained, which could be separated by ion-exchange chromatography (BioRex® 70). These investigators also suggest that the mechanism between a carbodiimide and a carboxyl group leads to the formation of an unstable acylisourea, which can either decompose to an acylurea derivative or react with a nucleophile. Work by Koshland and associates provided a major advance to this technology. Their approach used a water-soluble carbodiimide in the presence of excess amino acid nucleophile. The initial study by Hoare and Koshland[33] used N-benzyl-N'-3-dimethylaminopropylcarbodiimide. Subsequent studies used EDC[23,34] and resulted in the development of a quantitative method to measure carboxyl groups in proteins.[23] These initial studies also introduced the concept of using a unique nucleophile such as norleucine methyl ester, aminomethane-sulfonic acid, and norvaline. These initial studies used 0.1 M carbodiimide, pH 4.75, with 1.0 M glycine methyl ester in a pH-stat at 25°C. The possibility of a side reaction was discussed with reference to the possible modification of the phenolic hydroxyl of tyrosine to form the O-arylisourea.

This general approach continues to be used at present,[35] with most reactions being performed at pH 6 to 7 for reasons discussed later. Furthermore, few, if any, investigators perform the reaction in a pH-stat although, as noted later, there are issues with respect to the effect of buffer components. Most investigators use the "Good" buffers developed by Good and colleagues.[36,37]

There has been a useful amount of work on the characterization of EDC. Border and colleagues[38] evaluated the stability of EDC in aqueous solution. EDC has a t_{j} of 37 h (pH 7.0), 20 h (pH 6.0), and 3.9 h (pH 5.0) in 50 mM 2-(N-morpholino)ethanesulfonic acid at 25°C; in the presence of 100 mM glycine, the t_{j} values were 15.8 h (pH 7.0), 6.7 h (pH 6.0), and 0.73 h (pH 5.0). The authors suggest that this supports the use of EDC at pH 6.0 or 7.0, but at pH 5.0 stability would be an issue. This study also reported a major decrease in the stability of EDC under the previously given solvent conditions in the presence of 10 mM phosphate or 10 mM ATP. Lei and co-workers[39] reported kinetic studies on the hydrolysis of EDC in aqueous solution under acidic conditions. The conversion of EDC to the corresponding acylurea was measured by mass spectrometry and capillary electrophoresis. Consistent the Borders' results,[38] the rate of decomposition increased with increasing acidity (decreased pH). Mirsky and colleagues[40] used the decrease in absorbance at 214 nm ($\varepsilon = 6.3 \times 10^{-3}\ M^{-1}\ \text{cm}^{-1}$ to measure the stability of EDC. These investigators also report decreased stability with decreasing pH. The presence of citrate, acetate, or phosphate increased the rate of EDC decomposition. Sehgal and Vijay[41] optimized the conditions for EDC-mediated coupling of a carboxyl-containing compound to an amine matrix (Affi-Gel® 102). They noted that the presence of N-hydroxysuccinimide greatly improved the coupling of butyric acid to the matrix.

Modification of the active-site cysteinyl group in papain has been reported[42] as occurring under conditions (pH 4.75, 25°C) in which 6:14 carboxyl groups are modified together with 9/19 tyrosyl residues. Modification of the tyrosyl residues is reversed by 0.5 M hydroxylamine, pH 7.0 (5 h at 25°C) as first demonstrated by Carraway and Koshland.[43] Despite the problems with side reaction, modification of the carboxyl group in proteins with a water-soluble carbodiimide and an appropriate nucleophile (e.g., [14C]-glycine ethyl ester, norleucine methyl ester — easily detected by amino acid analysis — aminomethylsulfonic acid) has proven extremely useful.[44] Ammonium ions can be used as the attacking nucleophile to generate asparaginyl and glutaminyl residues from exposed carboxyl groups. Modification was accomplished in 5.5 M NH$_4$Cl, pH 4.75, for 3 h at 25°C. Under these conditions, ca. 11 of the 15 free carboxyl groups in chymotrypsinogen were converted to the corresponding amide.[45] 1,2-Diaminoethane or diaminomethane can be coupled to aspartic acid residues to produce a trypsin-sensitive bond.[46] The work of Lin and co-workers[47] provides another example of using this chemistry to introduce a new functional group into a protein. The first step involves the water-soluble carbodiimide-mediated coupling of cystamine to protein carboxyl groups. Reduction of the coupled cystamine with dithiothreitol results in 2-aminothiol functional groups bound to protein carboxyl groups.

Several studies have used nitrotyrosine ethyl ester as the modifying nucleophile with EDC.[48–52] Pho and co-workers[47] reported the inactivation of yeast hexokinase with 1-cyclohexyl-3-(2-morpholinoethyl)carbodiimide metho-p-sulfonate (CMC). The reaction was performed in 0.1 M phosphate, pH 6.0, and was accelerated by the addition of nitrotyrosine ethyl ester. The nitrotyrosine nucleophile provides a chromophoric label for the modified carboxyl group ($\varepsilon = 4.6 \times 10^3$ M^{-1} cm^{-1}, 0.1 M NaOH). The nitrotyrosine also provides for the isolation of modified peptides by immunoaffinity chromatography.[49]

A number of studies have used several different reagents for the site-specific modification of carboxyl groups in the same research study. Several compare a water-insoluble carbodiimide, N,N'-dicyclohexylcarbodiimide (DCC) and EDC. Sigrist-Nelson and Azzi[53] studied the effect of carboxyl-modifying reagents on proton translocation in a liposomal chloroplast ATPase complex. Proton transport was inhibited by DCC but not by EDC. Sussman and Slayman[54] reported similar differences with *Neurospora crassa* plasma membrane ATPase. This difference most likely reflects the hydrophobic nature of DCC, permitting it to pass through membranes (membrane-permeant reagent). Subsequent studies demonstrated that EDC can inhibit the purified F$_1$ ATPase,[55] but the reaction is complex and multiple products, including cross-linked derivatives, are obtained. Shoshan-Barmatz and co-workers[56] showed that voltage-dependent anion channel activity in skeletal muscle sarcoplasmic reticulum was inhibited by DCC (20 mM MES, 0.2 M NaCl, pH 6.4) but not by EDC or Woodward's reagent K. Pan and colleagues[57] modified purified vacuolar H+-pyrophosphatase (*Vigna radiate* L.) with several carboxyl reagents [50 mM MOPS, 20% (v/v) glycerol, 1 mM EGTA, and 1 mM DTT, pH 7.2, 33°C]. Rapid inactivation was observed with DCC and much slower inactivation by either EDC or Woodward's reagent K. Bjerrum and colleagues[58] used an impermeant, water-soluble carbodiimide, 1-ethyl-3-(1-azonia-4,4-dimethylpentyl) carbodiimide to inhibit anion exchange in human red blood cell membranes. The rate and extent of inactivation were increased by adding tyrosine ethyl ester.

It is not necessary to have a nucleophile present to study the modification of a carboxyl group with a carbodiimide. It does make identification of the modified residue less complicated, but, at the same time, can introduce kinetic ambiguity. Planas and colleagues[4] studied the modification of carboxyl groups in *Bacillus licheniformis* 1,3-1,4-β-glucanase. Modification of the native enzyme with EDC (20 mM MES, 20 mM HEPES, and 40 mM 4-hydroxy-1-methylpiperidine, pH 5.5, 25°C) resulted in the rapid loss of activity to a residual level of 23%; addition of glycine ethyl ester reduced the activity to 8%. It was not possible to use the analytical approach of Levy[59] to obtain useful kinetic data, and it was concluded that the loss of activity reflected the modification of two carboxyl groups at the active site. A pH dependence study showed a single titration curve with the most rapid inactivation at acid pH. It was possible to estimate a pK^a of 5.3 ± 0.2 for the modified groups. Table 5.1 gives examples of carboxyl group modification in proteins.

There are examples of carboxyl group modification with reagents expected to react far more effectively with other nucleophiles (Figure 5.3). An example of this is the reaction of iodoacetamide with ribonuclease T_1 to form the glycolic acid derivative of the glutamic acid residue, as elegantly shown by Takahashi and co-workers.[60] Another example is the modification of a specific carboxyl group in pepsin by *p*-bromophenacyl bromide.[61] (The use of *p*-bromophenacyl bromide in the specific modification of proteins is not uncommon, but is generally associated with the modification of cysteine, histidine, or methionine.) In the study of pepsin, optimal inactivation (ca. 12-fold molar excess of reagent, 3 h, 25°C) was obtained in the pH range of 1.5 to 4.0, with a rapid decrease in the extent of inactivation at pH 4.5 and above. (The effect of pH higher than 5.5 to 6.0 on the modification of pepsin cannot be studied because of irreversible denaturation of pepsin at pH 6.0 and above.) In studies with a 10% molar excess of *p*-bromophenacyl bromide, pH 2.8, at 37°C for 3 h, complete inactivation of the enzyme was obtained concomitant with the incorporation of 0.93 Mol of reagent per mole of pepsin (assessed by bromide analysis). Attempts to reactivate the modified enzyme with a potent nucleophile such as hydroxylamine were unsuccessful, but reactivation could be obtained with sulfhydryl-containing reagents (i.e., β-mercaptoethanol, 2,3-dimercaptopropanol, thiophenol). It has been subsequently established that reaction occurs at the β-carboxy group of an aspartic acid residue (formation of 2-*p*-bromophenyl-1-ethyl-2-one β-aspartate).[62] These investigators noted that the reduction of enzyme under somewhat harsh conditions (LiBH$_4$ in tetrahydrofuran) resulted in the formation of homoserine. Cury and colleagues[63] modified ASP49 in the phospholipase A from *Bothrops asper* snake venom. This modification eliminates enzymatic activity and also abolishes the hyperalgesia effect.

TABLE 5.1
Modification of Carboxyl Groups in Proteins

Protein	Reaction Conditions	Carbodiimide	Molar Excess	Nucleophile	Carboxyl Groups Modified	Other Functional Groups Modified	Ref.
Lysozyme	pH 4.75 at 25°C, pH-stat,[a] 0.1 M carbodiimide	BDC[b]	140	1.0 M glycine methyl ester	2.1/11 (5 min) 4.7/11 (60 min) 8.1/11 (5–6 h)	—	1
Lysozyme			140	0.1 M nitrotyrosine ethyl ester	1.4/11[c]	—	1
Trypsin			250	1.0 M glycine methyl ester	4.6/11 (5 min) 8.8/11 (60 min) 12.5/11 (5–6 h)[d]	—	1
Chymotrypsin			250	1.0 M glycine methyl ester	6.2/17 (5 min) 11.8/17 (60 min) 15.5/17 (5–6 h)	—	1
Trypsinogen	pH 4.5 at 25°C	EDC[f]	25	1.0 M glycine ethyl ester[e]	—	—	2
Chymotrypsinogen	pH 4.0 at 25°C, pH-stat	EDC		1.0 M glycine ethyl ester	13/14[g]	—	3
	pH 6.0 at 25°C, pH-stat				10/14	—	3
	pH 8.0 at 25°C, pH-stat				3/14	—	3
Chymotrypsin	pH 4.75 at 25°C, pH-stat	EDC		1.0 M glycine methyl ester	12.7[h] 15.6[i] 10.6 13.5[j]	—	4 4
Lysozyme	pH 4.75 at 25°C, pH-stat	BDC		0.25 M aminomethanesulfonic acid	8.5–9.5[j]	—	5
				10 M glycine methyl ester	6.5–7.5[j]	—	5

Protein	Conditions	Reagent	Concentration	Nucleophile	Ratio	Residues modified	Ref.
Trypsin	pH 4.75 at 25°C, pH-stat	EDC	—	1.0 M glycinamide	7.9[k] / 1.7[k]	3–5 Tyr[l]	6
Albumin	pH 4.75	EDC	—	Glycine methyl ester or L-argininamide	1[m]	—	7
Chymotrypsin	pH 4.0[n]	EDC	—	Glycine ethyl ester	15/15	—	8
Papain	pH 4.75 at 25°C	EDC	300–1200	Glycine ethyl ester	6/14	6–10 Tyr,[o] active-site cysteinyl residue	9
Yeast Hexokinase	0.1 M phosphate, pH 6.0 at 20°C[p]	CMC[q]	500–3000	Nitrotyrosine[r] ethyl ester	1	Cysteinyl (2)	10
Phosphorylase[b]	pH 5.1 at 25°C	CMC	—	Glycine[s] ethyl ester	3[t]	—	11
α-Mannosidase	pH 4.2, 0.1 M MES, 1.0 M NaCl	EDC	2000	Glycine[g] ethyl ester	8	—	12
Yeast enolase	0.050 M MES, 1 mM MgCl$_2$, 0.01 mM EDTA, pH 6.1	EDC, CMC[u]	2000	—	—	[v]	13
3-Phosphoglycerate kinase	0.1 M phosphate, pH 6.1 at 17°C	CMC	2000	Nitrotyrosine[v] ethyl ester	1	—	14
cAMP-dependent protein kinase	pH 6.5, 0.050 M MES, 23°C	EDC	—	Glycine ethyl ester	1.7[w]	[x]	15
Human Fc fragment	pH 4.75	EDC	—	Glycine ethyl ester	25	[y]	16
Pancreatic phospholipase A$_2$	0.25 M cacodylate, pH 5.5	EDC	—	Semicarbazide	13/15[z]	—[aa]	17
Spinach plastocyanin	pH 3.5	EDC	—	Semicarbazide	14/15[z]	—[aa]	17
	pH 6.0 at 23°C, borate	EDC	—	Ethylenediamine	4.3/16	—	18
Mitochondrial F$_1$-ATPase	0.05 M triethanolamine H$_2$SO$_4$, pH 7.0	DCC[cc]	6	—	2[bb]	—	19
Mitochondrial transhydrogenase	1 mM Tricine, pH 7.0 with 0.1 M choline chloride and 2% MeOH	DCC	—	—	—	—[dd]	20

(continued)

TABLE 5.1 (continued)
Modification of Carboxyl Groups in Proteins

Protein	Reaction Conditions	Carbodiimide	Molar Excess	Nucleophile	Carboxyl Groups Modified	Other Functional Groups Modified	Ref.
Lysozyme	pH 5.0	EDC	3.5[ee]	Ethylenediamine, ethanolamine, 4-(5)-(aminomethyl) imidazole, histamine, D-glucosamine, methylamine	1	—	21
Restriction endonuclease *Eco* RI	0.1 *M* triethanolamine pH 7.0, 2.0 *M* KCl at 20°C	CMC	—	—[ff]	—	—	22
Thylakoid membrane proteins	pH 7.5 (HEPES)	DDC	—	Glycine ethyl ester	—	—	23
		EDC	—	Glycine ethyl ester	—	—	

[a] It is generally necessary to add dilute HCl (0.2 *M*) during the course of the reaction to maintain the pH at 4.75.

[b] BCD, *N*-benzyl-*N'*-(3-dimethylaminopropyl)carbodiimide.

[c] Reaction time not given.

[d] Excess might result from autolysis of trypsin preparations, which would create "new" free carboxyl groups.

[e] Reactions are generally terminated by dilution into cold sodium acetate (1.0 *M*, pH 3.5 to 5.5).

[f] EDC, 1-ethyl-(3-dimethylaminopropyl)carbodiimide.

[g] Determined by incorporation of [^{14}C]-glycine ethyl ester.

[h] 1-H reaction terminated with 1.0 *M* acetate, pH 4.75.

[i] In 7.5 *M* urea.

[j] Note the interesting difference in the extent of modification, which is dependent on nucleophile used. There is also an interesting difference in the time course of modification.

k In the presence and absence of the competitive inhibitor benzamidine.

l Tyrosyl residues regenerated in 0.5 M hydroxylamine, pH 7.1, with no effect on the EDC/glycinamide changes in catalytic activity.

m Complete modification of the carboxyl groups was achieved in 6.0 M guanidine with either L-argininamide or glycine methyl ester. After reduction and carboxymethylation, ca. 20% of the carboxyl groups are unreactive with either nucleophile. There is a further decrease in modification with the reduced and cyanoethylated derivative.

n Reaction at pH 4.0 results in apparent quantitative modification of carboxyl groups.

o Tyrosine modification is acid stable, but reversed in 0.5 M hydroxylamine, pH 7.0, for 5 h at 25°C. Activated papain is irreversibly modified at active-site cysteinyl residue (Cys-25) by EDC, whereas mercuripapain is not.

p Optimal inactivation occurred at pH 5.5 to 6.0, with marked decrease in extent of inactivation at more alkaline pH.

q CMC, 1-cyclohexyl-3-(2-morpholinoethyl)carbodiimide metho p-toluenesulfonate.

r Isolated peptide containing modified glutamic acid residue, using affinity chromatography with antinitrotyrosyl γ-globulin.

s Inactivation is not dependent on addition of nucleophile, but rate is greatly enhanced.

t Determined by incorporation of [¹⁴C] from [Metho-¹⁴C] CMC. Also determined from the extent of incorporation of N-(2,4-dinitrophenyl)-ethylene diamine (spectrophometry); $\varepsilon = 15,000 \ M^{-1} \ cm^{-1}$. It was determined that the modification of one carboxyl group is critical for potential catalytic function.

u These investigators also used 1-methyl-(4-azonia-4,4-dimethylphenyl)-carbodiimide (1-ethyl-3-(3-dimethylaminopropyl)-carbodiimide). This reagent was more effective than either EDC or CMC (least active) in the inactivation of the enzyme under these conditions.

v Reaction at cysteine and tyrosine excluded by amino acid analysis after acid hydrolysis.

w Determined γ-Glu-Gly after proteolysis with trypsin (2X), pronase, carboxypeptidases A and B, and leucine aminopeptidase.

x Direct determination not performed. Reaction at pH 8.0 (carbodiimide is more specific for phenolic hydroxyl at alkaline pH) did not result in loss of catalytic activity.

y Tyrosine modification excluded by amino acid analysis after acid hydrolysis.

z Radiolabeled semicarbazide (synthesized from [¹⁴C] cyanate) incorporation.

aa Tyrosine modification occurred. Tyrosine regenerated with neutral hydroxylamine. These investigators stressed the need to keep hydroxylamine exposure as brief as possible to avoid side reactions such as peptide bond cleavage or deamidation.

bb From incorporation of [¹⁴C]-carbodiimide.

cc DCC, dicyclohexylcarbodiimide.

dd Interchain cross-linking of transhydrogenase dimer occurred under these conditions.

ee Obtained specific modification of Asp-101 by using low molar excess of carbodiimide. Extent of modification is somewhat independent of amine (nucleophile) used. These investigators speculate that increased specificity is a reflection of binding of carbodiimide to substrate binding site close to Asp-101. These investigators purified reaction products to obtain selectively modified protein derivatives.

ff Rate and extent of inactivation not changed by the addition of glycine ethyl ester.

(continued)

TABLE 5.1 (continued)
Modification of Carboxyl Groups in Proteins

References for Table 5.1

1. Hoare, D.G., and Koshland, D.E., Jr., A procedure for the selective modification of carboxyl groups in proteins, *J. Am. Chem. Soc.* 88, 2057, 1966.

2. Radhakrishnan, T.M., Walsh, K.A., and Neurath, H., Relief by modification of carboxylate groups of the calcium requirement for the activation of trypsinogen, *J. Am. Chem. Soc.* 89, 3059, 1967.

3. Abita, J.P., Maroux, S., Delaage, M., and Lazdunski, M., The reactivity of carboxyl groups in chymotrypsinogen, *FEBS Lett.* 4, 203, 1969.

4. Carraway, K.L., Spoerl, P., and Koshland, D.E., Jr., Carboxyl group modification in chymotrypsin and chymotrypsinogen, *J. Mol. Biol.* 42, 133, 1969.

5. Lin, T.-Y., and Koshland, D.E., Jr., Carboxyl group modification and the activity of lysozyme, *J. Biol. Chem.* 244, 505, 1969.

6. Eyl, A.W., Jr., and Inagami, T., Identification of essential carboxyl groups in the specific binding site of bovine trypsin by chemical modification, *J. Biol. Chem.* 246, 738, 1971.

7. Frater, R., Reactivity of carboxyl groups in modified proteins, *FEBS Lett.* 12, 186, 1971.

8. Johnson, P.E., Stewart, J.A., and Allen, K.G.D., Specificity of α-chymotrypsin with exposed carboxyl groups blocked, *J. Biol. Chem.* 251, 2353, 1976.

9. Perfetti, R.B., Anderson, C.D., and Hall, PL., The chemical modification of papain with 1-ethyl-3-(3-dimethylaminopropyl) carbodiimide, *Biochemistry* 15, 1735, 1976.

10. Pho, D.B., Roustan, C., Tot, A.N.T., and Pradel, L.-A., Evidence for an essential glutamyl residue in yeast hexokinase, *Biochemistry* 16, 4533, 1977.

11. Ariki, M., and Fukui, T., Modification of rabbit muscle phosphorylase b by a water-soluble carbodiimide, *J. Biochem. (Tokyo)* 83, 183, 1978.

12. Paus, E., Reaction of α-mannosidase from *Phaseolus vulgaris* with group-specific reagents. Essential carboxyl groups, *Biochim. Biophys. Acta* 526, 507, 1978.

13. George, A.L., Jr., and Borders, C.L., Jr., Essential carboxyl residues in yeast enolase, *Biochem. Biophys. Res. Commn,* 87, 59, 1979.

14. Desvages, G., Roustan, C., Fattoum, A., and Pradel, L.-A., Structural studies on yeast 3-phosphoglycerate kinase. Identification by immunoaffinity chromatography of one glutamyl residue essential for 3-phosphoglycerate kinase activity. Its location in the primary structure, *Eur. J. Biochem.* 105, 259, 1980.

15. Matsuo, M., Huang, C.-H., and Huang, L.C., Modification and identification of glutamate residues at the arginine recognition site in the catalytic subun.t of adenosine 3′,5′-cyclic monophosphate-dependent protein kinase of rabbit skeletal muscle, *Biochem. J.* 187, 371, 1980.

16. Vivanco-Martinez, F., Bragado, R., Albar, J.P., Juarez, C., and Ortiz-Masllorens, F., Chemical modification of carboxyl groups in human Fcg fragment: structural role and effect on the complement fixation, *Mol. Immunol.* 17, 327, 1980.

17. Fleer, E.A.M., Verheij, H.M., and de Haas, G.H., Modification of carboxylate groups in bovine pancreatic phospholipase A2. Identification of aspartate-49 as Ca^{2+}-binding ligand, *Eur. J. Biochem.* 113, 283, 1981.

18. Burkey, K.O., and Gross, E.L., Effect of carboxyl group modification on redox properties and electron donation capability of spinach plastocyanin, *Biochemistry* 20, 5495, 1981.

19. Pennington, R.M., and Fisher, R. R., Dicyclohexylcarbodiimide modification of bovine heart mitochondrial transhydrogenase, *J. Biol. Chem.* 256, 8963, 1981.

20. Esch, F.S., Böhlen, P., Otsuka, A.S., Yoshida, M., and Allison, W.S., Inactivation of the bovine mitochondrial F1-ATPase with dicyclohexyl [14C] carbodiimide leads to the modification of a specific glutamic acid residue in the β-subunit, *J. Biol. Chem.* 256, 9084, 1981.

21. Yamada, H., Imoto, T., Fujita, K., Okazaki, K., and Motomura, M., Selective modification of aspartic acid-101 in lysozyme by carbodiimide reaction, *Biochemistry* 20, 4836, 1981.

22. Woodhead, J.L., and Malcolms, D.B., The essential carboxyl group in restriction endonuclease *EcoR*I, *Eur. J. Biochem.* 120, 125, 1981.

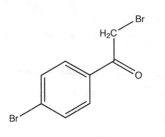

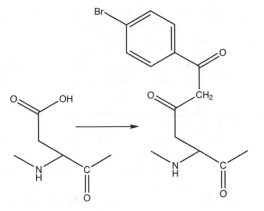

p-Bromophenacyl bromide

Aspartic Acid

2-*p*-bromophenyl-1-ethyl-2-one-
beta-aspartate

FIGURE 5.3 Modification of aspartic acid with *p*-bromophenacyl bromide.

REFERENCES

1. Kiss, T., Erdei, A., and Kiss, L., Investigation of the active site of the extracellular - D-xylosidase from *Aspergillus curonarius*, *Arch. Biochem. Biophys.* 399, 188, 2002.
2. Meier, T. et al., Evidence for structural integrity in the undecameric c-rings isolated from sodium ATP synthases, *J. Mol. Biol.* 325, 389, 2003.
3. Scholz, S.R. et al., Experimental evidence for a α- Me-finger nuclease motif to represent the active site of the capase-activaed DNase, *Biochemistry* 42, 9288, 2003.
4. Faijes, M. et al., Glycosynthase activity of *Bacillus licheniformis* 1,3-1,4-glucanase mutants: specificity, kinetics, and mechanism, *Biochemistry* 42, 13304, 2003.
5. Parsons, S.M., Jao, L., Dahlquist, F.W., Borders, C.L., Jr., Groff, T., Racs, J., and Raftery, M.A., The nature of amino acid side chains which are critical for the activity of lysozyme, *Biochemistry* 8, 700, 1969.
6. Parsons, S.M., and Raftery, M.A., The identification of aspartic acid residue 52 as being critical to lysozyme activity, *Biochemistry* 8, 4199, 1969.
7. Paterson, A.K., and Knowles, J.R., The number of catalytically essential carboxyl groups in pepsin. Modification of the enzyme by trimethyloxonium fluoroborate, *Eur. J. Biochem.* 31, 510, 1972.
8. Llewellyn, L.E., and Moczydlowski, E.G., Characterization of saxitoxin binding to saxiphilin, a relative of the transferrin family that displays pH-dependent ligand binding, *Biochemistry* 33, 12312, 1994.
9. Mackinannon, R., and Miller, C., Functional modification of a Ca^{2+}-activated K^+ channel by trimethyloxonium, *Biochemistry* 28, 8087, 1989.
10. Cherbavez, D.B., Trimethyoxonium modification of batrachotoxin-activated Na channels alters functionally important pore residues, *Biophys. J.* 68, 1337, 1995.
11. Gmez, M. et al., Chemical modification of -xylosidase from *Trichoderma reesei* QM 9414: pH-dependence of kinetic parameters, *Biochimie* 83, 961, 2001.
12. Woodward, R.B., Olofson, R.A., and Mayer, H., A new synthesis of peptides, *J. Am. Chem. Soc.* 83, 1010, 1961.

13. Dunn, B.M., Anfinsen, C.B., and Shrager, R.I., Kinetics of Woodward's Reagent K hydrolysis and reaction with staphylococcal nuclease, *J. Biol. Chem.* 249, 3717, 1974.

14. Bodlaender, P., Feinstein, G., and Shaw, E., The use of isoxazolium salts for carboxyl group modification in proteins. Trypsin, *Biochemistry* 8, 4941, 1969.

15. Saini, M.S., and Van Etten, R.L., An essential carboxylic acid group in human prostate acid phosphatase, *Biochim. Biophys. Acta* 568, 370, 1979.

16. Arana, J.L., and Vallejos, R.H., Two different types of essential carboxyl groups in chloroplast coupling factor, *FEBS Lett.* 123, 103, 1981.

17. Dinur, D., Kantrowitz, E.R., and Hajdu, J., Reaction of Woodward's Reagent K with pancreatic porcine phospholipase A2: modification of an essential carboxylate residue, *Biochem. Biophys. Res. Commn.* 100, 785, 1981.

18. Kooistra, C., and Sluyterman, L.A.A., Isosteric acid non-isosteric modification of carboxyl groups of papain, *Biochim. Biophys. Acta* 997, 115, 1989.

19. Johnson, A.R., and Dekker, E.E., Woodward's Reagent K inactivation of *Escherichia coli* L-threonine dehydrogenase: increased absorbance at 340-350 nm is due to modification of cysteine and histidine residues, not asparate or glutamate carboxyl groups, *Protein Sci.* 5, 382, 1996.

20. Puri, R.N., and Colman, R.W., A novel method for the chemical modification of functional groups other than a carboxyl group in proteins with *N*-ethyl-5-phenyl-isooxazolim-3-sulfonate (Woodward's Reagent K): inhibition of ADP-induced platelet responses involves covalent modification of aggregin, an ADP receptor, *Anal. Biochem.* 240, 251, 1996.

21. Bahar, S. et al., Persistence of external chloride and DIDS binding after chemical modification of Glu681 in human band 3, *Am. J. Physiol.* 277, C791, 1999.

22. Salhany, J.M., Sloan, R.L., and Cordes, K.S., The carboxyl side chain of glutamate 681 interacts with a chloride binding modifier site that allosterically modulates the dimeric conformational state of band 3 (AEI). Implications of the mechanism of anion/proton cotransport, *Biochemistry* 42, 1589, 2003.

23. Hoare, D.G., and Koshland, D.E., Jr., A method for the quantitative modification and estimation of carboxyl groups in proteins, *J. Biol. Chem.* 242, 2447, 1967.

24. George, A.L., Jr., and Border, C.L., Jr., Essential carboxyl groups in yeast enolase, *Biochem. Biophys. Res. Commn.* 87, 59, 1979.

25. Khorana, H.G., The chemistry of carbodiimides, *Chem. Rev.* 53, 145, 1953.

26. Kunkel, G.R., Mehrabian, M., and Martinson, H.G., Contact-site cross-linking agents, *Mol. Cell. Biochem.* 34, 2, 1981.

27. Iwamoto, H. et al., Staes of thin filament regulatory proteins as revealed by combining cross-linking/X-ray diffraction techniques, *J. Mol. Biol.* 317, 707, 2002.

28. Cook, G.M. et al., Purification and biochemical characterization of the F_1F_0 ATP synthase from thermophilic *Bacillus* sp. strain TA2.A1, *J. Bacteriol.* 185, 4442, 2003.

29. Das, A., and Ljungdahl, L.G., *Clostridium pasteurianum* F_1F_0 ATP synthase: operon, composition and some properties, *J. Bacteriol.* 185, 5527, 2003.

30. Sheehan, J.C., and Hlavka, J.J., The use of water-soluble and basic carbodiimides in peptide synthesis, *J. Org. Chem.* 21, 439, 1956.

31. Sheehan, J.C., and Hlavka, J.J., The cross-linking of gelatin using a water-soluble carbodiimide, *J. Am. Chem. Soc.* 79, 4528, 1957.

32. Riehm, J.P., and Scheraga, H.A., Structural studies on ribonuclease. XXI. The reaction between ribonuclease and a water-soluble carbodiimide, *Biochemistry* 5, 99, 1966.

33. Hoare, D.G., and Koshland, D.E., Jr., A procedure for the selective modification of carboxyl groups in proteins, *J. Am. Chem. Soc.* 88, 2057, 1966.

34. Lin, T.-Y., and Koshland, D.E., Jr., Carboxyl group modification and the activity of lysozyme, *J. Biol. Chem.* 244, 505, 1969.
35. Tohri, A. et al., Identification of the domains on the extrinsic 23 kD protein possibly involved in electrostatic interaction with the extrinsic 33 kD protein in spinach photosystem II, *Eur. J. Biochem.* 271, 962, 2004.
36. Good, N.E. et al., Hydrogen ion buffers for biological research, *Biochemistry* 5, 467, 1966.
37. Good, N.E., and Izawa, S., Hydrogen ion buffers, *Methods Enzymol.* 24, 53, 1972.
38. Gilles, M.A., Hudson, A.Q., and Borders, C.L, Jr., Stablity of water-soluble carbodiimides in aqueous solution, *Anal. Biochem.* 184, 244, 1990.
39. Lei, P.Q. et al., Kinetic studies on the rate of hydrolysis of N-ethyl-N'-(dimethylaminopropyl)carbodiimide I aqueous solution using mass spectrometry and capillary electrophoresis, *Anal. Biochem.* 310, 122, 2002.
40. Wrobel, N., Schinkinger, M., and Mirsky, V.M., A novel ultraviolet assay for testing side reactions of carbodiimide, *Anal. Biochem.* 305, 135, 2003.
41. Sehgal, D., and Vijay, I.K., A method for the high efficiency of water-soluble carbodiimide-mediated amidation, *Anal. Biochem.* 218, 87, 1994.
42. Perfetti, R.B., Anderson, C.D., and Hall, P.L., The chemical modification of papain with 1-ethyl-3(3-dimethylaminopropyl) carbodiimide, *Biochemistry* 15, 1735, 1976.
43. Carraway, K.L., and Koshland, D.E., Jr., Reaction of tyrosine residues in proteins with carbodiimide reagents, *Biochim. Biophys. Acta* 160, 272, 1968.
44. Carraway, K.L., and Koshland, D.E., Jr., Carbodiimide modification of proteins, *Methods Enzymol.* 25, 616, 1972.
45. Lewis, S.D., and Shafer, J.A., Conversion of exposed aspartyl and glutamyl residues in proteins to asparaginyl and glutaminyl residues, *Biochim. Biophys. Acta* 303, 284, 1973.
46. Wang. T.-T., and Young, N.M., Modification of aspartic acid residues to induce trypsin cleavage, *Anal. Biochem.* 91, 696, 1978.
47. Lin, C., Mihal, K.A., and Krueger, R.J., Introduction of sulfydryl groups into proteins at carboxyl sites, *Biochim. Biophys. Acta* 1038, 382, 1990.
48. Pho, D.B. et al., Evidence for an essential glutamyl residue in yeast hexokinase, *Biochemistry* 16, 4533, 1977.
49. Desvages, G. et al., Structural studies on yeast 3-phosphoglycerate kinase. Identification by immuno-affinity chromatography of one glutamyl residue essential for yeast 3-phosphoglycerate kinase activity. Its location in the primary structure, *Eur. J. Biochem.* 105, 259, 1980.
50. Lacombe, G., Van Thiem, N., and Swynghedauw, B., Modification of myosin subfragment 1 by carbodiimide in the presence of a nucleophile. Effect on adenosinetriphosphatase activity, *Biochemistry* 20, 3648, 1981.
51. Kerner, M. et al., Location of an essential carboxyl group along the heavy chain of cardiac and skeletal myosin subfragments 1, *Biochemistry* 22, 5843, 1983.
52. Hegde, S.S., and Blanchard, J.S., Kinetic and mechanistic characterization of recombinant *Lactobacillus viridencens* FemX (UDP-N-acetylmuramoyl pentapeptide-lysine N^6-alanine transferase, *J. Biol. Chem.* 278, 22861, 2003.
53. Sigrist-Nelson, K., and Azzi, A., The proteolipid subunit of the chloroplast adenosine triphosphatase complex. Reconstitution and demonstration of proton-conductive properties, *J. Biol. Chem.* 255, 10638, 1980.
54. Sussman, M.R., and Slayman, C.W., Modification of the *Neurospora crassa* plasma membrane (H+)-ATPase with N,N'-dicyclohexylcarbodiimide, *J. Biol. Chem.* 258, 1839, 1983.

55. Lotscher, H.R, deJay, C., and Capaldi, R.A., Inhibition of the adenosinetriphosphatase activity of *Escherichia coli* F_1 by the water-soluble carbodiimide, 1-ethyl-3-(3-dimethylaminopropyl)carbodiimide is due to modification of several carboxyls in the beta subunit, *Biochemistry* 23, 4134, 1984.

56. Shafir, I., Feng, W., and Shoshan-Barmatz, V., Dicyclohexylcarbodiimide interaction with the voltage-dependent anion channel from sarcoplasmic reticulum, *Eur. J. Biochem.* 253, 627, 1998.

57. Yang, S.J. et al., Localization of a carboxylic residue possibly involved in the inhibition of vacuolar H+-pyrophophatase by *N,N'*-dicyclohexylcarbodiimide, *Biochem. J.* 342, 641, 1999.

58. Bjerrum, P.J. et al., Functional carboxyl groups in the red blood cell anion exchange protein, *J. Gen. Physiol.* 93, 813, 1989.

59. Levy, H.M., Leber, P.D., and Ryan, E.A., Inactivation of myosin by 2,4-dinitrophenol and protection by adenosine triphosphate and other phosphate compounds, *J. Biol. Chem.* 238, 3654, 1963.

60. Takahashi, K., Stein, W.H., and Moore, S., The identification of a glutamic acid residue as part of the active site of ribonuclease T1, *J. Biol. Chem.* 242, 4682, 1967.

61. Erlanger, B.F., Vratsanos, S.M., Wassermann, M., and Cooper, A.G., Specific and reversible inactivation of pepsin, *J. Biol. Chem.* 240, PC3447, 1965.

62. Gross, E., and Morell, J.L., Evidence for an active carboxyl group in pepsin, *J. Biol. Chem.* 241, 3638, 1966.

63. Chacur, M. et al., Hyperalgesia induced by Asp49 and Lys49 phospholipases A2 from *Bothrops asper* snake venom: pharmacological mediation and molecular determinants, *Toxicon* 41, 667, 2003.

6 The Modification of Cysteine

Cysteine is usually the most powerful nucleophile in a protein and, as a result, is frequently the easiest to selectively modify with a variety of reagents. Cysteine is the sulfur analogue of serine in which the hydroxyl group is replaced with a sulfhydryl group. The reader is directed to an excellent review of thiols[1] for a more thorough discussion of both aliphatic and aromatic thiols. The bond dissociation energy for sulfhydryl groups is substantially less than that for the corresponding alcohol function, providing a basis for the increased acidity of sulfhydryl groups; for example, the pK_a for ethanethiol is 10.6 whereas it is 18 for ethanol. As a consequence, the reaction of cysteine with chloroacetate is slow (5.3×10^{-3} M^{-1} min^{-1}) whereas that with serine is nonexistent; the reaction of a cysteine residue at an enzyme active site (papain) is ca. 30,000 times faster (150 M^{-1} min^{-1}) than that of free cysteine at pH 6.0.[2]

Modification of cysteine residues most likely proceeds via a nucleophilic addition or displacement reaction, with the thiolate anion as the nucleophile (Figure 6.1). The reaction with the α-ketohaloalkyl compounds such as iodoacetate is an example of a nucleophilic displacement reaction, and the reaction of maleimide is a nucleophilic addition to an olefin. This reaction is an example of a Michael reaction. The chemistry of cysteine has been reviewed, and the reader is directed to reviews for further detail.[3-7] It is possible to group the reactions into general categories, based in part on the chemistry of the product formed and in part on the chemistry of the reagent used, as shown in Table 6.1. Oxidation and nitrosylation have been included. Although these reactions are not generally used for the site-specific modification of cysteine, they are included as they are of significant biological interest.[8-10] Table 6.2 presents a list of selected reagents frequently used for the modification of cysteine in proteins. This is not meant to be inclusive, but rather indicative of historical use, current trends, and the author's biases. Haloalkyl compounds, maleimides, and alkyl alkanethiosulfonates are also base structures for more complex molecules such as spectral probes and cross-linking agents. Other reagents such as Ellman's reagent [5,5′-dithiobis(2-nitrobenzoic acid)], cyanate, bromobimanes, and O-methylisourea are discussed later.

Local environment has a profound effect on the reactivity of cysteine residues in proteins. More than 30 years ago, Gerwin[12] demonstrated dramatic differences in the reaction of chloroacetic acid and chloroacetamide with the active-site cysteine of streptococcal proteinase. This difference in reactivity was due to the influence of the histidine residue at the active site. More recently, it has been shown[13] that local electrostatic potential modulates reactivity of individual cysteine residue in rat brain tubulin. Rat brain tubulin dimer contains 20 cysteine residues: 12 in the α-subunit

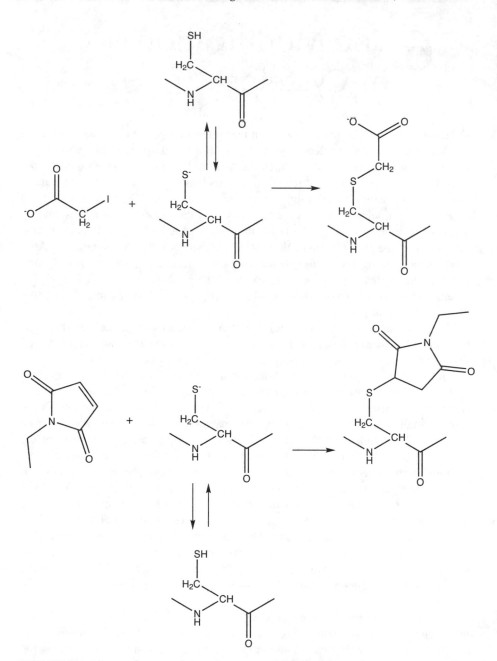

FIGURE 6.1 Reaction of cysteine with iodoacetate or *N*-ethylmaleimide. Also shown is the dissociation of a proton from the sulfhydryl group of cysteine to form the thiolate anion, which is the reactive species.

and 8 in the β-subunit. The rates of reaction of the cysteine residues in rat brain tubulin were determined with a variety of reagents in 0.3 *M* MES, pH 6.9, containing 1.0 m*M* EGTA and 1 m*M* MgCl$_2$ in the dark. The absence of light is critical, because

TABLE 6.1
Chemical Modification of Cysteine

Derivative	Example	Reagents
Thioether	S-Carboxymethyl cysteine	Iodoacetate, N-ethylmaleimide, bromobimane, bromoethylamine
Disulfide	S-5-Thio-2-nitrobenzoic cysteine, S-thio-2-ethylamine	5,5′-Dithiobis(2-nitrobenzoic acid); (2-ethylamino)methanethiosulfonate
Thioester	S-Acetyl cysteine	N-Acetylimidazole
Oxidation	Cysteic acid	Hydrogen peroxide, hydroxyl radicals
S-Nitrosylation	S-Nitrosyl cysteine	Nitric oxide and higher nitrogen oxides

haloalkane compounds such as monobromobimane undergo photolysis. The reagents evaluated included *syn*-monobromobimane, N-ethylmaleimide, iodoacetamide, and [5-((((2-iodoacetyl)amino) ethyl) amino)naphthalene-1-sulfonic acid] AEDANS. Of the 20 sulfhydryl groups, ca. 50% react equivalently with all reagents. Reaction is slower with iodoacetamide than with N-ethylmaleimide; more cysteine residues are modified with N-ethylmaleimide than with iodoacetamide, and the difference in the rates of reaction is ascribed to the differences in the chemistry of the reaction of the two compounds with the thiolate ion, with the reaction with iodoacetamide being a nucleophilic displacement whereas the reaction with N-ethylmaleimide an addition reaction. It was possible to identify a single highly reactive cysteine residue by reaction with chloroacetamide, which generally reacts with sulfhydryl groups more slowly than iodoacetamide does.[13]

The unique reactivity of cysteine has prompted investigators to use site-specific mutagenesis to place cysteine at particular points in a protein for the subsequent attachment of structural probes.[14–22] A cysteine residue is inserted into the sequence of a protein by using commercially available reagents and kits.[23] This technique allows the placement of a cysteine residue in a specific region of a protein, such as the extracellular domain or transmembrane domain of a membrane-associated protein. One of the early approaches was the substitution of a cysteine residue for a methionine residue at the S1′ binding site of carboxypeptidase Y.[17,22] This cysteine residue could be modified with a variety of alkylating or thioalkylating agents, leading to changes in the activity of the enzyme. For example, reaction with benzyl-methanethiosulfonate or phenacyl bromide caused a large decrease in activity toward substrates with a bulky leaving group (e.g., –OBz), whereas there was an increase in activity toward substrates with a small leaving group (e.g., –OMe). More recent studies on the alteration of enzyme activity by site-directed chemical modification after cysteine insertion have been reported by Desantis, Jones, and co-workers.[24,25] In the first study,[24] a cysteine residue was inserted in place of an asparagine residue (N52C) in the active site of subtilisin of *B. lentus*. This single cysteine residue was subsequently modified with a variety of alkylmethanethiosulfonate derivatives [70 mM CHES, 5 mM MES, 2 mM $CaCl_2$, pH 9.5, 20°C, 100-fold molar excess of methanethiosulfonate reagent in a poly(ethyleneglycol) (PEG-10000) polypropylene tube). There was a change in the pH dependence and specific activity with various derivatives

TABLE 6.2
Commonly Used Reagents for the Site-Specific Modification of Cysteine in Proteins

Reagent Type	Other Amino Acids Modified	Ref.
Haloalkyl[a]	Histidine, lysine, methionine, tyrosine	1–19
N-Alkyl maleimides	Lysine, tyrosine	20–39
5,5'-Ditgenreahiobis-(2-nitrobenzoic acid		40–53
p-Hydroxymercuribenzoate[b]		54–55
Alkylmethanethiosulfonates		56–66
Metal ions[c]		67–78

[a] Includes haloalkanes and α-haloketo compounds such as iodoacetic acid, iodoacetamide, and related derivatives.

[b] Includes related organic mercurial derivatives such as mercurinitrophenol derivatives and mercuriphenylsulfonate derivatives.

[c] Includes silver, gold, mercury, and zinc.

References for Table 6.2

1. Thompson, S.A., and Fiddes, J.C., Chemical characterization of the cysteines of basic fibroblast growth factor, *Ann. N.Y. Acad. Sci.* 638, 78, 1991.
2. Clark, S., and Konstantopoulos, N., Sulphydryl agents modulate insulin- and epidermal growth factor (EGF)-receptor kinase via reaction with intracellular receptor domains: differential effects on basal versus activated receptors, *Biochem. J.* 292, 217, 1993.
3. Kaslow, H.R., Schlotterbeck, J.D., Mar, V.L., and Burnette, W.N., Alkylation of cysteine 41, but not cysteine 200, decreases the ADP-ribosyltransferase activity of the S1 subunit of pertussis toxin, *J. Biol. Chem.* 264, 6386, 1989.
4. Park, S., and Burghardt, T.P., Isolating and localizing ATP-sensitive tryptophan emission in skeletal myosin subfragment 1, *Biochemistry* 39, 11732, 2000.
5. Castellano, F.N., Dattelbaum, J.D., and Lakowicz, J.R., Long-lifetime Ru(II) complexes as labeling reagents for sulfhydryl groups, *Anal. Biochem.* 255, 165, 1998.
6. Theuis, M., Loo, R.R.O., and Loo, J.A., In-gel derivatization of proteins for cysteine-specific cleavages and their analysis by mass spectrometry, *J. Proteome Res.* 2, 163, 2003.
7. Cheng, Y. et al., Ligand binding properties and structural studies of recombinant and chemically modified hemoglobins altered at 93 cysteine, *Biochemistry* 41, 11901, 2002.
8. Jahng, W.J. et al., A cleavable affinity biotinylating agent reveals a retinoid binding role for REP65, *Biochemistry* 42, 6159, 2003.
9. Corrie, J.E., Craik, J.S., and Monasinghe, V.R., A homobifunctional rhodamine for labeling proteins with defined orientations of a fluorophile. *Bioconj. Chem.* 9, 160, 1998.
10. Toutchkine, A., Nalbant, P., and Hahn, K.M., Facile synthesis of thiol-reactive Cy3 and Cy5 derivatives with enhanced water solubility, *Bioconj. Chem.* 13, 287, 2002.
11. Baja, E.S., and Fales, H.M., Overalkylation of a protein digest with iodoacetamide, *Anal. Chem.* 73, 3576, 2001.
12. Galvani, M. et al., Alkylation kinetics of protein in preparation for two-dimensional maps: a matrix assisted laser desorption/ionization-mass spectrometry investigation, *Electrophoresis* 22, 2058, 2001.
13. Lapko, V.N., Smith, D.L., and Smith, J.B., Identification of an artifact in the mass spectrometry of proteins derivatized with iodoacetamide, *J. Mass Spectrom.* 35, 572, 2000.
14. Sechi, S., and Chait, B.T., Modification of cysteine residues by alkylation. A tool in peptide mapping and protein identification, *Anal. Chem.* 70, 5150, 1998.
15. Lapko, V.N. et al., Surface topology of phytochrome A deduced from specific chemical modification with iodoacetamide, *Biochemistry* 37, 12526, 1998.

TABLE 6.2 (continued)
Commonly Used Reagents for the Site-Specific Modification of Cysteine in Proteins

16. Ramsier, V., and Chong, J.Y., Modification of cysteine residues with *N*-methyl iodoacetamide, *Anal. Biochem.* 221, 231, 1994.

17. Greimer, D.P. et al., Synthesis of the protein cutting reagent (*S*)-1-(*p*-bromoacetamidobenzyl)ethylenediaminetetraacetate and conjugation to side chains, *Bioconj. Chem.* 8, 44, 1997.

18. Pasquarello, C. et al., *N-t*-Butyliodoacetamide and iodoacetanilide: two new cysteine alkylating reagents for relative quantitation of proteins, *Rapid Commun. Mass Spectrom.* 18, 117, 2004.

19. Plowman, J.E. et al., The effect of oxidation or alkylation on the separation of wool keratin proteins by two-dimensional gel electrophoresis, *Proteomics* 3, 942, 2003.

20. Reyes, A.M., Bravo, N., Ludwig, H., Iriarte, A., and Slebe, J.C., Modification of Cys-128 of pig kidney fructose 1,6-bisphosphatase with different thiol reagents: size dependent effect on the substrate and fructose-2,6-bisphosphate interaction, *J. Protein Chem.* 12, 159, 1993.

21. May, J.M., Differential labeling of the erythrocyte hexose carrier by *N*-ethylmaleimide: correlation of transport inhibition with reactive carrier sulfhydryl groups, *Biochim. Biophys. Acta* 986, 207, 1989.

22. Ali, M.S., Roche, T.E., and Patel, M.S., Identification of the essential cysteine residue in the active site of bovine pyruvate dehydrogenase, *J. Biol. Chem.* 268, 22553, 1993.

23. Siksnys, V., and Pleckaityte, M., Role of the reactive cysteine residue in restriction endonuclease, *Biochim. Biophys. Acta* 1160, 199, 1992.

24. Juszczak, L.J. et al., UV resonance Raman study of 93-modified hemoglobin A: chemical modifier-specific effects and added influences of attached poly(ethylene glycol) chains, *Biochemistry* 41, 276, 2002.

25. Apuy, J.L. et al., Radiometric pulsed alkylation/mass spectrometry of the cysteine pairs in individual zinc fingers of MRG-binding transcription factor-1 (MTF-1) as a probe of zinc chelate stability, *Biochemistry* 40, 15164, 2001.

26. Ercel, N., Yang, P., and Aykin, M., Determination of biological thiols by high-performance liquid chromatography following derivatization by ThioGlo maleimide reagents, *J. Chromatogr. B* 753, 287, 2001.

27. Lucassen, R., Nishini, K., and Supattaporo, S., *In vitro* amplification of protease-resistant prion protein requires free sulfhydryl groups, *Biochemistry* 42, 4127, 2003.

28. Di Gleria, K., Hill, H.A.O., and Wang, L.L., *N*-(2-Ferrocene-ethyl)maleimide: a new electroactive sulfhydryl-specific reagent for cyteine-containing peptides and proteins, *FEBS Lett.* 390, 142, 1996.

29. Corriere, M. et al., Coupling of importin beta binding peptide on plasmid DNA: transfection efficiency is increased by modification of lipoplexes physico-chemical properties, *BMC Biotechnol.* 3, 14, 2003.

30. Baty, J.W., Hampton, M.B., and Winterbourn, C.C., Detection of oxidant sensitive thiol proteins by fluorescence labeling and two-dimensional electrophoresis, *Proteomics* 2, 1261, 2002.

31. van der Sluis, E.O., Nowen, N., and Driessen, A.J., SecY-SecY and SecY-SecG contacts revealed by site-specific crosslinking, *FEBS Lett.* 527, 159, 2002.

32. Indiveri, C. et al., Site-directed mutagenesis and chemical modification of the six native cysteine residues of the rat mitochondrial carnitine carrier: implications for the role of cysteine-136, *Biochemistry* 41, 8649, 2002.

33. Epps, D.E., and Vosters, A.F., The essential role of a free sulfhydryl group in blocking the cholesterol site of cholesterol ester transfer protein (CETP), *Chem. Phys. Lipids* 114, 113, 2002.

34. Niwayana, S., Kurano, S., and Matsumoto, H., Synthesis of D-labeled *N*-alkylmaleimides and application to quantitative peptide analysis by isotope differential mass spectrometry, *Bioorg. Med. Chem. Lett.* 11, 2257, 2001.

35. Phelps, K.K., and Walker, R.A., *N*-Ethylmaleimide inhibits Ncd motor function by modification of a cysteine in the stalk domain, *Biochemistry* 38, 10750, 1999.

(continued)

TABLE 6.2 (continued)
Commonly Used Reagents for the Site-Specific Modification of Cysteine in Proteins

36. Frillingos, S., and Kabak, H.R., Probing the conformation of the lactose permease of *Escherichia coli* by *in situ* site-directed sulfhydryl modification, *Biochemistry* 35, 3950, 1966.
37. Jones, P.C. et al., A method for determining transmembrane protein structure, *Mol. Membr. Biol.* 13, 53, 1996.
38. Yamaguchi, T. et al., Effects of chemical modification of a membrane thiol groups on hemolysis of human erythrocytes under hydrostatic pressure, *Biochim. Biophys. Acta* 1195, 205, 1994.
39. Jose, J., and Hendel, S., Monitoring the cellular surface display of recombinant proteins by cysteine labeling and flow cytometry, *ChemBioChem* 4, 396, 2003.
40. Mattsson, P., Pohjalainen, T., and Korpela, T., Chemical modification of cyclomaltodextrin glucanotransferase from *Bacillus circulans* var. *alkalophilus*, *Biochim. Biophys. Acta* 1122, 33, 1992.
41. Chang, G.-G., Satterlee, J., and Hsu, R.Y., Essential sulfhydryl group of malic enzyme from *Escherichia coli*, *J. Protein Chem.* 12, 7, 1993.
42. Nakayama, T., Tanabe, H., Deyashiki, Y., Shinoda, M., Hara, A., and Sawada, H., Chemical modification of cysteinyl, lysyl and histidyl residues of mouse liver 17β-hydroxysteroid dehydrogenase, *Biochim. Biophys. Acta* 1120, 144, 1992.
43. Okonjo, K.O., and Adejoro, I.A., Hemoglobins with multiple reactive sulfhydryl groups: the reaction of dog hemoglobin with 5,5′-dithiobis(2-nitrobenzoate), *J. Protein Chem.* 12, 33, 1993.
44. Karlstrom, A.R., Shames, B.D., and Levine, R.L., Reactivity of cysteine residues in the protease from human immunodeficiency virus: identification of a surface-exposed region which affects enzyme function, *Arch. Biochem. Biophys.* 304, 163, 1993.
45. Sheikh, S., and Katiyar, S.S., Chemical modification of octopine dehydrogenase by thiol-specific reagents: evidence for the presence of an essential cysteine at the catalytic site, *Biochim. Biophys. Acta* 1202, 251, 1993.
46. Gupta, K., and Panda, D., Perturbation of microtubule polymerization by quercetin through tubulin binding: a novel mechanism of its antiproliferative activity, *Biochemistry* 41, 13029, 2002.
47. Lee, J. et al., Identification of essential residues in 2′,3′-cyclic nucleotide 3′-phosphodiesterase. Chemical modification and site-directed mutagenesis to investigate the role of cysteine and histidine residues in enzymatic activity, *J. Biol. Chem.* 276, 14804, 2001.
48. Eyer, P. et al., Molar absorption coefficients for the reduced Ellman reagent: reassessment, *Anal. Biochem.* 312, 224, 2003.
49. Fernandez-Diaz, M.D. et al., Effects of pulsed electric fields on ovalbumin solutions and dialyzed egg white, *J. Agric. Food Chem.* 48, 2332, 2000.
50. Tremblay, J.M. et al., Modifications of cysteine residues in solution and membrane-associated conformations of phosphatidylinositol transfer protein have differential effects on lipid transfer activity, *Biochemistry* 40, 9151, 2001.
51. Mandelman, D., Jamal, J., and Poulos, T.L., Identification of two electron-transfer sites in ascorbate peroxidase using chemical modification, enzyme kinetics, and crystallography, *Biochemistry* 37, 17610, 1998.
52. Horn, M. et al., Free-thiol Cys 221 exposed during activation process in critical for native tetramer structure of cathepsin C (dipeptidyl peptidase I), *Protein Sci.* 11, 993, 2002.
53. Karim, C.B. et al., Cysteine reactivity and oligomeric structures of phospholamban and its mutants, *Biochemistry* 37, 12074, 1998.
54. Strange, R., Morante, S., Stefanini, S., Chiancone, E., and Desideri, A., Nucleation of the iron core occurs at the three-fold channels of horse spleen apoferritin: an EXAFS study on the native and chemically-modified protein, *Biochim. Biophys. Acta* 1164, 331, 1993.
55. Fann, M.C., Busch, A., and Maloney, P.C., Functional characterization of cysteine residues in G1pT, the glycerol 3-phosphate transporter of *Escherichia coli*, *J. Bacteriol.* 185, 2863, 2003.
56. Smith, D.J., Maggio, E.T., and Kenyon, G.L., Simple alkanethiol groups for temporary blocking of sulfhydryl groups of enzymes, *Biochemistry* 14, 766, 1975.

TABLE 6.2 (continued)
Commonly Used Reagents for the Site-Specific Modification of Cysteine in Proteins

57. Kenyon, G.L., and Bruice, T.W., Novel sulfhydryl reagents, *Methods Enzymol.* 47, 407, 1977.
58. Chang, H.-K., and Shieh, R.-C., Conformational changes in Kir2.1 channels during NH$_4$+-induced inactivation, *J. Biol. Chem.* 278, 908, 2003.
59. Akhand, A.A. et al., Evidence for both extra- and intracellular cysteine targets of protein modification for activation RET kinase, *Biochem. Biophys. Res. Commn.* 292, 826, 1001.
60. Pathak, R., Hendrickson, T.L., and Imperiali, B., Sulfhydryl modification of the yeast Wbplp inhibits oligosaccharide transferase activity, *Biochemistry* 34, 4179, 1995.
61. Holmgren, M. et al., On the use of thiol-modifying agents to determine channel topology, *Neuropharmacology* 35, 797, 1996.
62. Salleh, H.M., Patal, M.A., and Woodard, R.W., Essential cysteine in 3-deoxy-D-manno-octulosonic acid 8-phosphate synthase from *Escherichia coli*: analysis by chemical modification and site-directed mutagenesis, *Biochemistry* 35, 8942, 1996.
63. Orning, L., and Fitzpatrick, F.A., Modification of leukotriene A(4) hydrolase/aminopeptidase by sulfhydryl-blocking reagents: differential effects on dual enzyme activity by methyl-methanethiosulfonate, *Arch. Biochem. Biophys.* 368, 131, 1999.
64. Sacchi, V.F. et al., Glutamate 59 is critical for transport function of the amino acid cotransporter KAAT1, *Am. J. Physiol. Cell Physiol.* 285, C623, 2003.
65. Taylor, J.C., and Markham, G.D., Conformational dynamics of the active site loop of S-adenosylmethionine synthetase illuminated by site-directed spin labeling, *Arch. Biochem. Biophys.* 415, 164, 2003.
66. Loo, T.W., Bartlett, M.C., and Cherlee, D.M., Methanethiosulfonate derivatives of rhodamine and verapanil activate human P-glycoprotein at different sites, *J. Biol. Chem.* 278, 50136, 2003.
67. Hussain, S. et al., Potent and reversible interaction of silver with pure Na, K-ATPase and Na,K-ATPase-liposomes, *Biochim. Biophys. Acta* 1190, 402, 1994.
68. Bertini, I. et al., High resolution solution structure or the protein part of Cu7 metallothionein, *Eur. J. Biochem.* 267, 1008, 2000.
69. Myshkin, A.E., and Khromova, V.S., Peculiar features of the aggregation effect of silver (I) ion on hemoglobin, *Biochim. Biophys. Acta* 1651, 124, 2003.
70. Angevine, C.M., Herold, K.A., and Fillingame, R.H., Aqueous access pathways in subunit a of rotary ATP synthase extend to both sides of the membrane, *Proc. Natl. Acad. Sci.USA* 100, 13179, 2003.
71. McNeil, C.J., Athey, D., and Ho, W.O., Direct electron transfer bioelectronic interfaces: application to clinical analysis, *Biosens. Bioelectron.* 10, 75, 1995.
72. Handel, M.L. et al., Inhibition of AP-1 binding and transcription by gold and selenium involving conserved cysteine residues in Jun and Fos, *Proc. Natl. Acad. Sci.USA* 92, 4497, 1995.
73. Tian, Y. et al., Superoxide dismutase-based third-generation biosensor for superoxide anion, *Anal. Chem.* 74, 2428, 2002.
74. Safer, D., Undecagold cluster labeling of proteins at reactive cysteine residues, *J. Struct. Biol.* 127, 101, 1999.
75. Morby, A.P., Hobman, J.L., and Brown, N.L., The role of cysteine residues in the transport of mercuric ions by the Tn501 MerT and MerP mercury-resistance proteins, *Mol. Microbiol.* 17, 25, 1995.
76. Someya, Y., and Yamaguchi, A., Mercaptide formed between the residue Cys70 and Hg^{2+} or Co^{2+} behaves as a functional positively charged side chain operative in the Arg70 Cys mutant of the metal-tetracycline/H+ antiporter of *Escherichia coli*, Biochemistry 35, 9385, 1996.
77. Qian, H. et al., NMR solution structure or the oxidized form of MerP, a mercuric ion binding protein involved in bacterial ion resistance, *Biochemistry* 37, 9316, 1998.
78. Hisatome, I. et al., Block of sodium channels by divalent mercury: role of specific cysteinyl residues in the P-loop region, *Biophys. J.* 79, 1336, 2000.

TABLE 6.3
Modification of Site-Directed Modified Subtilisin

Modification	pK_a	k_{cat}/K_m(sec^{-1} mmol^{-1})
None (wild type)	7.01	157
None (N62C)	6.80	91
Ethylmethanethiosulfonate	6.79	200
Decylmethanethiosulfonate	6.29	256
2-Sulfonatoethylmethanethio-sulfonate	7.03	125

Source: Data from DeSantis, G., and Jones, J.B., *J. Am. Chem. Soc.* 120, 8582, 998. With permission.

(Table 6.3). The results obtained with the various derivatives are consistent with an increase in active-site hydrophobicity, stabilizing the unprotonated form of the active-site histidine. A plot of pH vs. k_{cat}/K_m for the wild-type enzyme showed a pH-dependent increase in k_{cat}/K_m, with a pK_a of ca. 7 with a plateau at pH 8 into the alkaline pH range. The derivative obtained from the modification of the N62C mutant enzyme with cyclohexyl-methane-thiosulfonate showed a bell-shaped curve, with a pK_a of ca. 6.45 for the ascending portion and 10 for the descending portion of the curve.

The major use of cysteine insertion/site-directed chemical modification has been directed toward elucidating topology, with emphasis on membrane proteins.[26-30] The approach involves the differential modification of inserted cysteine residues with membrane-permeant reagents (uncharged, e.g., *N*-ethylmaleimide, methylmethane-thiosulfonate) and membrane-impermeant reagents (charged, e.g., 2-sulfonatoethyl-methanethiosulfonate). Small (methylmethanethiosulfonate) and large (*N*-phenylma-leimide, benzyl-methanethiosulfonate) reagents are also used. It should be emphasized that topology or accessibility studies should be based on relative rates of reaction rather than all-or-none observations.

There are a number of studies on the combination of cysteine insertion and chemical modification. With the use of a single mutation, this can be defined as site-directed modification; this approach is not unique for cysteine but can be applied to insertion of lysine or tyrosine, followed by subsequent chemical modification. Table 6.4 lists some representative studies on the uses of cysteine insertion and subsequent chemical modification (site-directed modification), whereas some selected studies are described in more detail later.

Flitsch and Khorana prepared a series of bacteriorhodopsin mutants, each con-taining a single inserted cysteine residue that could be labeled with iodoacetate or 6-acryloyl-2-(dimethylamino)naphthalene[31] or a spin label.[32] Subsequent work by Hubbell and co-workers[33,34] increased the value of cysteine insertion followed by modification with a nitroxide spin label. There are several excellent recent reviews in this area.[35,36] Site-directed spin labeling of a transmembrane segment of the mitochondrial oxoglutarate carrier has been used to suggest that conformational changes occur during transport.[37] The cysteine residues in the native carrier protein were replaced with serine and 18 consecutive residues in the putative transmembrane

TABLE 6.4
Some Examples of Site-Directed Chemical Modification of Cysteine in Proteins

Protein	Reagents Used	Ref.
Human chloride channel protein (hClC-1)	2-Aminoethyl-MTS[a] [2-(Trimethylammonium)ethyl]-MTS 2-Sulfonatoethyl-MTS	1
Plasmid pI258CadC repressor	Methyl-MTS	2
Citrate transporter CitS (*Klebsiella pneumoniae*)	2-Sulfonatoethyl-MTS *N*-Ethylmaleimide [2-(Trimethylammonium)ethyl]-MTS	3
Erwin chrysanthemi oligogalacturonate porin KdgM	Biotin maleimide with polyethylene oxide spacer group	4
Membrane domain Nqo10 subunit of the proton-translocating NADH-quinone oxidoreductase of *Paracoccus denitrificans*	4-Acetamido-4′-[(iodoacetyl) amino]stilbene-2,2′-disulfonic acid *N*-Ethylmaleimide	5
Transmembrane region VIII of the *Rickettsia prowaekii* ATP/ADP translocase	2-Sulfonatoethyl-MTS [2-(Trimethylammonium)ethyl]-MTS 2-Aminoethyl-MTS	6
Kir6.2 subunit of the ATP-sensitive potassium channel	2-Aminoethyl-MTS 2-Sulfonato-MTS	7

[a] MTS, methanethiosulfonate.

References for Table 6.4

1. Fahlke, C. et al., Residues lining the inner pore vestibule of human muscle chloride channels, *J. Biol. Chem.* 276, 1759, 2001.
2. Sun, Y., Wong, M.D., and Rosen, B.P., Role of cysteinyl residues in sensing Pb (II), Cd (II), and Zn (II) by the plasmid pI258 CadC respressor, *J. Biol. Chem.* 276, 14955, 2001.
3. Subczak, I., and Lolkema, J.S., Accessibility of cysteine residues in a cytoplasmic loop of CitS of *Klebsiella pneumoniae* is controlled by the catalytic state of the transporter, *Biochemistry* 42, 9789, 2003.
4. Pellinen, T. et al., Topology of the *Erwinia chrysanthemi* oligogalacturonate porin KdgM, *Biochem. J.* 372, 329, 2003.
5. Kao, M.-C. et al., Characterization and topology of the membrane domain Nqo10 subunit of the proton-translocating NADH-quinone oxidoreductase of *Paracoccus denitrificans*, *Biochemistry* 42, 4534, 2003.
6. Winkler, H.H., Daugherty, R.M., and Audia, J.P., Cysteine-scanning mutagenesis and thiol modification of the *Rickettsia prowazekii* ATP/ADP translocase: evidence that the TM VIII faces an aqueous channel, *Biochemistry* 42, 12562, 2003.
7. Trapp, S. et al., Identification of residues contributing to the ATP binding site of Kir6.2, *EMBO J.* 22, 2903, 2003.

segment were individually replaced with cysteine, yielding 18 individual mutant proteins. The expressed proteins were reduced with dithioerythritol and taken into 0.17 M dodecylmaltoside, 1 mM EDTA, 10 mM Tris-HCl, pH 7.1, by gel filtration. The mutant proteins were subsequently modified with (1-oxyl-2,2,5.5-tetramethyl-Δ3-pyrroliine-3-methyl)methanethiosulfonate.

Kapanidis and Weiss reviewed the use of cysteine insertion for single-molecule fluorescence spectroscopy.[38] In addition to single-site modification with a fluorophore such a tetramethylrhodamine, there is intramolecular cross-linking of appropriately spaced single cysteine residues with a bis-functional fluorophore such as bis[(N-iodoacetyl-piperazinyl)-sulfonerhodamine. Single-cysteine substitution mutants within subunit 8 of yeast F110-ATP synthase were obtained and then modified with fluorescein-5-maleimide.[39] Study of the various derivatives provided a detailed topology of the protein.

Cysteine replacement of an arginine residue and an aspartate residue on the Na+/dicarboxylate cotransporter demonstrated that these residues are conformationally sensitive by reaction with methanethiosulfonate reagents.[40] The R349C mutant had reduced activity, which was increased on modification with (2-aminoethyl)methanethiosulfonate, but only the presence of sodium ions. The D373C mutant was inhibited by [2-(trimethylammonium)ethyl] methanethiosulfonate in the presence of sodium ions and was prevented by substrate. Three methanethiosulfonate derivatives were used in this study: (2-aminoethyl)methanethiosulfonate (membrane permeant), sodium(2-sulfonatoethyl)methanethiosulfonate, and [2-(trimethylammonium)ethyl] methane-thiosulfonate (both membrane impermeant).

In my opinion, the best example of the site-directed modification of cysteine is the work from Kabak's laboratory on the lactose permease protein from *Escherichia coli*.[41–45] In a recent study,[46] Kabak and colleagues used site-directed modification of cysteine to study the topology of the helix IX domain of lactose permease from *Escherichia coli*. A functional mutant protein in which the cysteine residues had been replaced was subjected to cysteine-scanning mutagenesis in the helix IX region. One mutant (G296C) was not expressed, and a second (R302C) was inactive. Three of the mutants (A291C, F308C, and T310C) were modified by N-ethylmaleimide at 25°C in 100 mM potassium phosphate, 10 mM MgSO$_4$, pH 7.5; reaction was moderately increased (A291C, I298C) or unaffected (T310C) by the presence of a ligand, β-D-galactopyranosyl-1-thio-β-D-galactopyranoside. Two mutant proteins reacted with N-ethylmaleimide only in the presence of β-D-galactopyranosyl-1-thio-β-D-galactopyranoside; the remaining 13 mutant proteins did not react with N-ethylmaleimide. When the modification reaction was performed at 0°C, reaction with N-ethylmaleimide was not detected with the F308C mutant or with the A295C mutant. Prior modification of the mutant proteins with 2-sulfonatoethyl-methanethiosulfonate precluded subsequent reaction with N-ethylmaleimide, demonstrating that the residues differentially modified by N-ethylmaleimide were equally exposed to solvent.

Spatial proximity of multiple cysteinyl residues could be evaluated by disulfide bond formation. In this approach (site-directed cross-linkage), multiple mutant proteins are expressed with cysteinyl residues placed at targeted locations in the sequence. Intramolecular disulfide formation is shown to trap structural fluctuations. Disulfide formation is initiated by cupric 1,10-phenanthroline as the oxidative catalyst at 22°C. The reaction can be terminated at various time points by the addition of EDTA and N-ethylmaleimide to block residual free cysteine residues. A number of recent studies have combined cysteine insertion and subsequent controlled oxidation to form disulfide bonds.[47–62]

   Another use of site-directed chemical modification at cysteine residues concerns the recent interest in chemical rescue. The term *chemical rescue* has been suggested to stem from the studies of Toney and Kirsch in 1989[63] on the restoration of the activity of a mutant (K238A) form of aspartate aminotransferase by addition of an exogenous amine or from the even earlier observations of Brooks and Benisek[64] on the activation of a Y14G mutant of delta 5,3-ketosteroid isomerase by phenols. It can be argued that these studies, in turn, benefited from the observations of Inagami[65,66] on the activation of trypsin by alkylguanidines in the hydrolysis of neutral substrates. Although this general approach is still used for mutant enzymes,[67] the term *chemical rescue* is also applied to the restoration of activity to a mutant enzyme in which, for example, a lysine residue is replaced with an alanine and activity is lost. Activity is then restored when, instead of an alanine residue, oligonucleotide-directed mutagenesis is used to insert a cysteine residue, which is then modified to produce a lysine analogue such as *S*-aminoethyl cysteine (Figure 6.2). The basic chemistry used for this reaction stems from the work of Cole and colleagues[68,69] on the modification of cysteine residues with ethylenimine (aziridine) to yield *S*-2-aminoethylcysteine in proteins to create novel sites for cleavage by trypsin. Modification was accomplished in 1.0 *M* Tris, pH 8.6, after reduction with 2-mercaptoethanol; conversion to *S*-2-aminoethylcysteine was also achieved with the use of 2-bromoethylamine (hydrobromide salt). *In situ* modification of proteins with 2-bromoethylamine has been used for cysteine-specific cleavages and subsequent analysis by mass spectrometry.[69,70] Ethylenimine was used to restore binding activity to P-selectin mutants (K111C).[71] Reaction with ethylenimine was performed in 0.5 *M* borate, pH 8.5, at 23°C. Modification with *N*-ethylmaleimide did not restore activity. A similar study showed restoration of activity of the K267C mutant of *Pseudomonas mevalonii* 3-hydroxy-3-methylglutaryl coenzyme A reductase by modification with 2-bromoethylamine (500 m*M* Tris, pH 8.0, 12 h, 16°C, dark; modification with denaturation required for activity restoration).[72] Modification with iodoacetamide to yield the *S*-carboxamidomethylcysteine derivative did not restore activity. 2-Bromoethylamine has also been used to restore activity to an active-site mutation (K107C) in aldolase.[73]

   2-Aminoethylmethanethiosulfonate has been used to restore activity to a mutant (R425C) of the citrate transporter CitP.[74] A related approach has been the conversion of cysteinyl residues to unnatural amino acid analogues in proteins.[75] Table 6.5 gives some examples. There are interesting derivatives. Also, note the information on the reaction rates of the various haloalkanes with cysteine.

   Alkyl alkanethiosulfonates (e.g., methyl methanethiosulfonate; Figure 6.3) have been extensively used in the past decade to modify cysteine residues in proteins. George Kenyon and colleagues in 1975 described methyl, ethyl, and trichloromethyl derivatives of methanethiosulfonate and propylpropanethiosulfonate.[76] These alkyl alkanethiosulfonates form mixed disulfides with the sulfhydryl group of cysteine, with the release of sulfinic acid. The mixed disulfides are of variable stability, more stable than thioesters but less stable than thioethers. The stability is influenced by the nature of substituents on the alkyl function. The modification is highly specific for sulfhydryl groups and is easily reversed by mild reduced agents such as 2-mercaptoethanol or dithiothreitol. Reaction could occur at sites other than cysteine in

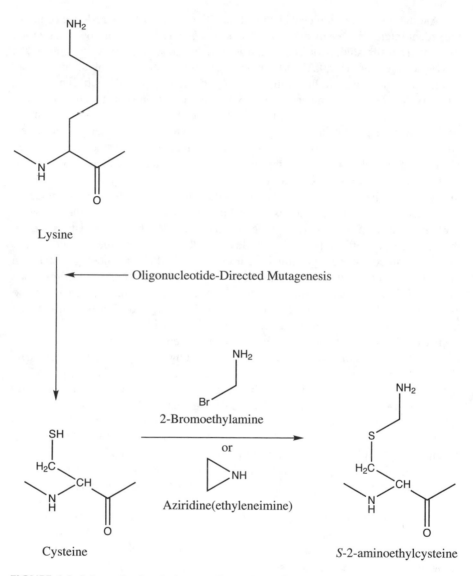

FIGURE 6.2 Scheme for chemical rescue. Shown is the change of a lysine residue to a cysteine residue by oligonucleotide-directed mutagenesis and the subsequent modification of the cysteine with 2-bromoethylamine or aziridine(ethyleneimine) to form S-aminoethylcysteine.

proteins.[77] Such reactions would most likely occur at amino groups, and these modifications would not be reduced by reducing agents such as dithiothreitol or 2-mercaptoethanol. To the best of my knowledge, there are no reports of the modification of amino groups in proteins with alkylthiosulfonates, but there are modifications with alkylthiosulfonates that are only partially reversed on reduction.[78] Modification of protein sulfhydryl groups is usually accomplished at pH 7 to 9, reflecting the importance of the thiolate anion as the reactive species.[79] In addition to pH dependence

TABLE 6.5
Alkylation of Cysteine to Produce Amino Acid Analogues

Reagent	Product	Amino Acid Analogue	Reaction Rate $(M^{-1}sec^{-1})^a$
2-Bromoethanol	$-CH_2CH_2OH$	Serine	1.1×10^5
Iodomethane	$-CH_3$	Methionine	1.5×10^{-1}
2-Bromoethylamine	$-CH_2CH_2NH_3^+$	Lysine	1.6×10^{-3}
Bromoethane	$-CH_2CH_3$	Methionine	5.6×10^{-4}
4-Chloromethyl-pyrazole	—	Histidine	5.3×10^{-1}
4-Bromoethyl-imidazole	—	Histidine	9.9×10^{-3}
4-Chloromethyl-imidazole	—	Histidine	5.1×10^{-1}

[a] Rates were determined at 25°C in 0.5 M borate, pH 9.5, for the reaction the reagent with N-acetylcystamine in the presence of 1.0. mM EDTA by measurement of sulfhydryl groups with 5,5'-dithiobis(2-nitrobenzoic acid).

Source: Data from Schindler, J.F., and Viola, R.E., *J. Protein Chem.* 15, 737–742, 1996. With permission.

studies that demonstrate a clear preference for reaction with the thiolate anion, this chapter presents a listing of the dependence of the second-order rate constant for the reaction on the pK_a of the individual thiol. The reader is also directed to a study by Stauffer and Karlin[80] on the effect of ionic strength on the reaction of alkylthiosulfonates with simple and protein-bound thiols. The extent of modification is best determined by the incorporation of radiolabeled reagent. In later work,[81,82] Kenyon and Bruice extended the understanding of the use of this class of reagents as part of a more general review of the modification of cysteine residues. Most work has used alkyl derivatives of methanethiosulfonate; such reagents are frequently referred to as MTS reagents.[83] Alkyl alkanethiosulfonates (MTS derivatives) have received much more use as reagents for the determination of cysteine accessibility[84] in proteins than as reagents for the modification of functionally important residues. In particular, there has been extensive use for the study of membrane proteins and other complex systems in which cysteine scanning approaches have been used for site-directed modification studies. As noted by Karlin and Akabas[84] and in Chapter 1, accessibility of a residue for chemical modification is relative and not absolute. Thus, the rates of reaction at the various sites should be determined and expressed as second-order rate constants. I am not aware of such studies on cysteine accessibility.

There is a wealth of information in the review by Karlin and Akabas,[84] which should be carefully considered by those interested in the use of MTS reagents. MTS reagents do undergo base-catalyzed hydrolysis and therefore should be prepared in neutral media immediately before use. Some investigators prepare stock solutions of reagent that are stored at −20°C. Stability under these conditions has not been validated and is not recommended. Contrary to early understanding,[81] there is an apparent dependence on the rate of reaction on the alkyl function. Manipulation of the alkyl function alters the membrane permeability of various MTS reagents. Table 6.6 summarizes these results.

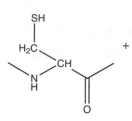
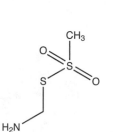
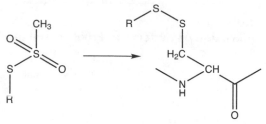

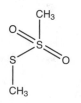

Methyl Methanethiosulfonate
Neutral

2-Aminoethyl Methanethiosulfonate
Charged

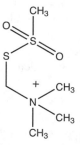

[2-(trimethylamonnium)ethyl] methanethiosulfonate

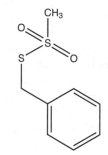

Benzyl Methanethiosulfonate

FIGURE 6.3 Some representative alkyl methanethiosulfonates.

Several specific studies on the use of alkyl alkanethiosulfonates are discussed in some detail later, and additional studies are listed in Table 6.7. Before considering this information, it is useful to consider the following comments. First, as noted previously, most of the work published in this area is not as useful as it could be, as few investigators are concerned with the measurement of reaction rate. Although most investigators appear to appreciate the issues with stability of the reagents in the preparation of the reaction mixture, little consideration is given to potential differences in the stability of MTS derivatives during the reaction. As shown by Karlin and Akabas, there are significant differences in the stability of various reagents. Finally, the investigators do not appear to appreciate issues with the potential lability of mixed disulfide derivatives. I was not a little discouraged to see at

TABLE 6.6
Some Properties of MTS Reagents

MTS Reagent	Rate Constant[a]	Partition Coefficient[b]	Hydrolysis Rate[c]
(2-Aminoethyl)	7.6	400	12
(2-Hydroxyethyl)	0.95	1.3	—
(2-Sulfonatoethyl)	1.7	2500	370
[2-(trimethyl-ammonium)ethyl]	21.2	690	11.2

[a] Reaction with 2-mercaptoethanol in 58 mM sodium phosphate, 0.1 mM EDTA, pH. 7.0, at 20°C; ($\times 10^{-4}$) M^{-1} sec^{-1}.

[b] Partition in water:n-octanol.

[c] $t_{\frac{1}{2}}$, 20°C, 190 mM sodium phosphate, pH 7.0, for hydrolysis to sulfenic acid and sulfinic acid derivatives.

Source: Data adapted from Karlin, A., and Akabas, M.H., *Methods Enzymol.* 293, 123, 1998. With permission.

least one study refer to the chemistry of modification by the methanethiosulfate reagents as an alkylation reaction.

Reaction with [2-(trimethylammonium)ethyl]methanethiosulfonate, a membrane-impermeant derivative, has been used to study a neuronal glutamate transporter.[85] An externally accessible cysteine was introduced into excitatory amino transporter-1 for the site-directed introduction of a fluorescent maleimide reporter group. The net current induced by aspartate in the oocyte system is blocked with [2-(trimethylammonium)ethyl] methanethio-sulfonate, (2-aminoethyl)methanethiosulfonate, and β-maleimidopropionic acid. Subsequent reaction with dithiothreitol restored current reduced by (2-aminoethyl)methane-thiosulfonate but not by β-maleimidopropionic acid. This is the behavior expected of a mixed disulfide formed by reaction of cysteine with (2-aminoethyl)methanethiosulfonate and for the thioether formed by the alkylation of cysteine with the maleimide derivative. The modification reactions were performed in Tris/HEPES buffer at pH 7.5.

Methanethiosulfonate derivatives have been used to study the mechanism of ion channel inhibition by chemical modification.[86] The question here is whether the inhibition is by blocking pore or by allosteric gating in Kir6.2 channels. Differential inhibition of channel function was observed with [2-(trimethylammonium)ethyl]methanethiosulfonate and [2-(triethylammonium)hexyl]methanethiosulfonate. The reagents were dissolved in dimethylsulfoxide and added to the bath solution (140 mM KCl, 2 mM EGTA, 0.2 mM MgCl$_2$, 5 mM HEPES, pH 7.4, with 1 or 10 μM ATP) via a gravity-driven perfusion pump.

Methanethiosulfonate derivatives have been used to study the mitochondrial ornithine/citrulline carrier.[87] Both the ornithine/ornithine (antiport) and ornithine/H$^+$ (unidirectional) transport modes were inhibited by the methanethiosulfonate derivatives and other sulfhydryl reagents. (2-aminoethyl)methanethiosulfonate was the most effective in antiport transport activity (IC$_{50}$ 0.17 μM) with [2-(trimethylammonium)ethyl]methanethiosulfonate (IC$_{50}$ 0.42 μM) and (2-sulfonatoethyl)methanethiosulfonate (IC$_{50}$ 200 μM). Reference to the partition coefficients obtained by Karlin

TABLE 6.7
Reaction of Methanethiosulfonate Reagents with Proteins

Protein	MTS Reagent	Conditions	Ref.
Nicotinic receptor[a]	(2-Aminoethyl)-]MTS [2-(Trimethylammonium) ethyl]-MTS (2-Sulfonatoethyl)-MTS	Sodium phosphate, pH 7.0[b]	1
Yeast oligosaccharyl transferase Wbplp	Methyl-MTS (N-Biotinylamino)ethyl-MTS	50 mM HEPES, 140 mM sucrose, pH 7.5, with 0.6% Nonidet P-40	2
Leukotriene A(4) hydrolase/aminopeptidase	Methyl-MTS	20 mM MOPS, pH 8.0, at 2°C	3
Lamb uterine estrogen receptor alpha (ER alpha)	Methyl-MTS	20 mM Tris-HCl, pH 8.5, 4 h, 0°C	4
Human sodium/bile acid cotransporter	(2-Sulfonatoethyl)-MTS [2-(Trimethylammonium) ethyl]-MTS	Modified Hank's balanced salt solution	5
Bovine heart mitochondrial ADP/ATP carrier	Methyl-MTS	250 mM sucrose, 0.2 mM EDTA, 10 mM PIPES, pH 7.2, 25°C	6
Nicotinic acetylcholine receptor	(2-Aminoethyl)-MTS [2-(Trimethylammonium) ethyl]-MTS [3-(Trimethylammonium) propyl]-MTS	50 mM HEPES, 96 mM NaCl, 2 mM KCl, 0.3 mM CaCl$_2$, 1 mM MgCl$_2$, pH 7.6	7
Oxalate-formate transporter of Oxalobacter foringenes	(2-Carboxyethyl)-MTS (2-Sulfonatoethyl)-MTS [2-(Trimethylammonium) ethyl]-MTS	20 mM potassium phosphate, pH 8.0, 23°C	8
Mitochondrial ornithine/citrulline carrier	(2-Aminoethyl)-]MTS [2-(Trimethylammonium) ethyl]-MTS (2-Sulfonatoethyl)-MTS	10 mM HEPES, 60 mM sucrose, pH 8.0	9

[a] This study contains a comparison of the rates of reaction of MTS compounds with simple thiols and protein-bound thiols.

[b] A variety of buffers were used in this study with varying concentrations of NaCl and detergent.

References for Table 6.7

1. Stauffer, D.A., and Karin, A., Electrostatic potential of the acetylcholine binding sites in the nicotinic receptor probed by reactions of binding-site cysteines with charged methanethiosulfonates, *Biochemistry* 33, 6840, 1994.
2. Pathak, R., Hendrickson, T.L., and Imperiali, B., Sulfhydryl modifications of the yeast Wbplp inhibits oligosaccharide transferase activity, *Biochemistry* 34, 419, 1995.
3. Orning, L., and Fitzpatrick, F.A., Modification of leukotriene A(4) hydrolase/aminopeptidase by sulfhydryl-blocking reagents: differential effects on dual enzyme activities by methyl methanethiosulfonate, *Arch. Biochem. Biophys.* 368, 131, 1999.
4. Aliau, S. et al., Steroidal affinity labels of the estrogen receptor alpha. 4. Electrophilic 11beta-aryl derivatives of estradiol, *J. Med. Chem.* 43, 613, 2000.

TABLE 6.7 (continued)
Reaction of Methanethiosulfonate Reagents with Proteins

5. Hallén, S., Frykland, J., and Sachs, G., Inhibition of the human sodium/bile acid cotransporter by site-specific methanethiosulfonate sulfhydryl reagents, *Biochemistry* 39, 6743, 2000.
6. Hashimoto, M. et al., Irreversible extrusion of the first loop facing the matrix of the bovine heart mitochondrial ADP/ATP carrier by labeling the Cys (56) residue with the SH-reagent methyl methanethiosulfonate, *J. Biochem.* 127, 443, 2000.
7. Sullivan, D.A., and Cohen, J.B., Mapping the agonist binding site of the nicotinic acetylcholine receptor. Orientation requirements for activation by covalent agonists, *J. Biol. Chem.* 275, 12651, 2000.
8. Fu, D. et al., Structure/function relationships in OXIT, the oxalate-formate transporter of *Oxalobacter firmigenes*. Assignment of transmembrane helix II to the translocation pathway, *J. Biol.Chem.* 276, 8753, 2001.
9. Tonazzi, A., and Indiveri, C., Chemical modification of the mitochondrial ornithine/citrulline carrier by SH reagents: effects on the transport activity and transition from carrier to pore-like function, *Biochim. Biophys. Acta* 1611, 123, 2003.

and Akabas[84] suggests a correlation between the relative effectiveness of the various methanethiosulfonate reagents as inhibitors in this complex system and their ability to cross a membrane as determined in the water – *n*-octanol system.

The reaction between alkyl alkanethiosulfonates and cysteine residues in proteins is an example of the formation of a mixed disulfide. There are other examples of the use of this reaction for the site-specific modification of cysteine residues in proteins. Cystine or cystamine have proven effective in the modification of guanylate cyclase.[88] This reaction involves the formation of a mixed disulfide and is easily reversed by adding dithiothreitol. There are a number of examples of the *in vivo* formation of mixed disulfides.[89–91] Another modification similar to that of alkanethiosulfonates is the reaction of sodium tetrathionate (Figure 6.4) with cysteine residues to yield the *S*-sulfenylsulfonate derivative,[92] which is primarily used for the protection of cysteinyl residues.[93,94]

Haloacetates such as chloroacetic acid and iodoacetic acid, the corresponding amides, and derivatives have been extremely useful reagents for the specific modification of cysteinyl residues. Haloacetates and haloacetamides react with cysteine

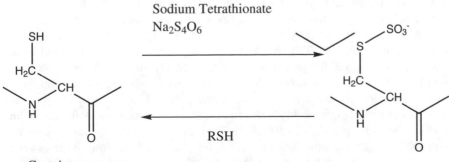

FIGURE 6.4 Reversible conversion of cysteine to *S*-sulfocysteine.

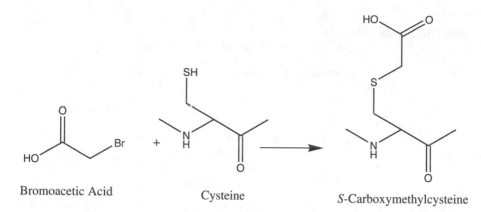

Bromoacetic Acid   Cysteine   S-Carboxymethylcysteine

FIGURE 6.5 Alkylation of cysteine with bromoacetic acid to form S-carboxymethylcysteine.

via an S_N2 reaction mechanism to give the corresponding carboxymethyl or carbox-
amidomethyl derivatives (Figure 6.5). When a rapid reaction is desired, iodine-
containing compounds are generally used. Dahl and McKinley-McKee[95] conducted
a rather detailed study of the reaction of alkyl halides with thiols. Reactivity of alkyl
halides depends not only on the halogen but also on the nature of the alkyl groups.
These investigators emphasized that the reactivity of an alkyl halide such as iodoacetate depends not only on the leaving potential of the halide substituent (I > Br >>>
Cl; 130:90:1), but also on the nature of the alkyl group. The rate of reaction of
2-bromoethanol with the sulfhydryl group of l-cysteine (pH 9.0) is ca. 1000 times
less than that observed with bromoacetic acid. The reactions are extremely pH
dependent, emphasizing the importance of the thiolate anion in the reaction.
Although these studies provide a useful framework, this is not always the situation.
Iodide is a highly polarizable and a good leaving group, but salvation can be
problematic. The reader is referred to an excellent review by Plapp[96] on affinity
labeling. With nitrogen as the nucleophile, the order of reaction of α-haloacetic acids
with 4-(p-nitrobenzyl) pyridine was I > Br >> Cl, consistent with the order of reaction
with thiols as described by Dahl and McKinley-McKee; the order of reactivity with
the histidine residues in RNase was Br > I > Cl and Br > I > Cl with DNase.
A series of steroid affinity labels (estradiol derivatives) for the estrogen receptor
were developed,[97] which contained either an epoxy function or a bromoacetyl function. Although both the epoxide derivative and the bromoacetyl derivatives were
stable in buffer (20 m*M* Tris-HCl, pH 7.0), they were rapidly hydrolyzed in the
cytosol to the vicinal diol or alcohol, respectively. Hydrolysis of the bromoacetyl
ester bond to the steroid moiety could not be prevented by adding esterase inhibitors
such as diisopropylphosphorofluoridate. Despite the problems of hydrolysis, bromoacetyl derivatives irreversibly bound to the estrogen receptor, blocking the binding
of radiolabeled ligand ([6,7-³H]estradiol). The reaction of the bromoacetyl derivative
with the estrogen receptor could be block by prior modification of the estrogen
receptor preparation with methyl methanethiosulfonate. Subsequent studies[98] used
chloroacetamido derivatives as well as bromoacetamido derivatives. Although

2-bromo-N-[2[4-3-17-cihydroxyestra-1,3,5(10)-triene 11-yl]phenoxy]ethyl]-N-methyl-acetanilide and 2-chloro-N-[2[4-3-17-cihydroxyestra-1,3,5(10)-triene 11-yl]phe-noxy]ethyl]-N-methylacetanilide bound to the cytosolic lamb uterine receptor with similar affinity, irreversible binding was obtained with the chloroacetamido derivative only. [Ru(2,2'bipyridine)$_2$(5-iodoacetamido-1,10-phenanthroline)(PF$_6$)$_2$ and [Ru(2,2'bi-pyridine)$_2$(5-chloroacetamido-1,10-phenanthroline) (PF$_6$)$_2$ have been synthesized and characterized as long-lifetime luminescent reagents.[99] The modification of pro-teins (human serum albumin or human immunoglobulin G) was accomplished with a 10-fold molar excess of reagent in 100 mM sodium phosphate, pH 7.1, The dye moiety binds to the protein via noncovalent interactions, but excess reagent is removed by dialysis (50 mM MOPS, pH 7.0) after completion of the modification of reaction (16 h, 4°C, dark). As expected, the chloroacetamido derivative was less reactive than the iodoacetamide derivative. The synthesis of an affinity-labeling reagent based on the structure of Vitamin A and containing a chloroacetyl function has been reported.[100] (3R-3-[boc-lys(biotinyl-O)-all-$trans$-retinol chloroacetate was synthesized. The chloroacetyl function reacts with a cysteine residue (10 mM Tris, 2 mM dithiothreitol, 1 mM EDTA, 1% Triton X-100, pH 9.0; 1 h, 4°C) and the labeled protein can be isolated by binding to an avidin matrix. The presence of two base-sensitive ester linkages in the reagent permits the release of the modified proteins from the avidin matrix in 200 mM sodium carbonate/bicarbonate, pH 11.

There are situations in which fast reaction rates are not necessarily desirable, such as the studies of Gerwin on streptococcal proteinase.[12] The reaction of strep-tococcal proteinase with iodoacetate or iodoacetamide[92] was too rapid to allow for accurate analysis of rates. This particular study was of considerable importance because it emphasized the importance of microenvironmental effects on the reaction of cysteine with α-halo acids and α-halo amides. Chloroacetic acid was far less effective than chloroacetamide. The sulfhydryl group at the active site of strepto-coccal proteinase has enhanced reactivity in that modification with iodoacetate readily occurred in the presence of a 100- to 1000-fold excess of 2-mercaptoethanol or cysteine. The enhanced reactivity of the active-site cysteine is also apparent from a comparison of the relative rates of modification of streptococcal proteinase and reduced glutathione. The rate of modification of streptococcal proteinase is 50 to 100 times more rapid than that of glutathione. The unique properties of this cysteine residue can be explained in part by the presence of an adjacent histidyl residue, which was demonstrated by an elegant series of studies by Liu.[92] Although histidine residues react with α-halo acids and amides, the presence of an adjacent cysteine residue precludes the use of this class of reagents to demonstrate the presence of a histidyl residue at the active site of streptococcal proteinase. Liu took advantage of the reversible modification of cysteinyl residues with sodium tetrathionite to modify the active-site histidine with iodoacetate.

The reaction of chloroacetic acid and chloroacetamide with papain has also yielded interesting results.[100,101] In studies with chloroacetamide, the active-site sulf-hydryl group of papain reacts at a rate more than 10-fold faster than free cysteine (5.78 M^{-1} sec^{-1} vs. 0.429 M^{-1} sec^{-1}).[20] As was the situation with streptococcal proteinase, there are dramatic differences in the rate of reaction of papain with chloroacetic acid and chloroacetamide. The reaction with chloroacetic acid has a pH optimum of ca. 7, whereas the optimum for reaction with chloroacetamide is at a pH 9. This investigation notes the influence of the neighboring histidyl residue, as has been discussed for streptococcal proteinase. These data further emphasize the importance of neighboring functional group effects on cysteinyl reactivity in proteins as well as the importance of rigorous evaluation of the effect of pH on the rate of the modification reaction.

Jörnvall and co-workers[102] used reaction with iodoacetate to probe differences in structure in wild-type β-galactosidase and various mutant forms of the enzyme. The modification reactions were performed in 0.1 M Tris, pH 8.1, under nitrogen in the dark. The reaction was terminated by adding excess 2-mercaptoethanol. Kalimi and Love[103] examined the reaction of the hepatic glucorticoid receptor with iodoacetamide in 0.010 M Tris, 0.25 M sucrose. Again, this reaction was performed in the dark. Kallis and Holmgren[104] examined the differences in reactivity of two sulfhydryl groups present at the active site of thioredoxin. The pH dependence of the reaction with iodoacetate suggested that one group had a pK_a of 6.7, whereas the second had a pK_a of 9.0. Iodoacetamide showed the same pH dependence, but the rate of reaction was ca. 20-fold higher than with iodoacetate. For example, at pH 7.2, the second-order rate constant for reaction with iodoacetate was 5.2 M^{-1} sec^{-1}, whereas it was 107.8 M^{-1} sec^{-1} for iodoacetamide. Figure 6.6 shows the results from this study. The low pK_a of one of the sulfhydryl groups was suggested to be a reflection of the presence of an adjacent lysine residue. Mikami and co-workers examined the inactivation of soybean α-amylase with iodoacetamide and iodoacetate.[105] Inactivation with iodoacetamide occurred ca. 60 times more rapidly than with iodoacetate at pH 8.6. Hempel and Pietruszko[106] showed that human liver alcohol dehydrogenase is inactivated by iodoacetamide but not by iodoacetic acid. These experiments were performed in 0.030 M sodium phosphate, pH 7.0, containing 0.001 M EDTA.

Reduction under denaturing conditions followed by alkylation with haloalkyl reagents or other alkylating agents is a common procedure in preparation of the protein for further analysis.[107–110] The success of analytical procedures such as two-dimensional electrophoresis[111,112] in proteomics depends on the effectiveness of this process. Historically, the reduction and alkylation step has had a important role in the determination of protein primary structure.[113] It is a useful step at present in the determination of protein modification by mass spectrometry.[114] In two-dimensional electrophoresis, reduction under denaturing conditions is critical in the preparation of the sample for analysis. Alkylation prevents the formation of spurious spots during the subsequent electrophoretic analysis.[115,116] Thus, all protocols for the separation of proteins by two-dimensional electrophoresis include a reduction and alkylation step at some point in the analysis. The reduction and alkylation step could be used in the preparation of the sample before the isoelectric focusing step[117,118] or between the isoelectric focusing step and the SDS/PAGE (polyacrylamide gel electrophoresis in the presence of sodium dodecyl sulfate) step. In the latter, this would involve the *in situ* treatment of the immobilized pH gradient (IPG) strip before the SDS/PAGE step. It is probably best to alkylate before the isoelectric focusing step to avoid the issue of disulfide bond reformation (disulfide scramble) during separation. Another consequence that can be avoided by alkylation is the β-elimination of cysteine in

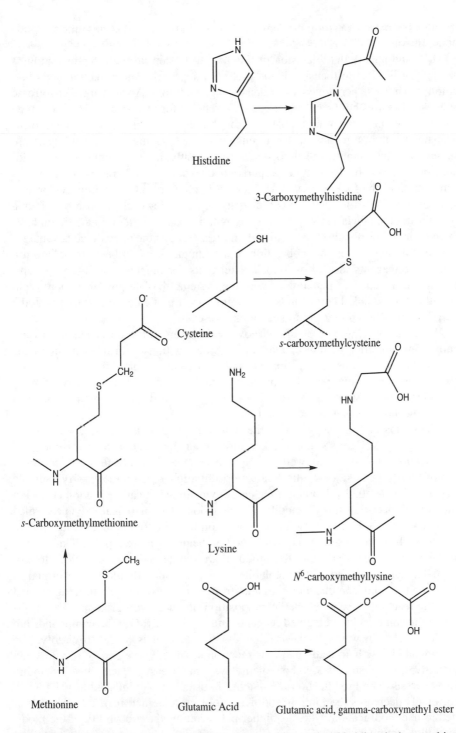

FIGURE 6.6 Potential reaction of an α-haloacetate with cysteine, histidine, lysine, methionine, or glutamic acid to yield various derivatives.

the alkaline pH range, forming dehydroalanine, and consequent unwanted peptide bond cleavage.[119]

It would appear that inclusion of the alkylation step during the preparation of sample for isoelectric focusing is the preferred approach. The reduction and alkylation step must be performed under denaturing conditions. An uncharged chaotropic agent such as urea has been the denaturant of choice for this process. However, urea has the potential of creating problems through a dismutation reaction, which forms cyanate, which in turn can react with nucleophiles including cysteine and lysine in subject proteins to increase heterogeneity.[117] Recently, it was recommend that thiourea be included in the sample preparation cocktail (9.0 M urea is the common concentration; the inclusion of 2 M thiourea is recommended) to improve protein solubilization[120] before the isoelectric focusing step. However, the use of thiourea complicates the alkylation step by reacting with iodoacetic acid or iodoacetamide.[121] Such inhibition is not observed with maleimides.[122] Given the operational advantages provided by thiourea in the preparation of the sample, it might be preferable to use maleimide reagents instead of iodoalkyl reagents for blocking sulfhydryl groups. The rate of reaction of maleimides with the cysteine thiolate groups is more rapid than that for iodoalkyl compounds. This might be of particular importance with the introduction of a reporter group such as fluorescein.

In studies with wool proteins, carboxymethylation with iodoacetate was essentially complete after 10 min of reaction.[123] Side reactions with other amino acids occurred at longer times of reaction. Other investigators reported that the alkylation process required a long period of time and was incomplete after 6 h of reaction.[125] It is clear from a study of the reaction of cysteine residues in proteins that local environmental factors influence reaction rate.

These observations suggest that the reduction and alkylation of proteins is not a trivial process. Care should be taken to monitor the reaction both for complete modification of cysteine as well as for side reactions (Figure 6.6), which could complicate the analysis of results. For example, it might be useful to modify a sample mixture with N-ethylmaleimide after carboxymethylation/carboxamidomethyation reactions. It is critical that all modification reactions with haloalkanes be performed in the dark; this might be a useful consideration for all cysteine modifications. The α-halo acids decompose in water, with the rate being far more rapid at alkaline pH. In the case of iodoacetic acid, the products are iodide and glycolic acid. Most of the haloalkyl compounds are photolabile. Thus, reactions should be performed in the dark to avoid complications arising from decomposition of the reagent and possible reaction of some of the decomposition products with proteins. We recrystallize the commercially obtained reagents and store over P_2O_5. The compounds are readily soluble in water. In the case of the free acid, it is useful to dissolve the compound in base before adding to the reaction mixture. In the case of α-haloacetyl derivatives, the resultant S-carboxymethylcysteine is easily quantitated by amino acid analysis. The modification of cysteine is usually accomplished at pH 8 to 9. Lower pH decreases the rate of modification of cysteine and increases the risk of methionine modification, whereas higher pH increases the probability of the modification of amino groups.

The development of isotope-coded affinity tags (ICATs)[126–127] as a method for studying differential protein expression[128,129] has fostered the development of novel reagents based on the iodoacetyl function. Examples include N-(13-iodoacetamido-4,7,10-trioxatridecanyl) biotinamide,[130] N-ethyliodoacetamide,[131,132] and N-t-butyl-iodoacetamide and iodoacetanilide.[133] These reagents are prepared as the light form (hydrogen) and the heavy form (deuterium). The biotin derivative is referred to as the ICAT reagent and is available with ^{13}C.[134] The biotin tag enables the rapid purification of peptides or proteins from mixtures after modification of cysteine residues with this reagent. Another approach to cysteine modification for proteomics is the use of ethyleneimine for an in-gel derivatization of proteins after electrophoretic separation[135] introducing a new site for cleavage by trypsin. Finally, vinyl pyridine is an alternative for the alkylation of cysteine residues in proteins[136] and deuterated derivatives have been prepared[137] for use in the study of differential protein expression. A thorough understanding of the chemistry of the alkylation reactions with haloalkanes is critical for the rigorous use of ICATs with modification chemistry based on haloalkane derivatives. The modification reactions for the attachment of such compounds to proteins are usually not well described and are poorly monitored. Based on the available data, it is not unreasonable to assume that iodine-based reagents would be the most useful. However, if there is a thought toward the use of such modification for quantitative proteomics, validation of the reaction will be required.

Although haloacetates and haloacetamides continue to be useful,[138–155] there has been far greater interest in the use of this chemistry as a mechanism for introducing a larger molecule that can serve as a structural probe. Examples include 5-iodoac-etamido-fluorescein[156] (Table 6.8), 5-[2-((iodoacetyl)amino)-ethyl]naphthalene-1-sulfonic acid (1,5-IAEDANS),[157–165] and 4-(2-iodoacetamido)-TEMPO,[166,167] cyanine dyes,[168] and some newly developed solvent-sensitive fluorescent dyes.[169] The iodoacetyl function group has been used for introducing a biotin probe in proteins[139,170–173] in addition to its use in the development of ICATs. A recent study[174] demonstrated the use of (4-iodobutyl)triphenylphosphonium for the modification of specific cysteine residues in membrane preparation. This reagent is accumulated by mitochondria, permitting the selective medication of cysteine residues inside the mitochondria. This reagent was used to monitor the thiol redox state of individual mitochondrial proteins. Inactivation of hamster arylamine N-acetyltransferase one by bromoacetamido derivatives of aniline and 2-aminofluorene[175] is worth some discussion, considering that the same active-site cysteine residue is modified by similar reagents by apparently kinetically different processes. Arylamine N-transferases are cytosolic enzymes that catalyze the acetylation of arylamines in an acetyl-coenzyme A-dependent reaction. In this study, (2-bromoacetylamino)fluorine demonstrates saturation kinetics consistent with an affinity reaction, whereas bromoacetanilide is an alkylating agent reacting by an S_N2 reaction similar to that for iodoacetic acid, which has been previously demonstrated to inhibit these enzymes.[176,177] A previous study suggested that bromoacetanilide demonstrated saturation kinetics with the rabbit enzyme.[178]

Reagents have been developed for the chemical cleavage of peptide bonds, using complexes of ferric ion and EDTA.[179–182] The effectiveness of these metal chelates

TABLE 6.8
Use of 5-Iodoacetamido-Fluorescein for the Modification of Proteins

Protein	Ref.
Alkaline protease from *Conidiobolus* sp.	1
Chaperone GroEL	2
Outer membrane protein, MopB, from *Methylococcus capsulatus* (Bath)	3
Skeletal myosin subfragment 1	4
Ca- and Mg-actin filaments	5
Oxidant-sensitive thiol proteins	6
Calf alpha-crystallin	7
Proteinase K	8
Creatine kinase	9
Acetylcholine receptor	10

References for Table 6.8

1. Tanksale, A.M. et al., Evidence for tryptophan in proximity to histidine and cysteine as essential to the active site of an alkaline protease, *Biochem. Biophys. Res. Commn.* 270, 910, 2000.
2. Hammerstrom, P. et al., Protein substrate binding induces conformational changes in the chaperonin GroEL. A suggested mechanism for unfoldase activity, *J. Biol. Chem.* 275, 22832, 2000.
3. Fjellbrikeland, A. et al., Molecular analysis of an outer membrane protein, MopB, of *Methylococcus capsulatus* (Bath) and structural comparisons with proteins of the OmpA family, *Arch. Microbiol.* 173, 346, 2000.
4. Park, S., and Burghardt, T.P., Isolating and localizing ATP-sensitive tryptophan emission in skeletal myosin subfragment 1, *Biochemistry* 39, 11732, 2000.
5. Hild, G., Nyitrai, M., and Somogyi, B., Intermonomer flexibility of Ca- and Mg-actin filaments at different pH values, *Eur. J. Biochem.* 347, 451, 2002.
6. Baty, J.W., Hampton, M.B., and Winterbourn, C.C., Detection of oxidant sensitive thiol proteins by fluorescence labeling and two-dimensional electrophoresis, *Proteomics* 2, 1261, 2002.
7. Putilina, T. et al., Subunit exchange demonstrates a differential chaperone activity of calf alpha-crystallin toward beta LOW and individual gamma crystallins, *J. Biol. Chem.* 278, 13747, 2003.
8. Pandhare, J. et al., Slow tight binding inhibition of proteinase K by a proteinaceous inhibitor: conformational alternations responsible for conferring irreversibility to the enzyme-inhibitor complex, *J. Biol. Chem.* 278, 48735, 2003.
9. Gregor, M. et al., Frequency-domain lifetime fluorometry of double-labeled creatine kinase, *Physiol. Res.* 52, 579, 2003.
10. Fairclough, R.H. et al., Agonist-induced transitions of the acetylcholine receptor, *Ann. N.Y. Acad. Sci.* 998, 101, 2003.

is based on proximity to the peptide bond rather that the amino acids participating in the scissile peptide bond. An artificial protease, iron (*S*)-1-[*p*-bromoaceta-mido]benzyl]-EDTA (Fe-BABE), was linked to a cysteine residues in subunit I and subunit II of the cytochrome *bd* quinol oxidase of *Escherichia coli*.[183] Cleavage is initiated by adding hydrogen peroxide and ascorbate.[180] Identification of the cleavage sites permits the determination of spatial relationships between the two subunits. This study allowed the division of cysteine residues in a target protein into three

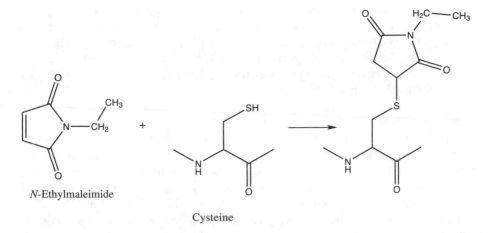

FIGURE 6.7 Reaction of *N*-ethylmaleimide with cysteine.

categories: (1) a group alkylated by the reagent but producing no cleavage products on the addition of ascorbate and hydrogen peroxide, (2) a group alkylated and producing cleavage products, and (3) some cysteine residues that do not react with the alkylation with the FE-BABE reagent. A more recent application showed the specific cleavage of the alpha subunit of *Escherichia coli* RNA polymerase following alkylation of the single cysteine residue[184] [55]Fe-BABE has been used as a reagent for the directed detection of cysteine residues on PAGE.[185]

Alkylation with *N*-iodoacetyl-3-iodotyrosine has proved useful in the study of vicinal cysteinyl residues in proteins.[186] Other approaches such as reaction with arsonous acid derivatives, such as melarson oxide, have also proved useful.[187,188]

N-Ethylmaleimide reacts with sulfhydryl groups in proteins (Figure 6.7) with considerable specificity.[189–191] This reaction is a Michael addition, which is a reaction between a nucleophile (thiolate anion) and an olefin (the maleimide ring). Bednar examined the chemistry of the reaction of *N*-ethylmaleimide with cysteine and other thiols in some detail.[192] The reaction of NEM with simple thiols can be described by the Brønsted equation. The second-order rate constant for the reaction of NEM with the thiolate anion of 2-mercaptoethanol is $10^7 \, M^{-1} \, min^{-1}$. This value is at least 5×10^{10} times higher than the reaction with the thiol. This study also reports data for the decomposition of NEM in several buffers and should be considered for the determination of truly accurate kinetic data. This reaction can be followed by the decrease in absorbance at 300 nm, the absorbance maximum of *N*-ethylmaleimide. The extinction coefficient of *N*-ethylmaleimide is $620 \, M^{-1} \, cm^{-1}$ at 302 nm.[189] The spectrophotometric assay is not sensitive and the modification is usually monitored by the incorporation of radiolabeled reagent. The alkylation product (*S*-succinyl cysteine) is stable and can be determined by amino acid analysis following acid hydrolysis. Although the reagent is reasonably specific for cysteine, reaction with other nucleophiles must be considered.[193] A diagonal procedure for the isolation of cysteine-containing peptides modified with *N*-ethylmaleimide has been reported.[194] This procedure is based on the hydrolysis of the reaction product of cysteine and

N-ethylmaleimide to cysteine-S-N-ethyl succinamic acid, generating a new negative charge. Various derivatives of maleimide provide the basis for the design of cross-linking reagents. Brown and Matthews[195,196] studied the reaction of lactose repressor protein with N-ethylmaleimide, two spin-label derivatives of N-ethylmaleimide, and a fluorophore derivative. The spin-labeled compounds showed the same pattern of reaction with the three cysteinyl residues as seen with N-ethylmaleimide. The fluorophore-derivative (N-(3-pyrene)maleimide) shows a slightly different reaction pattern. Other probes include 4-maleimido-2,2,6,6-tetramethylpiperidine-1-oxyl,[197] maleimidotetra-methylrhodamine,[197] N-(1-pyrenyl)maleimide,[198] 2-(4-maleimidoanilino) naphthalene-6-sulfonic acid,[199] 2,5-dimethoxy-4-stil-benylmaleimide,[199] rhodamine maleimide,[199] and eosin-5-maleimide.[200] Le-Quoc and colleagues examined the effect of the nature of the N-substituent groups on the rate of sulfhydryl group modification in succinate dehydrogenase.[201] The derivatives used were N-ethylmaleimide, N-butylmaleimide, and N-benzylmaleimide. The most reactive thiol groups in succinate dehydrogenase are probably located in an apolar environment, because the benzyl derivative reacted twice as fast as the ethyl derivative. (These modification reactions were performed in 0.050 M sodium phosphate, pH 7.6, at 37°C.) Modification with N-ethylmaleimide has found considerable use in the modification of inserted cysteine residues in the study of transmembrane proteins.[202] The synthesis of deuterated N-alkylmaleimides has been reported,[203] and these reagents have been used for quantitative peptide analysis in the study of differential protein expression. An electroactive probe has been developed that uses maleimide coupled to sulfhydryl groups in proteins.[204]

Some new fluorescent maleimide derivatives have been developed to measure thiols in biological fluids.[205] ThioGlo™ 3 (9-acetoxy-[4-(2,5-dihydro-1H-pyrrol-1-yl)phenyl]-3-oxo-3H-naphtho(2,1-b)pyran reacts with sulfhydryl groups via a maleimide function and the product is detected by fluorescence (excitation, 365 nm; emission, 445 nm). ThioGlo appears to be most useful for the precolumn derivatization of thiols in biological fluids before HPLC analysis. It is as sensitive (50 fMol) as N-(1-pyrenyl)maleimide for glutathione and somewhat more sensitive than monobromobimane. The reaction is performed in 100 mM Tris, 10 mM borate, 5 mM serine, 1 mM EDTA (pH 7.0) for tissue homogenates for 5 min at 23°C. The reaction is terminated by adding 2 N HCl to bring the pH to 2.5. This reagent has been used for the labeling of proteins in tissue extracts before electrophoretic analysis.[206]

Pulse-chase labeling of cysteine residues has been used to determine the relative reactivity of cysteine residues in the zinc finger regions of the MRE-binding domain of MRE-binding transcription factor-1 (MTF-121).[207] The results are used as a surrogate measurement of the zinc chelate stability of individual zinc fingers in MTF-1. In these experiments, the protein is first reacted with a 10-fold molar excess (to cysteine) of d_5-N-ethylmaleimide and at various time points, portions are removed and added to a solution containing H_5-N-ethylmaleimde (at a 100-fold molar excess to cysteine) and trypsin or chymotrypsin. Mass spectrometric analysis is used to determine the ratio of deuterium to hydrogen label at each cysteine residues, which, in turn, is a measure of reactivity and can be used to derive second-order rate constants for the alkylation reaction at each individual cysteine.

Maleimide poly(ethylene glycol) derivatives have been used to modify cys93 in human adult hemoglobin.[208] Subject maleimides include maleimide phenyl carbamyl (O″-methyl PEG 5000) and O′-(2-maleimidoethyl) –O′-methyl-PEG5000. Modification with the various N-alkylmaleimides was accomplished in phosphate-buffer saline, pH. 7.4, at various temperatures. A related study[209] describes the synthesis of maleimide derivatives of carbohydrates as chemoselective tags for site-specific glycosylation of proteins. Amino derivatives of carbohydrates were first coupled to 6-maleimidohexanoic acid N-hydroxysuccinimide ester to form the 6′-maleimdohex-anamidoethylglycosides which can then be coupled to a cysteinyl residue in the protein.

A novel maleimide reagent has been developed for the quantitative analysis of protein mixtures.[210] Acid-labile isotope-coded extractants (ALICE) are N-alkyl maleimide derivatives when there is an acid lability function between the alkylmaleimide and a polymer matrix. Reaction between a reduced (tributylphosphine) peptide mixtures and the reagent is accomplished in a 50/50 mixture (v/v) of acetonitrile and 200 mM sodium phosphate, pH 7.2. The bound polymer-peptide can be removed by filtration and the peptides released from the matrix with 5% trifluoroacetic acid in dichloromethane.

A variety of reagents have been used to define the role of cysteine in arginyl-tRNA synthase from *Escherichia coli*.[211] This protein contains four cysteine residues. Analysis with 5,5′-dithiobis(2-nitrobenzoic acid)(DTNB) suggested that two cysteine residues are located on the surface (reactive with DTNB) and two cysteine residues were buried (not reactive with DTNB). Iodoacetic acid reacted with one DTNB-sensitive residue with 50% loss of activity, whereas N-ethylmaleimide modified both DTNB-sensitive residues with a total loss of activity. However, replacement of the four cysteine residues with alanine via oligonucleotide-mediated mutagenesis had minimal effect on activity. It was concluded that the loss of activity observed with chemical modification was a result of steric hindrance rather than the loss of a specific chemical functional group.

The surface display of recombinant proteins in *Escherichia coli* can be assessed by reaction with maleimide reporter groups.[212] Cells expressing proteins with a single cysteine residue are incubated in phosphate-buffer saline and 2-mercaptoethanol at 23°C for 30 min, followed by washing with cold phosphate-buffered saline. N-biotin maleimide or N-fluorescein maleimide was then used to label the proteins expressed on the cell surface. The extent of modification was assessed by flow cytometry. Table 6.9 lists some recent studies on the use of N-alkylmaleimides for the site-directed modification of cysteine in various proteins.

Hydrophobic derivatives of N-ethylmaleimide (Figure 6.8) can be used as probes of the environment surrounding a sulfhydryl group in membrane anion channels.[213] Reaction with N-ethylmaleimide, N-benzylmaleimide, and N,N′-1,2-phenylenedimaleimide was evaluated and the reaction rate increased with increasing hydrophobicity. N-Phenylmaleimide has been used to modify a sulfhydryl group in the acetylcholine receptor.[214,215] Detergent was required for the modification reaction (10 mM MOPS to 100 mM NaCl to 0.1 mM EDTA with 0.02% sodium azide and 1% cholate). These studies identified cysteine residues potentially important in membrane function.

TABLE 6.9
N-Alkylmaleimides for the Site-Directed Modification of Cysteine in Proteins[a]

Protein	Reagent	Ref.
Tsr, *E.coli*, serine chemoreceptor	N-Ethylmaleimide	1
CDP-6-Deoxy-Δ 3,4-glucoseen reductase	N-Ethylmaleimide	2
E.coli F1-ATPase	N-Ethylmaleimide	3
Transmembrane protein segments	N-Fluorescein-5-maleimide[b], N-benzophenone-4-maleimide[c]	4
Lactose permease *E.coli*	N-Ethylmaleimide	5
Cytochrome b, *Rhobacter spheroids*	N-Ethylmaleimide	6
Acetate kinase, *Methanosarcina thermophilia*	N-Ethylmaleimide	7
Microtubule motor kinesin	N-Ethylmaleimide	8
Rat mitochondrial carnitine carrier[d]	N-Phenylmaleimide	9

[a] These studies are selected from the several in which cysteine residues have been inserted into a protein by oligonucleotide-directed site-specific mutagenesis either as a replacement for a specific amino acid residues or as a sterically compatible replacement for topology studies.

[b] Hydrophilic; modified residues in aqueous phase.

[c] Hydrophobic; modified residues in lipid phase.

[d] Also used N-ethylmaleimide, methanethiosulfonate reagents, p-chloromercuribenzoate, and mercuric chloride.

References for Table 6.9

1. Iwana, T. et al., *In vivo* sulfhydryl modification of the ligand-binding site of Tsr, the *Escherichia coli* serine chemoreceptor, *J. Bacteriol.* 177, 2218, 1995.
2. Ploux, O. et al., Mechanistic studies on CDP-6-deoxy-delta-3,4-glucoseen reductase: the role of cysteine residues as probed by chemical modification and site-directed mutagenesis, *Biochemistry* 334, 4159, 1995.
3. Haugton, M.A., and Capaldi, R.A., Asymmetry of *Escherichia coli* F1-ATPase as a function of the interaction of alpha-beta subunit pairs with the gamma and epsilon subunits, *J. Biol. Chem.* 270, 20568, 1995.
4. Jones, P.C. et al., A method for determining transmembrane protein structure, *Mol. Membr. Biol.* 13, 53, 1996.
5. Frillingos, S., and Kaback, H.R., Probing the conformation of the lactose permease of *Escherichia coli* by in situ site-directed sulfhydryl modification, *Biochemistry* 35, 390, 1996.
6. Tian, H. et al., The involvement of serine 175 and alanine 185 of cytochrome b of *Rhodobacter spheroids* cytochrome bcl complex in interaction with iron-sulfur proteins, *J. Biol Chem.* 272, 23722, 1997.
7. Singh-Wissmann, K. et al., Idenfication of essential glutamates in the acetate kinase from *Methanosarcina thermophlia*, *J. Bacteriol.* 180, 1129, 1998.
8. Phelps, K.K., and Walker, R.A., N-Ethylmaleimide inhibits Ncd motor function by modification of a cysteine in the stalk domain, *Biochemistry* 38, 10750, 1999.
9. Indiveri, C. et al., Site-directed mutagenesis and chemical modification of the six cysteine residues of the rat mitochondrial carnitine carrier: implication for the role of cysteine-136, *Biochemistry* 41, 8449, 2002.

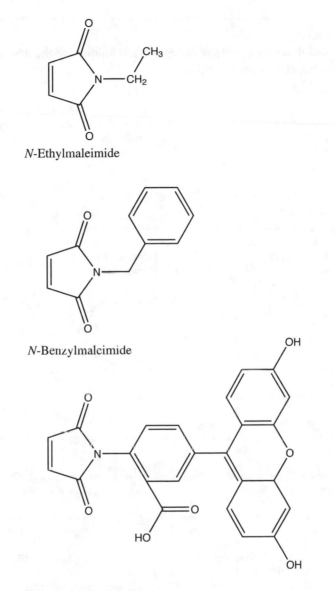

N-Ethylmaleimide

N-Benzylmalcimide

N-Fluorosceinmaleimide

FIGURE 6.8 Various N-alkylmaleimides.

Subsequent studies using site-specific mutagenesis have supported the importance of these cysteinyl residues.[215] Site-directed modification of cysteine has been demonstrated to be useful in the study of the relationship between structure and function in proteins. N-ethylmaleimide continues to be useful for this purpose and for the study of membrane proteins (Table 6.10). An equal number of proteins in which

TABLE 6.10
Some Examples of the Use of *N*-Ethylmaleimides for the Modification of Proteins at Cysteine Residues

Protein	Reagent	Conditions	Cys Insert	Ref.
Human erythrocyte membranes	*N*-Ethylmaleimide	Pressure	No	1
Lactose permease *E. coli*	*N*-Ethylmaleimide		Yes	2
Vascular smooth Muscle	*N*-Ethylmaleimide		No	3
Cystic fibrosis transmembrane conductance regulator	*N*-Ethylmaleimide		No	4
UDP-*N*-acetyl-muramoyl: L-alanine ligase *E. coli*	*N*-Ethylmaleimide		Yes	5
Glucose transporter GLUT1	*N*-Ethylmaleimide		Yes	6
Human melatonin MT(2) Receptor	*N*-Ethylmaleimide		Yes	7
Rotatory ATP synthase	*N*-Ethylmaleimide		Yes	8
ABC-type macrolide antibiotic exporter			Yes	9

References for Table 6.10

1. Yamaguchi, T. et al., Effects of chemical modification of membrane thiol groups on hemolysis of human erythrocytes under hydrostatic pressure, *Biochim. Biophys. Acta* 1195, 205, 1994.
2. Wu, J., Frillingos, S., and Kabak, H.R., Dynamics of lactose permease of *Escherichia coli* determined by site-directed chemical labeling and fluorescence spectroscopy, *Biochemistry* 34, 8257, 1995.
3. Tomera, J.F., Vascular chemical sulfhydryl alkylation *in vitro*: alterations in intracellular calcium and cAMP and cGMP metabolism, *Methods Find. Exp. Clin. Pharmacol.* 19, 113, 1997.
4. Cotton, J.F., and Welsh, M.J., Covalent modification of the regulatory domain irreversibly stimulates cystic fibrosis transmembrane conductance regulator, *J. Biol.Chem.* 272, 25617, 1997.
5. Nosal, L. et al., Site-directed mutagenesis and chemical modification of the two cysteine residues of the UDP-*N*-acetylmuramoyl: L-alanine ligase of *Escherichia coli*, *FEBS Lett.* 426, 309, 1998.
6. Olsowski, A. et al., Cysteine-scanning mutagenesis of helices 2 and y in GLUT1 identifies an exofacial cleft in both transmembrane segments, *Biochemistry* 39, 2469, 2000.
7. Mseeh, F., Gerdin, M.J., and Dubocovich, M.I., Idenfication of cysteines involved in ligand binding to the human melatonin MT(2) receptor, *Eur. J. Pharmacol.* 449, 29, 2002.
8. Angevine, C.M., and Fillingame, R.H., Aqueous access channels in subunit a of rotary ATP synthase, *J. Biol. Chem.* 278, 6066, 2003.
9. Kobayashi, N. et al., Membrane topology of ABC-type macrolide antibiotic exporter MacB in *Escherichia coli*, *FEBS Lett.* 546, 241, 2003.

endogenous cysteine residues have been modified are included in this tabulation for comparison of reaction conditions.

Localization of sulfhydryl groups within membranes has been achieved through comparison of the reaction with membrane-permeant and membrane-impermeant maleimide derivatives.[216–218] The use of these reagents can be traced back to the original observations of Abbott and Schachter in 1976.[219] The basic concept is to provide either a polar derivative or a derivative with steric considerations, which precludes passage through or into the membranes. Maleimide derivatives of glucosamine have been synthesized as affinity labels for the human erythrocyte hexose transport protein.[220] A particularly novel approach to this problem has been used by Falke and Koshland to analyze the structure of the aspartate receptor.[221] In this study, site-specific mutagenesis was used to place cysteinyl residues at six positions in the peptide chain. A novel membrane-impermeant maleimide derivative [N-(6-phosphonyl-n-hexyl)-maleimide] was used to study the reactivity of the individual sulfhydryl residues. From these studies, it was possible to map the domain structure of the receptor protein. Cysteinyl residues placed in the surface area could be modified by aqueous reagents, whereas transmembrane areas could be excluded by lack of reaction with membrane-impermeant reagents. Fluorescein-5-maleimide has been of particular value as a probe for the study of protein structure (Table 6.11). Another example of a Michael-type addition reaction involving cysteine is the reaction with acrylamide. Originally considered to be an unwanted complication of the electrophoresis of reduced proteins in acrylamide gel systems, modification with acrylamide is now considered to be a useful site-specific modification of cysteine in proteins.[222–225]

In 1958, George Ellman developed bis (p-nitrophenyl)disulfide (PNPD) to measure thiols.[226] This reagent was developed in response to a need for a more quantitative method to measure mercaptans in biological samples. This reagent was effective and sensitive. Reaction with a thiol-containing compound (p-nitrobenzylthiol, 2-mercaptoethanol, glutathione) yielded a mixed disulfide and p-nitrobenzylthiol. At pH values above 8, the nitrobenzylthiol anion was yellow, with a molar extinction coefficient of 13,600 $M^{-1}cm^{-1}$. Although PNPD was useful, solubility problems precluded extensive use. In 1959, Ellman developed 5,5'-dithiobis(2-nitrobenzoic acid) (DTNB; Ellman's reagent),[227] which at present is one of the more popular reagents for the modification and determination of the sulfhydryl group (Figure 6.9). Reaction with sulfhydryl groups in proteins results in the release of 2-nitro-5-mercaptobenzoic acid, which has a molar extinction coefficient of 13,600 M^{-1} cm^{-1} at 410 nm. The chemistry of DTNB was been the subject of continuing investigation after the original description, with most of the concern focusing on the extinction coefficient.[228–232] General consensus[232] would suggest a value of 14.15×10^3 M^{-1} cm^{-1} at 25°C and 13.8×10^3 M^{-1} cm^{-1} at 37°C (412 nm, 0.1 M sodium phosphate, pH 7.4). Although it important to have accurate values for extinction coefficients, the difference between the original value determined by Ellman in 1959 and the most recent value obtained in 2003 is of the order of 4%. An extremely precise extinction coefficient is more important when DTNB is used for the determination of thiols in biological fluids rather than for the absolute measurement of thiols in proteins, because most investigators do not use an accurate method for protein concentration analysis (see Appendix II).

Riddles and co-workers[230] studied the chemistry of DTNB in alkaline solutions as well as the reaction of DTNB with thiols. They concluded that the rate of reaction of DTNB was dependent on the pH and the pK_a of the thiol residue. The steric and electrostatic considerations discussed previously are also applicable to this reaction.

TABLE 6.11
Modification of Proteins with Fluorescein-5'-Maleimide

Protein	Ref.
G protein coupling domain of the beta-2-adrenergic receptor	1, 2
Lactose permease	3
Herpes simplex single-strand binding protein	4
Bovine heart cytochrome C oxidase	5
Spinach calmodulin	6
Membrane complex SeYEG	7
G protein coupled receptor	8
Ferric enterobactin	9
Aspartate receptor	10

References for Table 6.11

1. Ghanouni, P. et al., Functionally different agonists induce different conformations in the G protein coupling domain of the beta-2-adrenergic receptor, *J. Biol. Chem.* 276, 244330, 2001.
2. Ghanouni, P. et al., Agonist-induced conformational changes in the G protein-coupling domain of the beta 2 adrenergic receptor, *Proc. Natl. Acad. Sci. USA* 98, 5997, 2001.
3. Nachiel, E., and Gutman, M., Probing of the substrate binding domain of lactose permease by a proton pulse, *Biochim. Biophys. Acta* 1514, 33, 2001.
4. Dudas, K.C., Scouten, S.K., and Ruyechan, W.T., Conformational change in the herpes simplex single-strand binding protein induced by DNA, *Biochem. Biophys. Res. Commn.* 288, 184, 2001.
5. Marantz, Y. et al., Proton-collecting properties of bovine heart cytochrome C oxidase: kinetic and electrostatic analysis, *Biochemistry* 40, 15086, 2001.
6. Kogi, O. et al., Fluorescence dynamic anisotropy of spinach calmodulin labeled by a fluorescein chromophore at Cys-26, *Anal. Sci.* 18, 689, 2002.
7. van der Sluis, E.O., Nouwen, N., and Driessen, A.J., SecY-SecY and SecY-SecG contacts revealed by site-specific crosslinking, *FEBS Lett.* 527, 159, 2002.
8. Neumann, L. et al., Functional immobilization of a ligand-activated G-protein-coupled receptor, *ChemBioChem* 3, 993, 2002.
9. Cao, Z. et al., Spectroscopic observations of ferric enterobactin transport, *J. Biol. Chem.* 278, 1022, 2003.
10. Mehan, R.S., White, N.C., and Falke, J.J., Mapping out regions on the surface of the aspartate receptor that are essential for kinase activation, *Biochemistry* 42, 2952, 2003.

At pH 7.0 and 25°C, model thiols and some protein thiols react rapidly. There are examples of protein thiols that react more slowly ($t_j \geq 10$ min). An example of differences in site reactivity of inserted cysteine residues with DTNB is provided by Haas and colleagues with studies on mutant forms of adenylate cyclase.[233] Fluorescence resonance energy transfer (FRET) is a useful technique to study solution protein structure. The challenge in such studies is the development of suitable

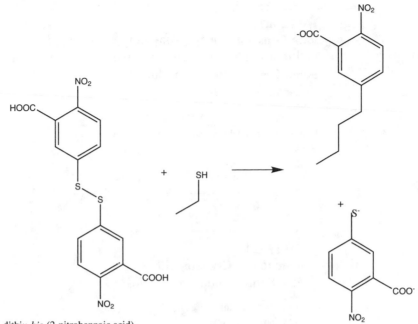

5,5'-dithio-*bis*-(2-nitrobenzoic acid)

2-nitro-5-mercaptobenzoic acid

FIGURE 6.9 Reaction of 5,5'-dithio-*bis*-(2-nitrobenzoic acid) with cysteine.

donor–recipient pairs of fluorescent probes. It is possible in some studies to use an intrinsic fluorophore such as tryptophan and 3-nitrotyrosine as a donor/recipient pair.[233] As noted previously, insertion of a cysteine residue provides a useful site of probe attachment. It would be a significant advantage to be able to introduce the two cysteine residues and differentially modify the individual residues with specific probes. In the adenylate cyclase study,[234] a series of single cysteine mutants was prepared and the rate of reaction of each mutant with DTNB determined. The goal was to identify the two mutants with the greatest difference in reactivity. The modification reactions were performed in 20 mM HEPES, pH 7.2, at 22.5°C and 40°C in the presence and absence of 100 mM NaCl. Change in absorbance at 440 nm was determined in a stopped-flow device. Table 6.12 gives some rate constants obtained in this study. The goal was to identify the two mutants with the greatest difference in reactivity. The studies shown in Table 6.12 permitted the identification of a double mutant (V169C/A188C) that demonstrated a minimum of a 106-fold difference in reactivity with DTNB between the two sites. The first modification used *N*-1-(pyrenylmethyl)iodoacetate (donor), and the single-substituted product (V169C-Py/A188C) was isolated by hydrophobic interaction chromatography on Phenyl Sepharose™ 6. V169C-Py/A188C was then modified with monobromobimane to yield the double-labeled product. Modification reactions were performed in 20 mM HEPES, pH 7.2, 2 mM EDTA containing 30 µM tris (2-carboxyethyl)phosphine, and 100 mM NaCl.

TABLE 6.12
Rate Constants for the Reaction of 5,5′-
Dithiobis(2-Nitrobenzoic Acid) with
Some Cysteine Insertion Mutants
of *Escherichia coli*

	K (mM^{-1} sec^{-1})	
Mutant	No Added Salt	With 100 mM NaCl
V169C	1.81	3.58
V142C	0.186	0.383
A188C	0.017	0.077

TABLE 6.13
Reaction of Cyanate with
Functional Groups in Proteins

Functional Group	pK_a	M^{-1} min^{-1}
α-Amino	7.8–8.2	1.4×10^{-1}
ε-Amino	10.5–10.8	2×10^{-3}
Sulfhydryl	8.3–8.5	4.0
Imidazole	7.0–7.2	1.8×10^{-1}

Source: From Stark, G.R., *Meth. Enzymol.* 11,
590, 1967. With permission.

Examples of the use of DTNB include studies on *Escherichia coli* citrate synthase[235] and D-amino acid transaminase[236] (0.1 *M* Tris, 0.002 *M* EDTA, pH 7.5). Other proteins studied with this reagent include rat brain nicotinic-like acetylcholine receptors[237] (calcium-containing Ringers solution, pH 7.4), lipophilin from human myelin[238] (0.001 *M* glycylglycine, 0.0001 *M* EDTA, pH 8.0), and human hemoglobin.[239] The last study followed the changes in absorbance at 450 nm to monitor the release of the 2-nitro-5-mercaptobenzoic acid. The molar extinction coefficients obtained at 450 nm were 5550 M^{-1} cm^{-1} (pH 6.0), 6510 M^{-1} cm^{-1} (pH 7.0), 6810 M^{-1} cm^{-1} (pH 8.0), 6940 M^{-1} cm^{-1} (pH 9.0), and 7010 M^{-1} cm^{-1} (pH 9.5). Feng and co-workers[240] also used the change in absorbance at 450 nm to monitor the reaction of DTNB with apomyoglobin. This study used the reaction of DTNB with inserted cysteine residues in ferric apomyoglobin to study intermediates in protein folding and to compare such results with those obtained from native-state amide hydrogen exchange. Protein folding–unfolding was studied as a function of guanidine hydrochloride concentration. At low denaturant concentration, there was a discrepancy between the results observed with hydrogen exchange and cysteine reactivity; this

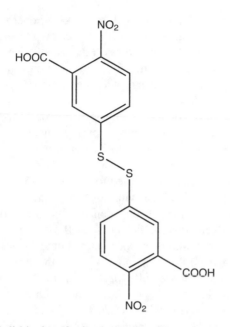
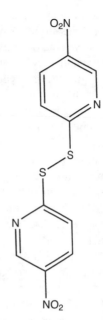

5,5'-dithio-*bis*-(2-nitrobenzoic acid)   2,2'-dithio-*bis*-(5-nitropyridine)

FIGURE 6.10 Structures of 5,5'-dithio-*bis*-(2-nitrobenzoid acid) and 2,2'-dithio-*bis*-(5-nitro-pyridine).

anomalous behavior was not observed when methyl methanethiosulfonate was used to modify the cysteine residues. Reaction with DTNB was also used to monitor conformational change in ovalbumin resulting from electric field pulses.[241]

It would appear that most recent studies with DTNB have used this reagent to measure cysteine residues in proteins rather than for the site-specific chemical modification of proteins. The reader is directed to several recent studies in which DTNB was used to measure the extent of cysteine modification by other sulfhydryl reagents.[242,243]

4,4'-Dithiodipyridine (Figure 6.10) is similar to 5,5'-dithiobis-(5-nitrobenzoic acid) in that a mixed disulfide is formed between a cysteinyl residue in the protein and the reagent, with the concomitant release of pyridine-4-thione.[235] The reaction of 4,4'-dithiodipyridine with protein sulfhydryl groups can be followed by spectroscopy ($\varepsilon_{324\,nm} = 19,800\ M^{-1}\ cm^{-1}$). The reaction is readily reversed by adding a reducing agent such as dithiothreitol.

The synthesis of a selenium analogue of this class of reagents, 6,6-diselenobis(3-nitrobenzoic acid), has been reported.[244] The selenium-containing reagent has the same reaction characteristics as those of the sulfur-containing compound in terms of specificity of reaction with cysteinyl residues in proteins. The reaction is monitored by spectroscopy following the release of 6-seleno-3-nitrobenzoate, which has a maximum at 432 nm. The extinction coefficient for the 6-seleno-3-nitrobenzoate anion varies slightly from 9532 (with excess reagent) to 10,200 M^{-1} cm^{-1} (with either excess cysteine or excess 2-mercaptoethanol). Other studies on the use of DTNB to modify sulfhydryl groups in proteins have included the modification of oncomodulin,[245] glutathione synthetase,[246] fibronectin,[247] and streptococcal NADH peroxidase.[248] In studies on the modification of ATP sulfurylase,[88] it was observed that DTNB was less potent than the more hydrophobic dithionitropyridine derivative.

Kimura and co-workers[249] introduced methyl 3-nitro-2-pyridyl disulfide and methyl 2-pyridyl disulfide. Both these reagents modify sulfhydryl groups forming the thiomethyl derivative. The spectrum of 3-nitro-2-pyridone is pH dependent. There is an isosbestic point at 310.4 nm, which can be used to determine the extent of the reaction of methyl-3-nitro-2-pyridyl disulfide with sulfhydryl groups. The difference in spectrum obtained does not show the pH dependence of the nitropyridyl derivative. At 343 nm, the change in extinction coefficient is 7060 M^{-1} cm^{-1}. The extinction coefficient (7600 M^{-1} cm^{-1}) of the 2-thiopyridinone at 343 nm is relatively stable from pH 3 to 8.0.[249] There is a marked decrease in absorbance above pH 8.0, reflecting the loss of a proton. Reaction with the sulfhydryl group in the protein clearly proceeds more rapidly at alkaline pH. In a related study, Drewes and Faulstich[250] prepared 2,4-dinitrophenyl-^{14}C-cysteinyl disulfide via a facile synthetic method as a means for introducing radiolabeled cysteine into proteins via disulfide exchange with free thiols. The reaction can be monitored by following the release of 2,4-dinitrophenol at 408 nm (ε = 12,700 M^{-1} cm^{-1}). The specificity of this reagent corresponded to that obtained with DTNB. Reaction with the sulfhydryl groups of papain was more rapid than that observed with DTNB. The resulting derivative can be easily reversed with thiols, but is stable to cyanogen bromide degradation and peptide purification.

p-Hydroxymercuribenzoate continues to be of use for the modification of sulfhydryl groups in proteins.[251–253] The reagent is obtained as p-chloromercuribenzoate, but is instantaneously converted to the hydroxy derivative in aqueous solution. This reagent was originally described by Boyer.[254] The absorbance change at 255 nm on modification is 6200 M^{-1} cm^{-1} at pH 4.6 and 7600 M^{-1} cm^{-1} at pH 7.0. Bai and Hayashi[255] have examined the reaction of organic mercurials with yeast carboxypeptidase (carboxypeptidase Y). Treatment of the modified enzyme with millimolar cysteine resulted in virtually complete recovery of catalytic activity. Bednar and co-workers[256] studied the inactivation of chalcone isomerase by p-chloromercuribenzoate and mercuric chloride. The modified protein could be readily reactivated by treatment with either thiols or KCN. The reactivation by KCN is based on the formation of a tight complex between cyanide and either organic or inorganic mercurials. The modification by mercuric chloride can be monitored by the increase in absorbance at 250 nm. Ojcius and Solomon[257] examined the inhibition of erythrocyte urea and water transport by p-chloromercuribenzoate. Other studies with this reagent have included dissociation of erythrocyte membrane proteins[258] and NADH peroxidase.[141]

2-Chloromercuri-4-nitrophenol is a compound related to the organic mercurial described previously. It has proved useful as a reporter group in the study of microenvironmental changes in the modified protein. An excellent example of this is provided from the studies of Marshall and Cohen[259] on the properties of ornithine transcarbamylase modified with 2-chloromercuri-4-nitrophenol. The enzyme from *Streptococcus faecalis* was modified in 0.1 M MOPS, 0.1 M KCl, pH 7.5, using changes in absorbance at 403 nm to follow the extent of modification. The bovine enzyme is carboxamidomethylated on a nonessential sulfhydryl group before reaction with the organic mercurial. Modification of the bovine enzyme with 2-chloromercuri-4-nitrophenol is performed in 0.020 M MOPS, 0.1 M KCl, pH 7.11, at 25°C. The modification was followed by the change in absorbance at 405 nm. Baines and Brocklehurst[260] reported the synthesis and characterization of 2-(2'-pyridylmercapto)-mercuri-4-nitrophenol, a reagent that has certain advantages.

A number of other modifications of sulfhydryl groups have proved useful. *O*-Methylisourea reacts with cysteinyl residues to form the *S*-methyl derivative.[261] *S*-methylation of proteins occurs *in vivo*.[262] Cyanate also can modify sulfhydryl groups.[263] Although this reaction is more rapid for sulfhydryl groups than for other nucleophiles such as amino groups, side reactions are a complication. Cleavage of peptide bonds at the amino terminus of cysteine residues is proving useful in peptide mapping of proteins.[264] The carbamoyl derivative of cysteine is stable at acid pH, but rapidly decomposes at alkaline pH. Specificity of modification has been obtained with some new reagents (Figure 6.11). 4-Chloro-7-nitrobenzo-2-oxa-1,3-diazole (4-chloro-7-nitrobenzofurazan; Nbf-Cl) is a reagent developed for the modification of amino groups.[265] It has also found application in the modification of sulfhydryl groups and is useful in that it introduces a fluorescent probe.[266-270] Nitta and coworkers[95] noted that there are other possible reaction products of Nbf-Cl, including the possibility of reaction products with sulfhydryl groups. The modification of the sulfhydryl group with concomitant reaction at the fourth position yields a derivative with a molar absorption coefficient of 13,000.[270] The reaction of Nbf-Cl with sulfhydryl groups in glutathione reductase and lipoamide dehydrogenase has also been reported.[271] Nitta and co-workers[95] have examined the chemistry of the reaction of Nbf-Cl with model sulfhydryl compounds in some detail.

4,4'-Diisothiocyanostilbene-2,2'-disulfonic acid has been used to study the importance of specific sulfhydryl groups in anion transport by membrane proteins.[272-275] Bimanes have proven to be useful structural probes for proteins.[278] Monobromobimane and monobromotrimethylammonium bimane are two examples of various derivatives available, with monobromobimane being considered a nonpolar or hydrophobic probe and monobromotrimethylammonium bimane being considered a polar probe. The reader is directed toward two early studies on the use of these reagents for the study of sulfhydryl group chemistry in proteins.[160,200]

The formation of 2-mercapto-5-nitrobenzoic acid, which occurs with the reaction of 2-nitrothiocyanobenzoic acid with thiols to form *S*-cyano derivatives, can be used for the quantitative determination of sulfhydryl groups. 2-Mercapto-5-nitrobenzoic acid has an absorbance maximum at 412 nm, with a molar extinction coefficient of 13,600 M^{-1} cm^{-1}.[227] Pecci and co-workers[279] characterized the reaction of rhodanese with 2-nitrothiocyanobenzoic acid. These investigators used a 1.3 molar excess of reagent in 0.050 M phosphate buffer, pH 8.0, at 18°C. The reaction was followed spectrophotometrically by the release of 2-mercapto-5-nitrobenzoic acid and was complete after 6 h.

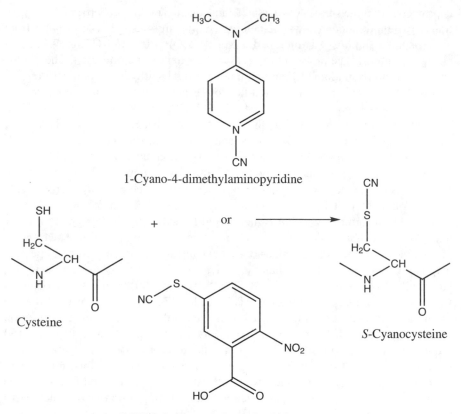

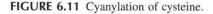

FIGURE 6.11 Cyanylation of cysteine.

    Cleavage at *S*-cyano-cysteinyl residues was first studied by Vanaman and Stark.[280] *S*-Cyanocysteine was first generated from the mixed disulfide of cysteine and 5-thio-2-nitrobenzoic acid by reaction with KCN (0.05 *M*) at pH 8.2 (0.2 *M* Tris-acetate with 20 m*M* EDTA).

    Toyo'oka and Imai[281] have reported the synthesis of a related compound, 4-(aminosulfonyl)-7-fluoro-2,1,3-benzoxadiazole, which is a fluorogenic reagent for sulfhydryl groups. In a subsequent study, Kirley[282] used this compound to label cysteinyl residues in proteins. Reaction readily occurred at pH 8.0 (100 m*M* borate, 2 m*M* EDTA, with 3% SDS).

    The modification of cysteinyl residues in proteins with 2-bromoethane sulfonate has been reported.[283] This derivatization procedure was developed in response to a need for a strongly hydrophilic substituent in samples for Edman degradation. The modification time is longer than for the corresponding carboxymethyl derivatives: 12 h for lysozyme, 24 h for insulin, and 48 h for glutathione. This derivative has considerable utility, because the *S*-sulfoethylated lysozyme derivative is soluble between pH 5.0 and 10.0 whereas the *S*-carboxymethylated derivative is not. This procedure has potential for primary structure analysis.

Mutus and co-workers[228] introduced 1-*p*-chorophenyl-4,4-dimethyl-5-diethy-lamino-1-penten-3-one hydrobromide as a reagent to modify thiol groups. This reagent readily and reversibly reacts with low-molecular-weight thiols such as cysteine or glutathione, with a large decrease in absorbance at 310 nM ($\varepsilon = 21,000\ M^{-1}$ cm^{-1}). The reaction with larger thiol-containing molecules such as proteins appears to be irreversible.

REFERENCES

1. Ohno A., and Oae, S., Thiols. In *Organic Chemistry of Sulfur* (Oae, S. Ed.). Plenum Press, New York, 1977, pp. 119, Chap. 4.
2. Sluyterman, L.A.A., The rate-limiting reaction in papain action as derived from the reaction of the enzyme with chloroacetic acid, *Biochim. Biophys. Acta* 151, 178, 1968.
3. Cecil, R., and McPhee, J.R., The sulfur chemistry of proteins, *Adv. Protein Chem.* 14, 255, 1959.
4. Liu, T.-Y., The role of sulfur in proteins. In *The Proteins*, Vol. 3, 3rd ed. (Neurath, H.H., and Hill, R.L., Eds.). Academic Press, New York, 1977, p. 240.
5. Torchinsky, Y.M., *Sulfur in Proteins*, Pergamon Press, Oxford, 1981.
6. Creighton, T.E., Chemical nature of polypeptides. In *Proteins, Structure and Molecular Principles*, W.H. Freeman, New York, Chap. 1.
7. Modena, G., Paradisi, C., and Scorrano, G., Solvation effects on basicity and nucleophilicity. In *Organic Sulfur Chemistry. Theoretic and Experimental Advances*, Bernardi, F., Csizmadia, I.G., and Mongini, A., Eds.). Elsevier, Amsterdam, 1985, pp. 569–597, Chap. 10.
8. Furuta, S. et al., Copper uptake is required for pyrrolidone dithiocarbamate-mediated oxidation and protein level increase of p53 in cells, *Biochem. J.* 365, 639, 2002.
9. Sharp, J.S., Becker, J.M., and Hettich, R.L., Protein surface mapping by chemical oxidation: structural analysis by mass spectrometry, *Anal. Biochem.* 313, 216, 2003.
10. Mannick, J.B., and Schonhoff, C.M., Nitrosylation: the next phosphorylation, *Arch. Biochem. Biophys.* 408, 1, 2002.
11. Tao, L., and English, A.M., Mechanism of *S*-nitrosylation of recombinant human brain calbindin D_{28K}, *Biochemistry* 42, 3326, 2003.
12. Gerwin, B.I., Properties of the single sulfhydryl group of streptococcal proteinase. A comparison of the rates of alkylation by chloroacetic acid and chloroacetamide, *J. Biol. Chem.* 242, 451, 1967.
13. Britto, P.J., Knipling, L., and Wolff, J., The local electrostatic environment determines cysteine reactivity of tubulin, *J. Biol. Chem.* 277, 29018, 2002.
14. Kanaya, S., Kimura, S., Katsuda, C., and Ikehara, M., Role of cysteine residues in ribonuclease H from *Escherichia coli*. Site-directed mutagenesis and chemical modification, *Biochem. J.* 271, 59, 1990.
15. Kato, H., Tanaka, T., Nishioka, T., Kimura, A., and Oda, J., Role of cysteine residues in glutathione synthetase from *Escherichia coli* B. Chemical modification and oligonucleotide site-directed mutagenesis, *J. Biol. Chem.* 263, 11646, 1988.
16. Altenbach, C., Flitsch, S.L., Khorana, H.G., and Hubbell, W.L., Structural studies on transmembrane proteins. II. Spin labeling of bacteriorhodopsin mutants at unique cysteines, *Biochemistry* 28, 7806, 1989.
17. Bech, L.M., and Breddam, K., Chemical modification of a cysteinyl residue introduced in the binding site of carboxypeptidase Y by site-directed mutagenesis [Abstract], *Carlsberg Res. Commn.* 53, 381, 1988.

18. Minard, P., Desmadril, M., Ballery, N., Perahia, D., Mouawad, L., Hall, L., and Yon, J.M., Study of the fast-reacting cysteines in phosphoglycerate kinase using chemical modification and site-directed mutagenesis, *Eur. J. Biochem.* 185, 419, 1989.
19. Kanaya, S., Kimura, S., Katsuda, C., and Ikehara, M., Role of cysteine residues in ribonuclease H from *Escherichia coli*. Site-directed mutagenesis and chemical modification, *Biochem. J.* 271, 59, 1990.
20. Ahmed, S.A., Kawasaki, H., Bauerle, R., Morita, H., and Miles, E.W., Site-directed mutagenesis of the alpha subunit of tryptophan synthase from *Salmonella typhimurium*, *Biochem. Biophys. Res. Commn.* 151, 672, 1988.
21. Thrower, A.R., Byrd, J., Tarbet, E.B., Mehra, R.K., Hamer, D.H., and Winge, D.R., Effect of mutation of cysteinyl residues in yeast Cu-metallothionein, *J. Biol. Chem.* 263, 7037, 1988.
22. Winther, J.R., and Breddam, K., The free sulfhydryl group (CYS341) of carboxypeptidase Y: functional effects of mutational substitutions, *Carlsberg Res. Commn.* 52, 263, 1987.
23. Sembrock, J., and Russell, D.W., Mutagenesis. In *Molecular Cloning: A Laboratory Manual*, Cold Spring Harbor Laboratory Press, Cold Spring Harbor, New York, 2001, pp. 13.11–13.105, Chap. 13. http://www.promega.com; http://www.stratagene.com.
24. DeSantis, G., and Jones, J.B., Chemical modifications at a single site can induce significant shifts in the pH profiles of a serine protease, *J. Am. Chem. Soc.* 120, 8882, 1998.
25. DeSantis, G., Shang, X., and Jones, J.B., Toward tailoring of the S_1 pocket of subtilisin *B. lentus*: Chemical modification of mutant enzymes as a strategy for removing specificity limitations, *Biochemistry* 28, 13391, 1999.
26. Karlin, B., and Akabas, M.H., Substituted-cysteine accessibility method, *Methods Enzymol.* 293, 123, 1998.
27. Toedt, G.H., Krishnan, R., and Friedhoff, P., Site-specific protein modification to identify the MutL interface of MutH, *Nucleic Acids Res.* 31, 819, 2003.
28. Venkatesan, P. et al., Site-directed sulfhydryl labeling of the lactose permease of *Escherichia coli*: N-ethylmaleimide-sensitive face of Helix II, *Biochemistry* 29, 10649, 2000.
29. Chan, B.S., Santriano, J.A., and Schuster, V.L., Mapping the substrate binding site of the prostaglandin transporter PGT by cysteine scanning mutagenesis, *J. Biol. Chem.* 274, 25564, 1999.
30. Frillingos, S. et al., Cys-scanning mutagenesis: a novel approach to structure-function in polytopic membrane proteins, *FASEB J.* 12, 1281, 1998.
31. Flitsch, S.L., and Khorana, H.G., Structural studies on transmembrane proteins. 1. Model study using bacteriorhodopsin mutants containing single cysteine residues, *Biochemistry* 28, 7800, 1989.
32. Alttenbach, C. et al., Structural studies on transmembrane proteins. 2. Spin labeling of bacteriorhodopsin mutants at unique cysteines, *Biochemistry* 28, 7806, 1989.
33. Todd, A.P. et al., Site-directed mutagenesis of colicin C1 provides specific attachment sites of spin labels whose spectra are sensitive to local conformation, *Proteins* 6, 294, 1989.
34. Barnakov, A. et al., Site-directed spin labeling of a bacterial chemoreceptor reveals a dynamic loosely packed transmembrane domain, *Protein Sci.* 11, 1472, 2002.
35. Tamm, L.K. et al., Structure, dynamics and function of the outer membrane protein A (Omp A) and influenza hemagglutinin fusion domain in detergene micelles by solution NMR, *FEBS Letters* 555, 139, 2003.

36. Hustedt, E.J., and Beth, A.H., Nitroxide spin-spin interactions: applications to protein structure and dynamics, *Annu. Rev. Biophys. Biomol. Struct.* 28, 129, 1999.
37. della Rocca, B.M. et al., The mitochondrial oxoglutarate carrier: Structural and dynamics properties of transmembrane segment IV studied by site-directed spin labeling, *Biochemistry* 42, 5493, 2003.
38. Kapanidis, A.N., and Weiss, S., Fluorescent probes and bioconjugation chemistries of single-molecule fluorescence analysis of biomolecules. *J. Chem. Phys.* 117, 10953, 2002.
39 Stephens, A.N. et al., The molecular neighborhood of subunit 8 or yeast mitochondria F1FO-ATP synthase probed by cysteine scanning mutagenesis and chemical modification, *J. Biol. Chem.* 278, 17867, 2003.
40. Yao, X., and Pajar, A.M., Arginine-349 and aspartate-373 of the Na$^+$/dicarboxylate cotransporter and conformationally sensitive residues, *Biochemistry* 42, 9789, 2003.
41. Kaback, H.R., Structure/function studies on the lactose permease of *Escherichia coli*, *Acta Physiol. Scand. Suppl.* 643, 21, 1998.
42. Abramson, J. et al., The lactose permease of *Escherichia coli*: overall structure, the sugar-binding site and the alternating model for transport, *FEBS Lett.* 555, 96, 2003.
43. Sondej, M. et al., Deduction of consensus binding sequences on proteins that bind IIAG1c of the phosphoenolpyruvate:sugar phosphotransferase system by cysteine scanning mutagenesis of *Escherichia coli* lactose permease, *Proc. Natl. Acad. Sci. USA* 96, 3525, 1999.
44. Kaback, H.R., Sahin-Toth, M., and Weinglass, A.B., The kamikaze approach to membrane transport, *Nat. Rev. Mol. Cell. Biol.* 2, 610, 2001.
45. Frillingos, S. et al., Binding of ligand or monoclonal antibody 4B1 induces discrete structural changes in the lactose permease of *Escherichia coli*, *Biochemistry* 36, 6408, 1997.
46. Zhong, W., Hu, Y., and Kaback, H.R., Site-directed sulfhydryl labeling of helix IX in the lactose permease of *Escherichia coli*, *Biochemistry* 42, 4904, 2003.
47. Loo, T.W., Bartlett, M.C., and Clarke, D.M., Disulfide cross-linking analysis shows that transmembrane segments 5 and 8 of human P-glycoprotein are close together on the cytoplasmic side of the membrane, *J. Biol. Chem.* 279, 7692, 2004.
48. Kawasaki-Nishi, S., Nishi, T., and Forgac, M., Interacting helical surfaces of the transmembrane segments of subunits a and c' of the yeast V-ATPase defined by disulfide-mediated cross-linking, *J. Biol. Chem.* 278, 41908, 2003.
49. Neale, E.J. et al., Evidence for intersubunit interactions between S4 and S5 transmembrane segments of the Shaker potassium channel, *J. Biol. Chem.* 278, 29079, 2003.
50. Samanta, S. et al., Disulfide cross-linking reveals a site of stable interaction between C-terminal regulatory domains of the two MalK subunits in the maltose transport complex, *J. Biol. Chem.* 278, 35265, 2003.
51. Satoh, Y. et al., Nearest neighbor analysis of the SecYEG complex. 1. Identification of a SecY-SecG interface, *Biochemistry* 42, 7434, 2003.
52. Miller, S. et al., The closed structure of the MscS mechanosensitive channel. Cross-linking of single cysteine mutants, *J. Biol. Chem.* 278, 32246, 2003.
53. Kawate, T., and Gouaux, E., Arresting and releasing staphylococcal alpha-hemolysin at intermediate stages of pore formation by engineered disulfide bonds, *Protein Sci.* 12, 997, 2003.
54. Shapovalov, G. et al., Open-state disulfide crosslinking between *Mycobacterium tuberculosis* mechanosensitive channel subunits, *Biophys. J.* 84, 2357, 2003.

55. Hemmi, H. et al., Inhibitory specificity change of the ovomucoid third domain of the silver pheasant upon introduction of an engineered Cys14-Cys39 bond, *Biochemistry* 42, 2524, 2003.

56. Martensson. L.G., Karlsson, M., and Carlsson, U., Dramatic stabilization of the native state of human carbonic anhydrase II by an engineered disulfide bond, *Biochemistry* 41, 15867, 2002.

57. Tian, S., and Jonas, A., Structural and functional properties of apolipoprotein A-I mutants containing disulfide-linked cysteines at positions 124 or 232, *Biochim. Biophys. Acta* 1599, 56, 2002.

58. Lima, W.F. et al., Human RNase H1 activity is regulated by a unique redox switch formed between adjacent cysteines, *J. Biol. Chem.* 278, 14906, 2003.

59. Struthers, M., Yu, H., and Oprian, D.D., Protein coupled receptor activation: Analysis of a highly constrained, "straitjacketed" rhodopsin, *Biochemistry* 39, 7938, 2000.

60. Yu, H., Kono, H., and Oprian, D.D., A general method for mapping tertiary contacts between amino acid residues in membrane-embedded proteins, *Biochemistry* 34, 14963, 1995.

61. Nagy, J.K. et al., Mapping the oligomeric interface of diacylglycerol kinase by engineered thiol cross-linking: homologous sites in the transmembrane domain, *Biochemistry* 39, 4154, 2000.

62. Yu, H., Kono, H., and Oprian, D.D., State-dependent disulfide cross-linking in rhodopsin, *Biochemistry* 38, 12028, 1999.

63. Toney, M.D., and Kirsch, J.F., Direct Bronsted analysis of the restoration of activity to a mutant enzyme by exogenous amines, *Science* 243, 1485, 1989.

64. Brooks, B., and Benisek, W.F., Specific activation of a tyrosine-glycine mutant of delta 5,3-ketosteroid isomerase by phenols, *Biochem. Biophys. Res. Commn.* 184, 1386, 1992.

65 Inagami, T., and Murachi, T., Induced activation of the catalytic site of trypsin, *J. Biol. Chem.* 238, PC1905, 1963.

66. Inagami, T., and Hatano, H., Effect of alkylguanidines on the inactivation of trypsin by alkylation and phosphorylation, *J. Biol. Chem.* 244, 1176, 1969.

67. Zheng, R., and Blanchard, J.S., Identification of active site residues in *E. coli* ketopantoate reductase by mutagenesis and chemical rescue, *Biochemistry* 39, 16244, 2000.

68. Raftery, M.A., and Cole, R.D., On the aminoethylation of proteins, *J. Biol. Chem.* 241, 3457, 1966.

69. Thevis, M., Loo, R.R.O., and Loo, J.A., In-gel derivatization of proteins for cysteine-specific cleavages and their analysis by mass spectrometry, *J. Proteome Res.* 2, 163, 2003.

70. Göransson, U., Broussalis, A.M., and Claeson, P., Expression of *Viola* cyclotides by liquid chromatography-mass spectrometry and tandem mass spectrometry sequencing of intercysteine loops after introduction of charges and cleavage sites by amino-ethylation, *Anal. Biochem.* 318, 107, 2003.

71. Hollenbaugh, D., Aruffo, A., and Senter, P.D., Effects of chemical modification on the binding activities of P-selectin mutants, *Biochemistry* 34, 5678, 1995.

72. Bochar, D.A. et al., Aminoethylcysteine can replace the function of the essential active site lysine of *Pseudomonas mevalonii* 3-hydroxy-3-methylglutaryl coenzyme A reductase, *Biochemistry* 38, 8879, 1999.

73. Hopkins, C.E. et al., Chemical-modification rescue assessed by mass spectrometry demonstrates that -thia-lysine yields the same activity as lysine in aldolase, *Protein Sci.* 11, 1591, 2002.

74. Bandell, M., and Lolkema, J.S., Arg-425 of the citrate transporter CitP is responsible high affinity binding of di- and tricarboxylates, *J. Biol. Chem.* 275, 39130, 2000.
75. Schindler, J.F., and Viola, R.E., Conversion of cysteinyl residues to unnatural amino acid analogues, *J. Protein Chem.* 15, 737, 1996.
76. Smith, D.J., Maggio, E.T., and Kenyon, G.L., Simple alkane thiol groups for temporary blocking of sulfhydryl groups of enzymes, *Biochemistry* 14, 766, 1975.
77. Kluger, R., and Tsue, W.-C., Amino group reactions of the sulfhydryl reagent methyl methanethiosulfonate. Inactivation of D-3-hydroxybutyrate and reaction with amines in water, *Can. J. Biochem.* 58, 629, 1980.
78. Pathak, R., Hendrickson, T.L., and Imperiali, B., Sulfhydryl modification of the yeast Wbp1p inhibits oligosaccharide transferase activity, *Biochemistry* 34, 4179, 1995.
79. Roberts, D.D. et al., Reactivity of small thiolate anions and cysteine-25 in papain toward methyl methanethiosulfonate, *Biochemistry* 25, 5595, 1986.
80. Stauffers, D.A., and Karlin, A. Electrostatic potential of the acetylcholine binding site in the nicotinic receptor probed by reaction of binding-site cysteines with charged methanethiosulfonates, *Biochemistry* 33, 6840, 1994.
81. Kenyon, G.L., and Bruice, T.W., Novel sulfhydryl reagents, *Methods Enzymol.* 47, 407, 1977.
82. Bruice, T.W., and Kenyon, G.L., Novel alkyl alkanethiolsulfonate sulfhydryl reagents. Modification of derivatives of L-cysteine, *J. Protein Chem.* 1, 47, 1982.
83. Toronto Research Chemicals, Inc. Methanethiosulfonate reagents: application to the study of protein topology and ion channels. Available at http://www.trc-canada.com/white_papers.lasso. 2004.
84. Karlin, A., and Akabas, M.H., Substituted-cysteine accessibility method, *Methods Enzymol.* 293, 123, 1998.
85 Barre, L., Kavanaugh, M.P., and Kanner, B.I., Dynamic equilibrium between coupled and uncoupled modes of neuronal glutamate transporter, *J. Biol. Chem.* 277, 13501, 2002.
86. Cui, Y., and Fan, Z., Mechanism of Kir6.2 channel inhibition by sulfhydryl modification: pore block or allosteric gating, *J. Physiol.* 540, 731, 2002.
87. Tonazzi, A., and Indiveri, C., Chemical modification of the mitochondrial ornithine/citrulline carrier by SH reagents: effects on the transport activity and transition from carrier to pore-like function, *Biochim. Biophys. Acta* 1611, 123, 2003.
88. Brandwein, H.J., Lewicki, J.A., and Murad, F., Reversible inactivation of guanylate cyclase by mixed disulfide formation, *J. Biol. Chem.* 256, 2958, 1981.
89. Thomas, J.A., Chai, Y.-C., and Jung, C.-H., Protein S-thiolation and dethiolation, *Methods Enzymol.* 233, 385, 1994.
90. Lim, A. et al., Identification of S-sulfonation and S-thiolation of a novel transthyretin Phe33Cys variant from a patient diagnosed with familial transthyretin amyloidosis, *Protein Sci.* 12, 1775, 2003.
91. Shenton, D., and Grant, C.M., Protein S-thiolation targets glycolysis and protein synthesis in response to oxidative stress in the yeast *Saccharomyces cerervesiae*, *Biochem. J.* 374, 513, 2003.
92. Liu, T.Y., Demonstration of the presence of a histidine residue at the active site of streptococcal proteinase, *J. Biol. Chem.* 242, 4029, 1967.
93. Mukhopadhyay, A., Reversible protection of disulfide bonds followed by oxidative folding render recombinant hCGbeta highly immunogenic, *Vaccine* 18, 1802, 2000.
94. Tikhonov, R.V. et al., Recombinant human insulin. VIII. Isolation of fusion proteins — S-sulfonate, biotechnological precursor of human insulin, from the biomass of transformed *Escherichia coli* cells, *Protein Expr. Purif.* 21, 176, 2001.

95. Dahl, K.S., and McKinley-McKee, J.S., The reactivity of affinity labels: a kinetic study of the reaction of alkyl halides with thiolate anions: a model reaction for protein alkylation, *Bioorg. Chem.* 10, 329, 1981.

96. Plapp, B.V., Application of affinity labeling for studying structure and function of enzymes, *Methods Enzymol.* 87, 469, 1982.

97. El Garrouj, D., Aumeles, A., and Borgna, J.-L., Steroidal affinity labels of the estrogen receptor. 1. 17-α(bromoacetoxy)alkyl.alkynylestradiols, *J. Med. Chem.* 36, 2973, 1993.

98. Aliau, S. et al., Steroidal affinity labels of the estrogen receptor α. 4. Electrophilic 11-aryl derivatives of estradiol, *J. Med. Chem.* 43, 613, 2000.

99. Castellano, F.N., Dattelbaum, J.D., and Lakowicz, J.R., Long-lifetime Ru(II) complexes as labeling reagents for sulfydryl groups, *Anal. Biochem.* 255, 165, 1998.

100a. Jahng, W.J. et al., A cleavable biotinylating agent reveals a retinoid binding role for RPE65, *Biochemistry* 42, 6159, 2003.

100b. Chaiken, I.M., and Smith, E.L., Reaction of chloroacetamide with the sulfhydryl group of papain, *J. Biol. Chem.* 244, 5087, 1969.

101. Chaiken, I.M., and Smith, E.L., Reaction of the sulfhydryl group of papain with chloroacetic acid, *J. Biol. Chem.* 244, 5095, 1969.

102. Jörnvall, H., Fowler, A.V., and Zabin, I., Probe of β-galactosidase structure with iodoacetate. Differential reactivity of thiol groups in wild-type and mutant forms of β-galactosidase, *Biochemistry* 17, 5160, 1978.

103. Kalimi, M., and Love, K., Role of chemical reagents in the activation of rat hepatic glucocorticoid-receptor complex, *J. Biol. Chem.* 255, 4687, 1980.

104. Kallis, G.-B., and Holmgren, A., Differential reactivity of the functional sulfhydryl groups of cysteine-32 and cysteine-35 present in the reduced form of thioredoxin from *Escherichia coli*, *J. Biol. Chem.* 255, 10261, 1980.

105. Mikami, B., Aibara, S., and Morita, Y., Chemical modification of sulfhydryl groups in soybean α-amylase, *J. Biochem.* 88, 103, 1980.

106. Hempel, J.D., and Pietruszko, R., Selective chemical modification of human liver aldehyde and dehydrogenases E₁ and E₂ by iodoacetamide, *J. Biol. Chem.* 256, 10889, 1981.

107. Crestfield, A.M., Moore, S., and Stein, W.H., The preparation and enzymatic hydrolysis of reduced and S-carboxymethylated proteins, *J. Biol. Chem.* 238, 622, 1963.

108. Friedman, M., Krull, L.H., and Cavins, J.F., The chromatographic determination of cystine and cysteine residues in proteins as *S*-β-(4-pyridyl-ethyl) cysteine, *J. Biol. Chem.* 245, 3868, 1970.

109. Mak, A.S., and Jones, B.L., Application of *S*-pyridylethylation of cysteine to the sequence analysis of proteins, *Anal. Biochem.* 84, 432, 1978.

110. Pflouq, M., Stoffer, B., and Jensen, A.L., In situ alkylation of cysteine residues in a hydrophobic membrane protein immobilized on polyvinylidene difluoride membranes by electroblotting prior to microsequence and amino acid analysis, *Electrophoresis* 13, 148, 1992.

111. Lundell, N., and Schreitmüller, T., Sample preparation of peptide-mapping: a pharmaceutical quality-control perspective, *Anal. Biochem.* 266, 31, 1999.

112. Görg, A. et al., The current state of two-dimensional electrophoresis with immobilized pH gradients, *Electrophoresis* 21, 1037, 2000.

113. Jahn, O. et al., The use of multiple ion chromatograms in on-line HPLC-MS for the characterization of post-translational and chemical modification of proteins, *Int. J. Mass. Spectrom.* 214, 37, 2002.

114. Sechi, S., and Chait, B.T., Modification of cysteine residues by alkylation. A tool in peptide mapping and protein identification, *Anal. Chem.* 70, 5150, 1998.

115. Herbert, B. et al., Reduction and alkylation of proteins in preparation of two-dimensional map analysis. Why, when, and how? *Electrophoresis* 22, 2046, 2001.

116. Hoving, S. et al., Preparative two-dimensional gel electrophoresis at alkaline pH using narrow range immobilized pH gradients, *Proteomics* 2, 127, 2002.

117. Shaw, M.M., and Riederer, B.M., Sample preparation for two-dimensional gel electrophoresis, *Proteomics* 3, 1408, 2003.

118. Malloy, M.P. et al., Overcoming technical variation and biological variation in quantitiative proteomics, *Proteomics* 3, 1912, 2003.

119. Herbert, B. et al., β-elimination: An unexpected artifact in proteome analysis, *Proteomics* 3, 826, 2003.

120. Rabilloud, T. et al., Improvement of the solubilization of proteins in two-dimensional electrophoresis with immobilized pH gradients, *Electrophoresis* 18, 307, 1997.

121. Galvani, M. et al., Protein alkylation in the presence/absence of thiourea in proteome analysis: a matrix assisted laser desorption/ionization-time of flight-mass spectrometry investigation, *Electrophoresis* 22, 2066, 2001.

122. Shaw, J., Rowlinson, R., Nickson, J., Stone, T., Sweet, A., Williams, K., and Tonge, R., Evaluation of saturation labelling two-dimensional gel electrophoresis fluorescent dyes, *Proteomics* 3, 1181–1195, 2003.

123. Tyagarajon, K., Pretzer, E., and Wiktorowicz, J.E., Thiol-reactive dyes for fluorescence labeling of proteomic samples, *Electrophoresis* 24, 2348, 2003.

124. Plowman, J.E. et al., The effect of oxidation or alkylation on the separation of wool keratin proteins by two-dimensional gel electrophoresis, *Proteomics* 3, 942, 2003.

125. Galvani, M. et al., Alkylation kinetics of proteins in preparation for two-dimensional maps. A matrix, *Electrophoresis* 22, 2058, 2001.

126. Jeffrey, D.A., and Bogyo, M., Chemical proteomics and its application to drug discover,. *Curr. Opin. Biotechnol.* 14, 87, 2003.

127. Bottari, P. et al., Design and synthesis of visible isotope-coded affinity tags for the absolute quantitation of specific proteins in complex mixtures, *Bioconjugate Chem.* 15, 280, 2004.

128. Mann, M , Quantitative proteomics? *Nature Biotechnol.* 17, 954, 1999.

129. Smolka, M., Zhou, H., and Aebersold, R., Quantitative protein profiling using two-dimensional gel electrophoresis, isotope-coded affinity tag labeling, and mass spectrometry, *Mol. Cell. Proteom.* 1, 19, 2002.

130. Gygi, P. et al., Quantitative analysis of complex protein mixtures using isotope-coded affinity tags, *Nature Biotechnol.* 17, 994, 1999.

131. Tao, W.A., and Aebersold, R., Advances in quantitative proteomics via stable isotope tagging and mass spectrometry, *Curr. Opin. Biotechnol.* 14, 100, 2003.

132. Shen, M. et al., Isolation and isotope labeling of cysteine- and methionine-containing tryptic peptides: application to the study of cell surface proteolysis, *Mol. Cell. Proteom.* 2, 315, 2003.

133. Pasquarello, C. et al., *N-t*-butylidoacetamide and iodoacetanilide: two new cysteine alkylating reagents for relative quantitation of proteins, *Rapid Commn. Mass Spectrom.* 18, 117, 2004,

134. Hansen, K.C. et al., Mass spectrometric analysis of protein mixtures at low levels using cleavable ^{13}C-isotope-coded affinity tag and multidimensional chromatography, *Mol. Cell. Proteom.* 2, 299, 2003.

135. Thevis, M., Loo, R.R.O., and Loo, J.A., In-gel derivatization of proteins for cysteine-specific cleavages and their analysis by mass spectrometry, *J. Proteom. Res.* 2, 163, 2003.

136. Friedman, M., Application of the *S*-pyridylethylation reaction to the elucidation of the structure and function of proteins, *J. Protein Chem.* 20, 431, 2001.

137. Sebastiano, R. et al., A new deuterated alkylating agent for quantitative proteomics, *Rapid Commn. Mass Spectrom.* 17, 2380, 2003.

138. Soper, T.S., Ueno, H., and Manning, J.M., Substrate-induced changes in sulfhydryl reactivity of bacterial D-amino acid transaminase, *Arch. Biochem. Biophys.* 240, 1, 1985.

139. Kaslow, H.R., Schlotterbeck, J.D., Mar, V.L., and Burnette, W.N., Alkylation of cysteine 41, but not cysteine 200, decreases the ADP-ribosyltransferase activity of the S1 subunit of pertussis toxin, *J. Biol. Chem.* 264, 6386, 1989.

140. Makinen, A.L., and Nowak, T., A reactive cysteine in avian liver phosphoenolpyruvate carboxykinase, *J. Biol. Chem.* 264, 12148, 1989.

141. Poole, L.B., and Claiborne, A., The non-flavin redox center of the streptococcal NADH peroxidase. I. Thiol reactivity and redox behavior in the presence of urea, *J. Biol. Chem.* 264, 12322, 1989.

142. Wu, H., Yao, Q.-Z., and Tsou, C.-L., Creatine kinase is modified by 2-chloromercuri-4-nitrophenol at the active site thiols with complete inactivation, *Biochim. Biophys. Acta* 997, 78, 1989.

143. Wang, Z.-X., Preiss, B., and Tsou, C.-L., Kinetics of inactivation of creatine kinase during modification of its thiol groups, *Biochemistry* 27, 5095, 1988.

144. Ramseur, U., and Chang, J.Y., Modification of cysteine residues with *N*-methyl iodoacetamide, *Anal. Biochem.* 221, 231, 1994.

145. Sather, J.K., and King, J., Intracellular trapping of a cytoplasmic folding intermediate of the phage P22 tailspike using iodoacetamide, *J. Biol. Chem.* 269, 25268, 1994.

146. Khan, A.R. et al., Accessibility and dynamics of Cys residues in bacteriophage Ike and M13 major coat protein mutants, *Biochemistry* 34, 12388, 1995.

147. Connolly, D.T., Heuvelman, D., and Glenn, K., Inactivation of cholesterol ester transfer protein by cysteine modification, *Biochem. Biophys. Res. Commn.* 223, 42, 1996.

148. Zeng, F.Y., and Weigel, P.H., Fatty acylation of the rat and human asialoglycoprotein receptors. A conserved cytoplasmic cysteine residue is acylated in all receptor subunits, *J. Biol. Chem.* 271, 32454, 1996.

149. Lapko, V.N. et al., Surface topography of phytochrome A deduced from specific chemical modification with iodoacetamide, *Biochemistry* 37, 12536, 1998.

150. Powlowski, J., and Sahlman, L., Reactivity of the two essential cysteine residues of the periplasmic mercuric ion-binding protein, MerP, *J. Biol. Chem.* 274, 33320, 1999.

151. Shingledecker, K., Jiang, S., and Paulus, H., Reactivity of the cysteine residues in the protein splicing active center of the *Myobacterium tuberculosis* RecA intein, *Arch. Biochem. Biophys.* 375, 138, 2000.

152. Dyllick-Brenzinger, M. et al., The role of cysteine residues in tellurite resistance medicated by the TehAB determinant, *Biochem. Biophys. Res. Commn.* 277, 394, 2000.

153. Firla, B. et al., Extracellular cysteines define ectopeptidase (APN, CD13) expression and function, *Free Radic. Biol. Med.* 32, 584, 2002..

154. Pitts, K.E., and Summers, A.O., The roles of thiols in the bacterial organomercurial lyase (MerB), *Biochemistry* 41, 10287, 2002.

155. Tamarit, J. et al., Biochemical characterization of yeast mitochondrial Grx5 monothiol glutaredoxin, *J. Biol. Chem.* 278, 25745, 2003.

156. Seifried, S.E., Wang, Y., and Von Hippel, P.H., Fluorescent modification of the cysteine 202 residue of *Escherichia coli* transcription termination factor rho, *J. Biol. Chem.* 263, 13511, 1988.

157. Pardo, J.P., and Slayman, C.W., Cysteine 532 and cysteine 545 are the *N*-ethylmaleimide-reactive residues of the *Neurospora* plasma membrane H+-ATPase, *J. Biol. Chem.* 264, 9373, 1989.

158. Miyanishi, T., and Borejdo, J., Differential behavior of two cysteine residues on the myosin head in muscle fibers, *Biochemistry* 28, 1287, 1989.

159. Bishop, J.E., Squier, T.C., Bigelow, D.J., and Inesi, G., (Iodoacetamido) fluorescein labels a pair of proximal cysteines on the Ca^{2+}-ATPase of sarcoplasmic reticulum, *Biochemistry* 27, 5233, 1988.

160. First, E.A., and Taylor, S.S., Selective modification of the catalytic subunit of cAMP-dependent protein kinase with sulfhydryl-specific fluorescent probes, *Biochemistry* 28, 3598, 1989.

161. Ramalingam, T.S., Das, P.K., and Podder, S.K., Ricin-membrane interaction: membrane penetration depth by fluorescence quenching and resonance energy transfer, *Biochemistry* 33, 12247, 1994.

162. Hamman, B.D. et al., Rotational and conformational dynamics of *Escherichia coli* ribosomal protein L7/L12, *Biochemistry* 35, 16672, 1996.

163. Karlstrom, A., and Nygren, P.A., Dual labeling of a binding protein allows for specific fluorescence detection of native protein, *Anal. Biochem.* 295, 22, 2001.

164. Teilum, K. et al., Early kinetic intermediate in the folding of acyl-CoA binding protein detected by fluorescence labeling and ultrarapid mixing, *Proc. Natl. Acad. Sci. USA* 99, 9807, 2002.

165. Tripet, B. et al., Kinetic analysis of the interactions between troponin C and the C-terminal troponin I regulatory region and validation of a new peptide delivery/capture system used for surface plasmon resonance, *J. Mol. Biol.* 323, 345, 2002.

166. Xie, L. et al., Graphical evaluation of alkylation of myosin's SH1 and SH2: the *N*-phenylmaleimide reaction, *Biophys. J.* 72, 858, 1997.

167. Holyoak, T., and Nowak, T., Structural investigation of the binding of nucleotide to phosphoenolpyruvate carboxykinase by NMR, *Biochemistry* 40, 11037, 2001.

168. Toutchkine, A., Nalbant, P., and Hahn, K.M., Facile synthesis of thiol-reactive Cy3 and Cy5 derivatives with enhanced water solubility, *Bioconj. Chem.* 13, 387, 2002.

169. Toutchkine, A., Kraynov, V., and Hahn, K., Solvent-sensitive dyes to report protein conformational changes in living cells, *J. Am. Chem. Soc.* 125, 4132, 2003.

170. Chu, Q., and Fukui, Y., *In vivo* dynamics of myosin II in *Dictyostelium* by fluorescent analogue cytochemistry, *Cell. Motil.Cytoskel.* 35, 254, 1996.

171. Szalecki, W., Synthesis of norbiotinamine and its derivatives, *Bioconjug. Chem.* 7, 271, 1996.

172. Thomas, T.H. et al., Erythrocyte membrane thiol proteins associated with changes in the kinetics of Na/Li countertransport: a possible molecular explanation of changes in disease, *Eur. J. Clin. Invest.* 28, 259, 1998.

173. Kim, J.R. et al., Identification of proteins containing cysteine residues that are sensitive to oxidation by hydrogen peroxide at neutral pH, *Anal. Biochem.* 283, 214, 2000.

174. Lin, T.-K. et al., Specific modification of mitochondrial protein thiols in response to oxidative stress. A proteomics approach. *J. Biol. Chem.* 277, 17048, 2002.

175. Guo, Z. et al., Arylamine *N*-acetyltransferases: covalent modification and inactivation of hamster NAT1 by bromoacetamido derivatives of aniline and 2-aminofluorene, *J. Protein Chem.* 22, 631, 2003.

176. Chung, J.G. et al., Evidence for arylamine *N*-acetyltransferase in *Hymenolepis nana*, *J. Microbiol. Immunol. Infect.* 30, 1, 1997.

177. Chung, J.G., Purification and characterization of an arylamine *N*-acetyltransferase in the nematode *Enterobius vermicularis*, *Microbios* 98, 15, 1999.

178. Andres, H.H. et al., On the active site of liver acetyl-CoA arylamine *N*-acetyltransferase from rapid acetylator rabbits (III/J), *J. Biol. Chem.* 263, 7521, 1988.

179. Bateman, R.C. et al., Nonenzymatic peptide α-amidation. Implications for a novel enzyme mechanism, *J. Biol. Chem.* 260, 9088, 1985.

180. Rana, T.M., and Meares, C.F., Specific cleavage of a protein by an attached iron chelate, *J. Am. Chem. Soc.* 112, 2457, 1990.

181. Rana, T.M., and Meares, C.F., Transfer of oxygen from an artificial protease to peptide carbon during proteolysis, *Proc. Natl. Acad. Sci.USA* 88, 10578, 1991.

182. Greiner, D.P. et al., Binding of the σ70 protein to the core subunits of *Escherichia coli* RNA polymerase, studied by iron-EDTA protein footprinting, *Proc. Natl. Acad. Sci. USA* 93, 71, 1996.

183. Ghaim, J.B. et al., Proximity mapping the surface of a membrane protein using an artificial protease: demonstration that the quinine-binding domain of subunit I is near the N-terminal region of subunit II of cytochrome *bd*, *Biochemistry* 34,11311, 1995.

184. Greiner, D.P. et al., Synthesis of the protein cutting reagent iron (*S*)-1-(*p*-bromoace-tamidobenzyl)ethylenediaminetetraacetate and conjugation to cysteine side chains, *Bioconjug. Chem.* 8, 44, 1997.

185. Johnson, L., and Gershan, P.D., Direct detection of protein thiol derivatization by PAGE, *BioTechniques* 33, 1292, 2002.

186. Gitler, C., Mogyoros, M., and Kalef, E., Labeling of protein vicinal dithiols: role of protein-S_2 to protein-$(SH)_2$ conversion in metabolic regulation and oxidative stress, *Methods Enzymol.* 233, 403, 1994.

187. Happersberger, H.P., Przybylski, M., and Glocker, M.O., Selective bridging of bis-cysteinyl residues by arsonous acid derivatives as an approach to the characterization of protein tertiary structure and folding pathways by mass spectrometry, *Anal. Biochem.* 24, 237, 1998.

188. Happersberger, H.P., Cowgill, C., and Glocker, M.O., Structural characterization of monomeric folding intermediates of recombinant human macrophage-colony stimulating factor (rhM-CSF) by chemical trapping, chromatographic separation and mass spectrometric peptide mapping, *J. Chromatogr. B.* 782, 393, 2002.

189. Gregory, J.D., The stability of N-ethylmaleimide and its reaction with sulfhydryl groups, *J. Am. Chem. Soc.* 77, 3922, 1955.

190. Leslie, J., Spectral shifts in the reaction of N-ethylmaleimide with proteins, *Anal. Biochem.* 10, 162, 1965.

191. Gorin, G., Martic, P.A., and Doughty, G., Kinetics of the reaction of N-ethylmaleimide with cysteine and some congeners, *Arch. Biochem. Biophys.* 115, 593, 1966.

192. Bednar, R.A., Reactivity and pH dependence of thiol conjugation to N-ethylmaleimide: detection of a conformational change in chalcone isomerase, *Biochemistry* 29, 3684, 1990.

193. Smyth, D.G., Blumenfeld, O.O., and Konigsberg, W., Reaction of N-ethylmaleimide with peptides and amino acids, *Biochem. J.* 91, 589, 1964.

194. Gehring, H., and Christen, P., A diagonal procedure for isolating sulfhydryl peptides alkylated with N-ethylmaleimide, *Anal. Biochem.* 107, 358, 1980.

195. Brown, R.D., and Matthews, K.S., Chemical modification of lactose repressor proteins using N-substituted maleimides, *J. Biol. Chem.* 254, 5128, 1979.

196. Brown, R.D., and Matthews, K.S., Spectral studies on Lac repressor modified with N-substituted maleimide probes, *J. Biol. Chem.*, 254, 5135, 1979.

197. Perussi, J.R., Tinto, M.H., Nascimento, O.R., and Tabak, M., Characterization of protein spin labeling by maleimide: evidence for nitroxide reduction, *Anal. Biochem.* 173, 289, 1988.

198. Marquez, J., Iriarte, A., and Martinez-Carrion, M., Covalent modification of a critical sulfhydryl group in the acetylcholine receptor: cysteine-222 of the α-subunit, *Biochemistry* 28, 7433, 1989.

199. Mills, J.S., Walsh, M.P., Nemcek, K., and Johnson, J.D., Biologically active fluorescent derivatives of spinach calmodulin that report calmodulin target protein binding, *Biochemistry* 27, 991, 1988.

200. Jezek, P., and Drahota, Z., Sulfhydryl groups of the uncoupling protein of brown adipose tissue mitochondria: distinction between sulfhydryl groups of the H^+-channel and the nucleotide binding site, *Eur. J. Biochem.* 183, 89, 1989.

201. Le-Quoc, K., Le-Quoc, D., and Gaudemer, Y., Evidence for the existence of two classes of sulfhydryl groups essential for membrane-bound succinate dehydrogenase activity, *Biochemistry* 20, 1705, 1981.

202. Niwayama, S., Kurano, S., and Matsumoto, N., Synthesis of D-labeled N-alkylmaleimides and application to quantitative peptide analysis by isotope differential mass spectrometry, *Bioorg. Med. Chem. Lett.* 11, 2257, 2001.

203. Jones, P.C. et al., A method for determining transmembrane protein structure, *Mol. Membr. Biol.* 13, 53, 1996.

204. Di Gleria, K. et al., N-(2-ferrocene-ethyl)maleimide: a new electroactive sulfhydryl-specific reagent for cysteine-containing peptides and proteins, *FEBS Lett.* 390, 142, 1996.

205. Ercal, N., Yang, P., and Aykin, N., Determination of biological thiols by high-performance liquid chromatography following derivatization by ThioGlo maleimide reagents, *J. Chromatogr. B* 752, 287, 2001.

206. Tyagarajan, K., Pretzer, A., and Wiktorowicz, J.E., Thio-reactive dyes for fluorescence labeling of proteomic samples, *Electrophoresis* 24, 2348, 2003.

207. Apuy, J.L. et al., Radiometric pulsed alkylation/mass spectrometry of the cysteine pairs in individual zinc fingers of MRE-binding transcription factor-1 (MTF-1) as a probe of zinc chelate stability, *Biochemistry* 40, 15164, 2001.

208. Juszczak, L.J. et al., UV resonance Raman study of 93 modified hemoglobin A: chemical modification-specific effects and added influences of attached poly(ethylene glycol) chains, *Biochemistry* 41, 376, 2002.

209. Ni, J., Singh, S., and Wang, L.-X., Synthesis of maleimide-activated carbohydrates as chemoselective tags for site-specific glycosylation of peptides and proteins, *Bioconj. Chem.* 14, 232, 2003.

210. Ziu, Y. et al., Acid-labile isotope-coded extractants: a class of reagents for quantitative mass spectrometric analysis of complex protein mixtures, *Anal. Chem.* 74, 4969, 2002.

211. Liu, M. et al., Effect of cysteine residues on the activity of arginyl-tRNA synthetase from *Escherichia coli*, *Biochemistry* 38, 11006, 1999.

212. Jose, J., and Handel, S., Monitoring the cellular surface display of recombinant proteins by cysteine-labeling and flow cytometry, *ChemBioChem* 4, 296, 2003.

213. Rial, E., Aréchaga, I., Sainz-de-la-Maza, E., and Nicholls, D.G., Effect of hydrophobic sulphydryl reagents on the uncoupling protein and inner-membrane anion channel of brown-adipose-tissue mitochondria, *Eur. J. Biochem.* 182, 187, 1989.

214. Yee, A.S., Corley, D.E., and McNamee, M.G., Thiol-group modification of *Torpedo californica* acetylcholine receptor: subunit localization and effects on function, *Biochemistry* 25, 2110, 1986.

215. Pradier, L., Yee, A.S., and McNamee, M.G., Use of chemical modifications and site-directed mutagenesis to probe the functional role of thiol groups on the gamma subunit of *Torpedo californica* acetylcholine receptor, *Biochemistry* 28, 6562, 1989.

216. Abbott, R.E., and Schachter, D., Topography and functions of sulfhydryl groups of the human erythrocyte glucose transport mechanism, *Mol. Cell. Biochem.* 82, 85, 1988.

217. May, J.M., Reaction of an exofacial sulfhydryl group on the erythrocyte hexose carrier with an impermeant maleimide. Relevance to the mechanism of hexose transport, *J. Biol. Chem.* 263, 13635, 1988.

218. May, J.M., Interaction of a permeant maleimide derivative of cysteine with the erythrocyte glucose carrier. Differential labelling of an exofacial carrier thiol group and its role in the transport mechanism, *Biochem. J.* 263, 875, 1989.

219. Abbott, R.E., and Schachter, D., Impermeant maleimides. Oriented probes of erythrocyte membrane proteins, *J. Biol. Chem.* 251, 7176, 1976.

220. May, J.M., Selective labeling of the erythrocyte hexose carrier with a maleimide derivative of glucosamine: relationship of an exofacial sulfhydryl to carrier conformation and structure, *Biochemistry* 28, 1718, 1989.

221. Falke, J.J., Dernburg, A.F., Sternberg, D.A., Zalkin, N., Milligan, D.L., and Koshland, D.E., Jr., Structure of a bacterial sensory receptor. A site-directed sulfhydryl study, *J. Biol. Chem.* 263, 14850, 1988.

222. Yan, J.X. et al., Identification and quantitation of cysteine in proteins separated by gel electrophoresis, *J. Chromatog. A* 813, 187, 1998.

223. Bordini, E., Hamdan, M., and Righetti, P.G., Probing acrylamide alkylation sites in cysteine-free proteins by matrix-assisted laser desorption/ionization time-of-flight, *Rapid Commn. Mass Spectrom.* 14, 840, 2000.

224. Mineki, R. et al., *In situ* alkylation with acrylamide for identification of cysteinyl residues in proteins during one- and two-dimensional sodium dodecyl sulphate-polyacrylamide gel electrophoresis, *Proteomics* 2, 1672, 2002.

225. Cahill, M.A. et al., Analysis of relative isotopologue abundances for quantitative profiling of complex protein mixtures labelled with acrylamide/D-3-acrylamide alkylation tag system, *Rapid Commn. Mass Spectrom.* 17, 1283, 2003.

226. Ellman, G.L., A colorimetric method for determining low concentrations of mercaptans, *Arch. Biochem. Biophys.* 74, 443, 1958.

227. Ellman, G.L., Tissue sulfhydryl groups, *Arch. Biochem. Biophys.* 82, 70, 1959.

228. Habeeb, A.F.S.A., Reaction of protein sulfhydryl groups with Ellman's reagent, *Methods Enzymol.* 25, 457, 1972.

229. Collier, H.B., A note on the molar absorptivity of reduced Ellman's reagent, 3-carboxylato-4-nitrothiophenolate, *Anal. Biochem.* 56, 310, 1973.

230. Riddles, P.W., Blakeley, R.L., and Zerner, B., Ellman's reagent: 5,5′-dithiobis(2-nitrobenzoic acid): a reexamination, *Anal. Biochem.* 94, 75, 1979.

231. Riddles, P.W., Blakeley, R.L., and Zerner, B., Reassessment of Ellman's reagent, *Methods Enzymol.* 91, 49, 1983.

232. Eyer, P. et al., Molar absorption coefficients for the reduced Ellman reagent: reassessment, *Anal. Biochem.* 312, 224, 2003.

233. Rischel, C., and Poulsen, F.M., Modification of a specific tyrosine enables tracing of end-to-end distance during apomyoglobin folding, *FEBS Lett.* 374, 105, 1995.

234. Ratner, V. et al., A general strategy for site-specific double labeling of globular proteins for kinetic FRET studies, *Bioconj. Chem.* 13, 1163, 2002.

235. Talgoy, M.M., Bell, A.W., and Duckworth, H.W., The reactions of *Escherichia coli* citrate synthase with the sulfhydryl reagents 5,5¢-dithiobis-(2-nitrobenzoic acid) and 4,4′-dithiodipyridine, *Can. J. Biochem.* 57, 822, 1979.

236. Soper, T.S., Jones, W.M., and Manning, J.M., Effects of substrates on the selective modification of the cysteinyl residues of D-amino acid transaminase, *J. Biol. Chem.* 254, 10901, 1979.

237. Lukas, R.J., and Bennett, E.L., Chemical modification and reactivity of sulfhydryls and disulfides of rat brain nicotinic-like acetylcholine receptors, *J. Biol. Chem.* 255, 5573, 1980.

238. Cockle, S.A., Epand, R.M., Stollery, J.G., and Moscarello, M.A., Nature of the cysteinyl residues in lipophilin from human myelin, *J. Biol. Chem.* 255, 9182, 1980.

239. Hallaway, B.E., Hedlund, B.E., and Benson, E.S., Studies of the effect of reagent and protein charges on reactivity of the 93 sulfhydryl group of human hemoglobin using selected mutations, *Arch. Biochem. Biophys.* 203, 332, 1980.

240. Feng, Z. et al., On the nature of conformational openings: native and unfolded-state hydrogen and thiol-disulfide exchange studies of ferric aquomyoglobin, *J. Mol. Biol.* 314, 153, 2001.

241. Fernandez-Diaz, M.D. et al., Effects of electric fields on ovalbumin solutions and dialyzed egg white, *J. Agric. Food Chem.* 48, 2332, 2000.

242. Liu, M. et al., Effect of cysteine residues on the activity of arginyl-tRNA synthetase from *Escherichia coli*, *Biochemistry* 38, 11006, 1999.

243. Tremblay, J.M. et al., Modifications of cysteine residues in the solution and membrane-associated conformations of phosphatidylinositol transfer protein have differential effects on lipid transfer activity, *Biochemistry* 40, 9151, 2001.

244. Luthra, M.P., Dunlap, R.B., and Odom, J.D., Characterization of a new sulfhydryl group reagent: 6,6'-diselenobis-(3-nitrobenzoic acid), a selenium analog of Ellman's reagent, *Anal. Biochem.* 117, 94, 1981.

245. Clayshulte, T.M., Taylor, D.F., and Henzl, M.T., Reactivity of cysteine 18 in oncomodulin, *J. Biol. Chem.* 265, 1800, 1990.

246. Kato, H., Tanaka, T., Nishioka, T., Kimura, A., and Oda, J., Role of cysteine residues in glutathione synthetase from *Escherichia coli* B. Chemical modification and oligonucleotide site-directed mutagenesis, *J. Biol. Chem.* 263, 11646, 1988.

247. Narasimhan, C., Lai, C.-S., Haas, A., and McCarthy, J., One free sulfhydryl group of plasma fibronectin becomes titratable upon binding of the protein to solid substrates, *Biochemistry* 27, 4970, 1988.

248. Poole, L.B., and Claiborne, A., Evidence for a single active-site cysteinyl residue in the streptococcal NADH peroxidase, *Biochem. Biophys. Res. Commn.* 153, 261, 1988.

249. Kimura, T., Matsueda, R., Nakagawa, Y., and Kaiser, E.T., New reagents of the introduction of the thiomethyl group at sulfhydryl residues of proteins with concomitant spectrophotometric titration of the sulfhydryl: methyl 3-nitro-2-pyridyl disulfide and methyl 2-pyridyl disulfide, *Anal. Biochem.* 122, 274, 1982.

250. Drewes, G., and Faulstich, H., 2,4-Dinitrophenyl[14C]cysteinyl disulfide allows selective radiolabeling of protein thiols under spectrophotometric control, *Anal. Biochem.* 188, 109, 1990.

251. Yao, S.Y.M. et al., Identification of Cys140 in helix 4 as an exofacial cysteine residue within the substrate-translocation channel of rat equilibrative nitrobenzylthioinosine (NBMPR)-insensitive nucleoside transporter rENT2, *Biochem. J.* 353, 387, 2001.

252. Fann, M.C., Busch, A., and Maloney, P.C., Functional characterization of cysteine residues in G1pT, the glycerol 3-phosphate transporter of *Escherichia coli*, *J. Bacteriol.* 185, 3863, 2003.

253. Ding, Z. et al., Inactivation of the human P2Y$_{12}$ receptor by thiol reagents requires interaction with both extracellular cysteine residues, Cys17 and Cys 270, *Blood* 101, 3908, 2003.

254. Boyer, P.D., Spectrophotometric study of the reaction of protein sulfhydryl groups with organic mercurials, *J. Am. Chem. Soc.* 76, 4331, 1954.

255. Bai, Y., and Hayashi, R., Properties of the single sulfhydryl group of carboxypeptidase Y. Effects of alkyl and aromatic mercurials on activities toward various synthetic substrates, *J. Biol. Chem.* 254, 8473, 1979.
256. Bednar, R.A., Fried, W.B., Lock, Y.W., and Pramanik, B., Chemical modification of chalcone isomerase by mercurials and tetrathionate. Evidence for a single cysteine residue in the active site, *J. Biol. Chem.* 264, 14272, 1989.
257. Ojcius, D.M., and Solomon, A.K., Sites of *p*-chloromercuribenzenesulfonate inhibition of red cell urea and water transport, *Biochim. Biophys. Acta*, 942, 73, 1988.
258. Clark, S.J., and Ralston, G.B., The dissociation of peripheral proteins from erythrocyte membranes brought about by *p*-mercuribenzenesulfonate, *Biochim. Biophys. Acta* 1021, 141, 1990.
259. Marshall, M., and Cohen, P.P., The essential sulfhydryl group of ornithine transcarbamylases-pH dependence of the spectra of its 2-mercuri-4-nitrophenol derivative, *J. Biol. Chem.* 255, 7296, 1980.
260. Baines, B.S., and Brocklehurst, K., A thiol-labelling reagent and reactivity probe containing electrophilic mercury and a chromophoric leaving group, *Biochem. J.* 179, 701, 1979.
261. Banks, T.E., and Shafer, J.A., Inactivation of papain by *S*-methylation of its cysteinyl residue with *O*-methylisourea, *Biochemistry* 11, 110, 1972.
262. Lapko, V.N., Smith, D.L., and Smith, J.B., Methylation and carbamylation of human γ-crystallins, *Protein Sci.* 12, 1762, 2003.
263. Stark, G., Modification of proteins with cyanate, *Methods Enzymol.* 11, 590, 1967.
264. Wu, J., and Watson, J.T., Optimization of the cleavage reaction for cyanylated cysteinyl peptides for efficient and simplified mass mapping, *Anal. Biochem.* 258, 268, 1998.
265. Ghosh, P.B., and Whitehouse, M.W., 7-Chloro-4-nitrobenzo-2-oxa-1,3-diazole: a new fluorigenic reagent for amino acids and other amines, *Biochem. J.* 108, 155, 1968.
266. Birkett, D.J., Price, N.D., Radda, G.K., and Salmon, A.G., The reactivity of SH groups with a fluorogenic reagent, *FEBS Lett.* 6, 346, 1970.
267. Birkett, D.J., Dwek, R.A., Radda, G.K., Richards, R.E., and Salmon, A.G., Probes for the conformational transitions of phosphorylase b. Effect of ligands studied by proton relaxation enhancement, fluorescence and chemical reactivities, *Eur. J. Biochem.* 20, 494, 1971.
268. Lad, P.M., Wolfman, N.M., and Hammes, G.G., Properties of rabbit muscle phosphofructokinase modified with 7-chloro-4-nitrobenzo-2-oxa-1,3-diazole, *Biochemistry* 16, 4802, 1977.
269. Nitta, K., Bratcher, S.C., and Kronman, M.J., Anomalous reaction of 4-chloro-7-nitrobenzofurazan with thiol compounds, *Biochem. J.* 177, 385, 1979.
270. Dwek, R.A., Radda, G.A., Richards, R.E., and Salmon, A.G., Probes for the conformational transitions of phosphorylase a. Effect of ligands studied by proton-relaxation enhancement, and chemical reactivities, *Eur. J. Biochem.* 29, 509, 1972.
271. Carlberg, I., and Mannervik, B., Interaction of 2,4,6-trinitrobenzenesulfonate and 4-chloro-7-nitrobenzo-2-oxa-1,3-diazole with the active sites of glutathione reductase and lipoamide dehydrogenase, *Acta Chem. Scand.* B34, 144, 1980.
272. Lepke, S., Fasold, H., Pring, M., and Passow, H., A study of the relationship between inhibition of anion exchange and binding to the red blood cell membrane of 4,4′-dithiocyano stilbene-2,2′-disulfonic acid (DIDS) and its dihydro derivative (H₂DIDS), *J. Membr. Biol.* 29, 147, 1976.

273. Bettendorff, L., Wins, P., and Schoffeniels, E., Thiamine triphosphatase from electrophorus electric organ is anion-dependent and irreversibly inhibited by 4,4'-diisothiocyanostilbene-2,2'-disulfonic acid, *Biochem. Biophys. Res. Commn.* 154, 942, 1988.

274. Speth, M., and Schulze, H.-U., On the nature of the interaction between 4,4'-diisothiocyanostilbene 2,2'-disulfonic acid and microsomal glucose-6-phosphatase. Evidence for the involvement of sulfhydryl groups of the phosphohydrolase, *Eur. J. Biochem.* 174, 111, 1988.

275. Tang, X.B., and Casey, J.R., Trapping of inhibitor-induced conformational changes in the erythrocyte membrane anion exchanger AE1, *Biochemistry* 38, 14565, 1999.

276. Yano, H. et al., A strategy for the identification of proteins targeted by thioredoxin, *Proc. Natl. Acad. Sci. USA* 98, 4794, 2001.

277. Yano, H., Kuroda, S., and Buchanan, B.B., Disulfide proteome in the analysis of protein function and structure, *Proteomics* 2, 1090, 2002.

278. Ralat, L.A., and Colman, R.F., Monobromobimane occupies a distinct xenobiotic substrate binding site in glutathione-*S*-transferase, *Protein Sci.* 12, 2575, 2003.

279. Pecci, L., Cannella, C., Pensa, B., Costa, M., and Cavallini, D., Cyanylation of rhodanese by 2-nitro-5-thiocyanobenzoic acid, *Biochim. Biophys. Acta* 623, 348, 1980.

280. Vanaman, T.C., and Stark, G.C., A study of the sulfhydryl groups of the catalytic subunit of *Escherichia coli* aspartate transcarbamylase. The use of enzyme-5-thio-2-nitrobenzoate mixed disulfides as intermediates in modifying enzyme sulfhydryl groups, *J. Biol. Chem.* 245, 3565, 1970.

281. Toyo'oka, T. and Imai, K., Isolation and characterization of cysteine-containing regions of proteins using 4-(aminosulfonyl)-7-fluoro-2,1,3-benzoxadiazole and high-performance liquid chromatography, *Anal. Chem.* 57, 1931, 1985.

282. Kirley, T.L., Reduction and fluorescent labeling of cyst(e)ine-containing proteins for subsequent structural analyses, *Anal. Biochem.* 180, 231, 1989.

283. Niketic, V., Thomsen, J., and Kristiansen, K., Modification of cysteine residues with 2-bromoethane-sulfonate. The application of *S*-sulfoethylated peptides in automatic Edman degradation, *Eur. J. Biochem.* 46, 547, 1974.

284. Mutus, B., Wagner, J.D., Talpas, C.J., Dimmock, J.R., Phillips, O.A., and Reid, R.S., 1-*p*-Chlorophenyl-4,4-dimethyl-5-diethylamino-1-penten-3-one hydrobromide, a sulfhydryl-specific compound which reacts irreversibly with protein thiols but reversibly with small molecular weight thiols, *Anal. Biochem.* 177, 237, 1989.

7 The Modification of Cystine

Cystine residues are generally considered critical for maintaining the native structure of a protein. The extent to which this concept is universally accurate is not clear. In support of the critical importance of disulfide bonds is the example that the ABA-1 allergen of the nematode *Ascaris* has a single cystine. Reduction of this cystine results in the loss of structural integrity and function.[1] Further support is obtained from studies in which the insertion of an additional disulfide into a protein increases stability.[2] However, there are also proteins such as tubulin that contain a relatively large number of cysteine residues without a single disulfide bond.[3] It is clear that the study of disulfide bonds is of value to the problem of protein folding.[4-11] Studies involving cysteine insertion via oligonucleotide-directed mutagenesis and subsequent formation of cystine have been of importance.[7-11] Disulfide bonds can also be formed in proteins containing cysteine during the heat treatment of biotherapeutics derived from blood plasma.[12] Several early studies[13,14] and review articles[15,16] should be considered before studying cystine residues in proteins. The review by Gilbert is of remarkable value for the wealth of solid chemistry and thermodynamics and is of particular importance for disulfide interchange studies.[17-20] Thiol–disulfide exchange refers to a reversible association process resulting either in cystine formation (disulfide cross-linking) or mixed disulfide formation. The formation of cysteine–glutathione is an example of a mixed disulfide. Reduced glutathione is included in many disulfide interchange studies. The reader is referred to one particularly useful study on membrane protein stability.[19] The experiments were performed in 0.1 M Tris-HCl, 0.2 M KCl, 1 mM EDTA, pH 8.6, with catalytic amounts of glutathione (1.5 mM total thiol, mixture of reduced and oxidized glutathione). The protein disulfide reductase system (thioredoxin, thioredoxin reductase, NADPH) has a major role in the intracellular processing of disulfide bonds.[20]

Most of the laboratory work on cystine involves reduction in the presence of denaturants and subsequent modification with an alkylating agent as part of the structural analysis of a protein. The reaction of cysteine with alkylating agents such as α-haloacetates/α-haloacetamides or 4-vinylpyridine is discussed in Chapter 6. Cleavage with various reducing agents is described later. The differential modification of cystine residues in proteins can provide useful information and several different approaches can be used for this purpose.

Cleavage of cystine can also be accomplished by oxidation. Gorin and Godwin[21] reported that cystine can be quantitatively converted to cysteic acid by reaction with iodate in 0.1 to 1.0 M HCl. This reaction has been applied to insulin. The reaction product was not completely characterized, but given the relationship between iodate consumption and the cystine residues in insulin, the primary reaction is the oxidation

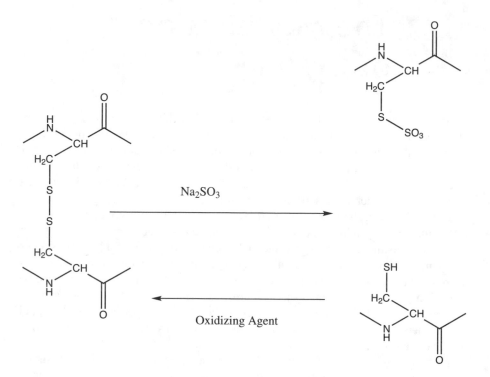

FIGURE 7.1 Oxidative sulfitolysis of cystine.

of disulfide bonds. This reaction was complete in 15 to 30 min. After longer periods of reaction, the iodination of tyrosine residues occurred. Oxidation of cystine can be accomplished under more vigorous conditions with reagents such as performic acid.[22] The *in vivo* oxidation of proteins has been of great interest in the past decade, as discussed in Chapter 1. Cystine is one of the targets for oxidation,[23] but carbonyl formation,[24] methionine oxidation,[25,26] and tyrosine modification[27] have been of greater interest. Oxidation of cysteine to cystine and mixed disulfide formation have been of interest in protein oxidation.[28] The analysis of cystine oxidation in proteins is complicated and the reader is referred to an excellent review by Underberg and co-workers[29] for a discussion of the issues.

The reaction of sulfite with cystine yields the *S*-sulfo derivative of cystine and cysteine (Figure 7.1). Cole[30] has reviewed the earlier literature and chemistry of this reaction. The reaction proceeds optimally at alkaline pH (pH 9.0). An oxidizing agent such as cupric ions or *o*-iodosobenzoate can be included to ensure effective conversion (oxidative sulfitolysis) of all cystine residues to the corresponding *S*-sulfocysteine derivatives. Another approach involves the inclusion of sodium tetrathionate in the reaction to convert the cysteine to *S*-sulfocysteine.[31] Scheraga and colleagues[32] developed an ingenious approach to the sulfitolysis reaction. They included 2-nitro-5-sulfothiobenzoate in the reaction, which resulted in the conversion of the cysteine to S-sulfocysteine, with the concomitant formation of 2-nitro-5-thiobenzoate. 2-Nitro-5-thiobenzoate can be measured at 412 nm and is proportional to the cystine residues in the protein. The reaction is reversible to form cysteine on treatment with a thiol such as 2-mercaptoethanol or dithiothreitol.

This reaction has been adapted to the controlled reduction of disulfide bonds in proteins in the absence of denaturing agents.[33] In the example presented, the disulfide bonds of bovine serum albumin were cleaved at pH 7.0 (0.1 M phosphate) at 40°C in the presence of 0.1 M sodium sulfite and simultaneously converted to the S-sulfo derivatives with oxygen and 0.4 mM cupric sulfate. The rate of reaction decreased markedly above pH 7.0.

Sulfitolysis has seen continuing use of the selective and reversible cleavage of disulfide bonds. Würfel and colleagues studied the reaction of sodium sulfite with a number of bacterial and plant thioredoxins.[34] The sulfitolysis reaction (0.1 M Tris-HCl, pH 8.0, 30°C, 4 h, 4 nM protein, 5 to 10 mM sodium sulfite) was performed in the presence or absence of guanidine hydrochloride. The process of sulfitolysis was measured by the reaction of the liberated thiol group with radiolabeled iodoacetate. The extent of sulfitolysis either in the presence or absence of denaturing agent is dependent on the primary structure of the thioredoxin. This study demonstrates that sulfitolysis can be used for the study of the effect of conformation on disulfide stability. This group extended these studies by use of fluorescence spectroscopy to study the structural requirements for the sulfitolysis of disulfide bonds in proteins.[35] There is an increase in fluorescence at 345 nm (excitation of tryptophan at 280 nm), reflecting the cleavage of disulfide bonds and the concomitant loss of the quenching effect on tryptophan fluorescence. These studies suggest that the presence of a charged amino acid near the disulfide bond influences the susceptibility to sulfitolysis. The influence of primary structure on the susceptibility of cystine to sulfitolysis is provided by a study comparing a subtilisin-like proteinase from a psychrotrophic *Vibrio* species, proteinase K, and aqualysin I.[36] The disulfide bonds of either proteinase K or aqualysin were not cleaved by sulfitolysis (0.2 M Tris-HCl, 20 mM EDTA, 0.1 M Na$_2$SO$_3$, pH 9.5, in the presence of 2-nitro-5-sultothiobenzoate[32]). The disulfide bonds of the *Vibrio* protease were cleaved under those conditions. The inclusion of the 2-nitro-5-sulfothiobenzoate permits the conversion of the thiol to the S-sulfo derivative, whereas the formation of 2-nitro-5-thiobenzoate permits measurement of disulfide cleavage. 2-Nitro-5-sulfothiobenzoate with sodium sulfite (pH 9.0, 200-fold molar excess of sodium sulfite) was also used for the processing of a fusion protein containing the extracellular domain of the P$_2$X$_2$ ion channel.[37] Sulfitolysis in the presence of sodium tetrathionate has proven useful for the quantitative conversion of cystine to S-sulfocysteine in the processing of biotherapeutics.[38-40]

Reduction of cystine is usually accomplished with a mild reducing agent such as -mercaptoethanol, dithiothreitol, or cysteine. Gorin and co-workers[41] have examined the rate of reaction of lysozyme with various thiols. At pH 10.0 (0.025 M borate), the relative rates of reaction were 2-mercaptoethanol, 0.2; dithiothreitol, 1.0; 3-mercaptopropionate, 0.4; and 2-aminoethanol, 0.01. The results with aminoethanethiol were somewhat surprising because the reaction (disulfide exchange) involves the thiolate anion and 2-aminoethanethiol would be ionized more extensively than the other mercaptans. Dithiothreitol has been a useful reagent in the reduction of disulfide bonds in proteins[42] as introduced by Cleland. Dithiothreitol and the isomeric form, dithioerythritol, are each capable of the quantitative reduction of disulfide bonds in proteins. Related reducing agents including bis(2-mercaptoethyl) sulfone and N,N'-dimethyl-N,N'-bis(mercaptoacetyl)hydrazine have been described.[43] Furthermore, the oxidized form of dithiothreitol has an absorbance maximum at 283 nm ($\varepsilon = 273$), which can be used to determine the extent of disulfide bond cleavage.[44] Homandberg and Wai[45] demonstrated that the reduction of urokinase by dithiothreitol in the presence of arginine allows the selective reduction of a disulfide bond joining the catalytically active chain to a nonessential 13-amino-acid peptide. A synthetic peptide can then be coupled to the free sulfhydryl group. Alliegro[46] reported that dithiothreitol has an effect on protein function unrelated to its effect on thiol-disulfide exchange. Insolubilized dihydrolipoic acid has also been proposed for use in the quantitative reduction of disulfide bonds.[47]

In most proteins, the free sulfhydryl groups (cysteine) derived from the reduction of cystine will, at alkaline pH, fairly rapidly undergo reoxidation to form the original disulfide bonds. This process can be accelerated by the sulfhydryl-disulfide interchange enzyme[48–50] or sulfhydryl oxidase.[51–53] Reoxidation of cysteine to cystine proceeds via the thiolate anion[13–16] and is minimized at pH values below 7. Although infrequently used, phosphorothioate had been demonstrated to effectively cleave disulfide bonds in proteins forming the S-phosphorothioate derivative.[54,55]

Light and co-workers examined the susceptibility of disulfide bonds in trypsinogen to reduction.[56] At pH 9.0 (0.1 M sodium borate), a single disulfide bond (Cys_{179}–Cys_{203}) is cleaved in trypsinogen by 0.1 M $NaBH_4$. The resulting sulfhydryl groups are blocked by alkylation. The characterization of the modified protein has been performed by the same group.[57] The disulfide bond modified under these conditions is critical in establishing the structure of the primary specificity site in trypsin.

From these studies, there is little doubt that the various disulfide bonds in a protein show different reactivity toward reducing agents. These differences in reactivity can be explored with various reagents and can be used with the aid of partial reduction followed by alkylation with radiolabeled iodoacetate to determine the position of disulfide bonds in proteins.[58]

A unique study of the reduction of a disulfide bond is provided by the study of the Rieske [2Fe-2S] center of *Thermos thermophilus*.[59] It is possible to reduce this disulfide bond by protein-film voltammetry. Reversible oxidation and reduction of the soluble Riscke domain of *Thermos themophilus* were observed with direct absorption of the protein (the soluble Rieske protein fragment obtained by heterologous expression in *Escherichia coli*) on a pyrolytic graphite-edge electrode.

The reduction of cystine to cysteine involves the addition of an electron to the sulfur with the proton coming along for the ride (depends on pH). An exception is direct hydride transfer. It has been demonstrated that disulfide bridges in α-lactalbumin are reduced by electron transfer from an excited tryptophan residue.[60,61] In studies with human α-lactalbumin,[61] ultraviolet light (270 to 290 nm, 1mW cm^{-2}, 2 to 4 h, 50 mM HEPES, 150 mM KCl, pH 7.8, 20°C) irradiation resulted in a 10-nm red shift of its tryptophan fluorescence emission spectrum (324 to 334 nm).

Reaction of the irradiated protein with 5,5'-dithiobis(2-nitrobenzoic acid) demonstrated the presence of free sulfhydryl groups. It is possible that a similar phenomena is involved in the changes in a cyclic-nucleotide-gated ion channel by ultraviolet irradiation.[62] The cleavage of a disulfide bond mediated by ultraviolet light has also been reported for bovine somatotropin.[63] Lyophilized recombinant bovine somatotropin was photolyzed by ultraviolet light (305 to 410 nm, λ_{max} = 350 nm). The protein had been lyophilized from carbonate buffer and the cake contained 6% moisture. Unlike the other examples of disulfide bond reduction mediated via tryptophan, the cysteine residues in photolyzed somatotropin appear to donate electrons back to tryptophan, leaving a pair of thiyl radicals that subsequently add oxygen to form the sulfonate. Disulfide bonds can also be cleaved by X-radiation (synchrotron radiation).[64] This study reported the reduction of a redox-active disulfide in a tryporedoxin. The radiation dose was less than that required to break a structural disulfide in lysozyme.[65]

Leach and co-workers[66] explored the electrolytic reduction of proteins. These investigators recognized that although small peptides containing disulfide bonds could be reduced by cathodic reduction, there would likely be problems with proteins because of size and tertiary structure considerations. Therefore, a small thiol was used as a catalyst for the reduction.

The use of trivalent phosphorus nucleophiles to reduce organic disulfides has been known for some time.[67] Tri-n-butylphosphine reduces S-sulfocysteine to cysteine[68] and also reduces disulfide bonds in proteins.[69] The reaction is performed under alkaline conditions (pH 8.0, 0.1 M Tris or 0.5 M bicarbonate) with n-propanol added (50/50, v/v) to dissolve tri-n-butylphosphine, which is insoluble in strictly aqueous solutions. This procedure has recently been used to reduce disulfide bonds in various proteins before reaction with 4-(aminosulfonyl)-7-fluoro-2,1,3-benzodiazole.[70–72] The reader is recommended to a review by Grayson and Farley[73] and Overman and co-workers[67] for additional references to early work on the trialkylphosphines and triarylphosphines.

The extensive application of trialkylphosphines/triarylphosphines for the modification of cystine residues in proteins was hampered by insolubility of reagents such as tri-n-butylphosphine. The synthesis of a water-soluble phosphine, Tris(2-carboxyethyl)phosphine (TCEP), was a significant advance.[74,75] The early development of this reagent[75] described the properties of the reagent. It is quite soluble in water (310 g l^{-1}). Dilute solutions (5 mM) are reasonably stable at acid pH values; at pH values above 7, the rate of conversion of the reagent to the oxide is significant. The reduction of disulfides proceeds very rapidly at pH 4.5 and below. Kinetic selectivity in the reduction of disulfides could be demonstrated.

Gray[76] extended these early observations to permit the use of TCEP reduction to establish the position of disulfide bonds in proteins. Because the reduction is performed at low pH (stock solution of 20 mM TCEP in 0.17 M citrate, pH 3.0, is stable for weeks at 23°C; the reduction is performed in 0.1% triflUoroacetic acid with 1 to 10 μM TCEP), it is possible to obtain partially reduced peptides by HPLC separation. Alkylation of the free thiols in the isolated peptides with 4-vinylpyridine permitted subsequent structural analysis of the peptide and disulfide bond assignment.

It is possible that some disulfide exchange can occur during the alkylation step, because alkylation and reoxidation to form cystine are competing reactions. In a subsequent study,[77] Gray applied the use of TCEP to assign disulfide bonds in echistatin.

There is continuing interest in the use of trialkylphosphines for the reduction of disulfide bonds the preparation of samples for separation by electrophoresis in proteomic research.[78–80] Komives and co-workers[81] were able to use the approach of Gray[76] to map the disulfide bonds in the fourth and fifth EGF-like domains of thrombomodulin. Watson and colleagues[82,83] used TCEP reduction followed by modification with cyanate[84] and subsequent peptide bond cleavage in ammonia[85] to assign disulfide bonds in proteins. Tetenbaum and Miller[86] used TCEP to reduce disulfide bonds in soybean trypsin inhibitor. The use of TCEP permitted the use of sulfur X-ray absorption spectroscopy to characterize the disulfides. Although TCEP is considered unstable at pH 7.0 and above,[75] useful studies with this reagent have been performed at pH 7.0 to 8.0.[86–88]

Disulfide bonds are somewhat unstable at alkaline pH (pH 13.0). Donovan[89] has examined this in some detail. With protein-bound cystine, there is change in the spectrum, with an increase in absorbance at 300 nm. Florence[90] has more recently studied this problem. This investigation presented evidence to suggest that cleavage of disulfide bonds in proteins by base proceeds via β-elimination from dehydroalanine and a persulfide intermediate that can decompose to form several products. Appendix I is included partially for the chemistry and partially for historic reasons.

APPENDIX I: MICROPROCEDURE FOR THE REDUCTION AND CARBOXYMETHYLATION OF PROTEINS

GENERAL COMMENTS

The following procedure has been developed in our laboratory to prepare proteins for tryptic hydrolysis. The procedure has been successfully used for 25 to 1000 µg amounts. The procedure has been adapted from previous work.[91,92]

MATERIALS

Use urea recrystallized. Prepare saturated solutions of urea at 50°C and add 10 g Amberlite MB-1 cation/anion exchange resin. Filter rapidly through a sintered glass filter. Place warm filtrate in the cold and allow to stand overnight. Obtain crystalline material by filtration (Whatman #1) and dry by lyophilization. Store in a brown bottle. Urea solutions must be freshly prepared for each procedure. Quality urea or guanidine hydrochloride/thiocyanate can be obtained from most first-tier suppliers and can be used without further purification. The problem of cyanate formation must be taken into consideration.[93] Mercaptoethanol can be obtained from the usual commercial sources (i.e., Eastman or Aldrich). With the trypsin-TPCK-treated solution, a 1.0 mg ml⁻¹ solution in 0.001 M HCl is prepared 2 to 4 h before use. For iodoacetic acid, the sodium salt is preferred, but this is frequently difficult to obtain. We usually use iodoacetic acid in NaOH. Dissolve 268 mg of iodoacetic acid (recrystallized from petroleum ether) in 1.0 ml 1.0 M NaOH. Note that 4-vinyl-pyridine[94] can also be used for blocking sulfhydryl groups.

PROCEDURE

1. Dialyze the protein sample in 0.1 M ammonium bicarbonate and lyophilized (Speed-Vac, Savant Instruments, Farmingdale, NY) in a 12 × 75 polypropylene tube.
2. Place 200 μl 8.0 M urea, 0.2% EDTA (1.80 g recrystallized urea) in a glass tube marked for 3.25 ml. Add 1.5 ml 1.4 M Tris, pH 8.6, and 0.15 ml 5% EDTA. Heat the suspension under the hot water tap until most of the urea is in the solution. Take the volume to the 3.25 ml mark by adding deionized water. Flush the tube with with nitrogen and add 5 μl mercaptoethanol. Flush the tube again with nitrogen and place at 37°C. Run a solvent blank tube with each series. This usually involves the initial processing of a volume of dialysis solvent equal to that of the protein samples.
3. Transfer the samples to microcentrifuge tubes and wash the initial tubes with 800 μl ETOH-HCl (98:2 ETOH: conc. HCl), with the washes being transferred to the microcentrifuge tubes. Place the samples overnight at 20°C. This step results in the precipitation of the reduced protein.
4. After centrifugation, dissolve the pellet fractions in 200 μl l^{-1} Tris-urea containing 500 μg iodoacetic acid. Place the sodium salt and the reactions at 37°C for 30 min. Prepare the Tris-urea-iodoacetate by adding 10 μl iodoacetic acid/NaOH (see previously) per milliliter of the Tris-urea-EDTA (see previously). After the incubation, precipitate the reduced carboxymethylated protein as described previously with ETOH-HCl. Add 800 μl of ETOH-HCl, and place the reactions at 20°C overnight.
5. Centrifuge the tubes and dry by lyophilization.
6. Suspend the reduced carboxymethylated proteins in 1000 μl 0.1 M ammonium bicarbonate and add 10 μl of trypsin. Place the tubes at 37°C, and after 4 h add an additional 10 μl of trypsin. Allow the digestion to continue for 16 h (overnight). Freeze the samples subsequent to chromatographic analysis.

NOTES

1. The temperature for the two precipitation steps is critical. We have found that ice bath temperatures are not adequate.
2. The urea solutions must be freshly prepared.
3. Avoid multiple transfers of samples. In this procedure, only two tubes are used with the transfer step, including a wash step as well.

REFERENCES

1. McDermott, L. et al., Mutagenic and chemical modification of the ABA-1 allergen of the nematode *Ascaris*: consequences for structure and lipid binding properties, *Biochemistry* 40, 9918, 2001.
2. Ikegaya, K. et al., Kinetic analysis of enhanced thermal stability of an alkaline protease with engineered twin disulfide bridges and calcium-dependent stability, *Biotechnol. Bioeng.* 81, 187, 2003.
3. Britto, P.J., Knipling, L., and Wolff, J., The local electrostatic environment determines cysteine reactivity of tubulin, *J. Biol. Chem.* 277, 29018, 2002.
4. Creighton, T.E., Disulfide bonds as probes of protein folding pathways, *Methods Enzymol.* 131, 83, 1986.
5. Wedemeyer, W.J. et al., Disulfide bonds and protein folding, *Biochemistry* 39, 4207, 2000.
6. Scheraga, H.A., Konishi, Y., and Ooi, T., Multiple pathways for regenerating ribonuclease A, *Adv. Biophys.* 18, 21, 1984.
7. Falke, J.J., and Koshland, D.E., Global flexibility in a sensory receptor: a site-directed cross-linking approach, *Science* 237, 1596, 1987.
8. Cai, S.J. et al., Probing catalytically essential domain interactions in histidine kinase EnvZ by targeted disulfide crosslinking, *J. Mol. Biol.* 328, 409, 2003.
9. Bunn, M.W., and Ordal, G.W., Transmembrane organization of the *Bacillus subtilis* chemoreceptor MCpB deduced by cysteine disulfide crosslinking, *J. Mol. Biol.* 331, 941, 2003.
10. Klco, J.M., Lassere, T.B., and Baranski, T.J., C5a receptor oligomerization. I. Disulfide trapping reveals oligomers and potential contact surfaces in a G protein-coupled receptor, *J. Biol. Chem.* 278, 35345, 2003.
11. van der Sluis, E.O., Nouwen, N., and Driessen, A.J.M., SecY-SecY and SecY-SecG contacts revealed by site-specific crosslinkage, *FEBS Lett.* 527, 159, 2002.
12. Smales, C.M., Pepper, D.S., and James, D.C., Protein modification during anti-viral heat-treatment bioprocessing of factor VIII concentrates, factor IX concentrates, and model proteins in the presence of sucrose, *Biotechnol. Bioeng.* 77, 37, 2002.
13. Eldjarn, L., and Pihl, A., The equilibrium constants and oxidation-reduction potentials of some thiol-disulfide systems, *J. Am. Chem. Soc.* 79, 4589, 1957.
14. Eldjarn, L., and Pihl, A., On the mode of action of x-ray protective agents II. Interaction between biologically important thiols and disulfides, *J. Biol. Chem.* 225, 499, 1957.
15. Cecil, R., and McPhee, J.R., The sulfur chemistry of proteins, *Adv. Protein Chem.* 14, 255, 1959.
16. Gilbert, H.F., Molecular and cellular aspects of thiol-disulfide exchange, *Adv. Enzymol.* 63, 69, 1990.
17. Baldwin, R.L., Intermediates in protein folding reactions and the mechanism of protein folding, *Annu. Rev. Biochem.* 44, 453, 1975.
18. Lehle, K., Wrba, A., and Jaenicke, R., *Erythrina caffia* trypsin inhibitor retains its native structure and function after reducing its disulfide bonds, *J. Mol. Biol.* 239, 276, 1994.
19. Cristian, L., Lear, J.D., and DeGrado, W.F., Determination of membrane protein stability via thermodynamic coupling of folding to thiol-disulfide interchange, *Protein Sci.* 12, 1731, 2003.
20. Kern, P. et al., Chaperone properties of *Escherichia coli* thioredoxin and thioredoxin reductase, *Biochem. J.* 371, 965, 2003.

21. Gorin, G., and Godwin, W.E., The reaction of iodate with cystine and with insulin, *Biochem. Biophys. Res. Commn.* 25, 227, 1966.
22. Moore, S., On the determination of cystine as cysteic acid, *J. Biol. Chem.* 239, 235, 1963.
23. Pattison, D.I., and Davies, M.J., Absolute rates constants for the reaction of hypochlorous acid with protein side chains and peptides bonds, *Chem. Res. Toxicol.* 14, 1453, 2001.
24. Levine, R.L. et al., Carbonyl assays for determination of oxidatively modified proteins, *Methods Enzymol.* 233, 247, 1994.
25. Johnson, D., and Travis, J., The oxidative inactivation of human alpha-1-antiproteinase inhibitor. Further evidence for methionine at the reactive center, *J. Biol. Chem.* 254, 4022, 1979.
26. Dong, J. et al., Metal binding and oxidation of amyloid-β within isolated senile plaque cores: Raman microscopic evidence, *Biochemistry* 42, 2768, 2003.
27. Giulivi, C., and Davies, K.J.A., Dityrosine: a marker for oxidatively modified proteins, *Methods Enzymol.* 233, 363, 1994.
28. Ghezzi, P., and Bonetto, V., Redox proteomics: identification of oxidatively modified proteins, *Proteomics* 3, 1145, 2003.
29. Reubsaet, J.L.E. et al., Analytical techniques used to study the degradation of proteins and peptides: chemical instability, *J. Pharm. Biomed. Anal.* 17, 955, 1998.
30. Cole, R.D., Sulfitolysis, *Methods Enzymol.* 11, 206, 1967.
31. Pihl, A., and Lange, R., The interaction of oxidized glutathione, cystamine monosulfoxide, and tetrathionate with the –SH groups of rabbit muscle D-glyceraldehyde 3-phosphate dehydrogenase, *J. Biol. Chem.* 237, 1356, 1962.
32. Thannhauser, T.W., Konishi, Y., and Scheraga, H.A., Sensitive quantitative analysis of disulfide bonds in polypeptides and proteins, *Anal. Biochem.* 138, 181, 1984.
33. Kella, N.K.D., and Kinsella, J.E., A method for the controlled cleavage of disulfide bonds in proteins in the absence of denaturants, *J. Biochem. Biophys. Methods* 11, 251, 1985.
34. Würfel, M., Häberlein, I., and Follman, H., Facile sulfitolysis of the disulfide bonds in oxidized thioredoxin and glutaredoxin, *Eur. J. Biochem.* 211, 609, 1993.
35. Häberlein, I., Structure requirements for disulfide bridge sulfitolysis of oxidized *Escherichia coli* thioredoxin studied by fluorescence spectroscopy, *Eur. J. Biochem.* 223, 473, 1994.
36. Kristjánsson, M.M. et al., Properties of a subtilisin-like proteinase from a psychrotropic *Vibrio* speciies. Comparison with proteinase K and aqualysin I, *Eur. J. Biochem.* 260, 752, 1999.
37. Kim, M., Yoo, O.J., and Choe, S., Molecular assembly of the extracellular domain of P2X$_2$, an ATP-gated ion channel, *Biochem. Biophys. Res. Commun.* 240, 618, 1997.
38. Nilsson, J. et al., Integrated production of human insulin and its C-peptide, *J. Biotechnol.* 48, 241, 1996.
39. Mukhopadhyay, A., Reversible protection of disulfide bonds followed by oxidative folding render recombinant hCGβ highly immunogenic, *Vaccine* 18, 1802, 2000.
40. Tikhonov, R.V. et al., Recombinant human insulin. VII. Isolation of fusion protein-S-sulfonate, biotechnological precursor of human insulin from the biomass of transformed *Escherichia coli* cells, *Protein Expr. Purif.* 21, 176, 2001.
41. Gorin, G., Fulford, R., and Deonier, R.C., Reaction of lysozyme with dithiothreitol and with other mercaptans, *Experientia* 24, 26, 1968.
42. Cleland, W.W., Dithiothreitol, a new protective reagent for SH groups, *Biochemistry* 3, 480, 1964.

43. Singh, R., and Whitesides, G.M., Reagents for rapid reduction of disulfide bonds in proteins. In *Techniques in Protein Chemistry VI* (Crabb, J.W., Ed.). Academic Press, San Diego, 1996, pp. 269–266.

44. Iyer, K.S., and Klee, W.A., Direct spectrophotometric measurement of the rate of reduction of disulfide bonds. The reactivity of the disulfide bonds of bovine α-lactalbumin, *J. Biol. Chem.* 248, 707, 1973.

45. Homandberg, G.A., and Wai, T., Reduction of disulfides in urokinase and insertion of a synthetic peptide, *Biochim. Biophys. Acta* 1038, 209, 1990.

46. Alliegro, M.C., Effects of dithiothreitol on protein activity unrelated to thiol-disulfide exchange: for consideration in the analysis of protein function with Cleland's reagent, *Anal. Biochem.* 282, 102, 2000.

47. Gorecki, M., and Patchornik, A., Polymer-bound dihydrolipoic acid: a new insoluble reducing agent for disulfides, *Biochim. Biophys. Acta* 303, 36, 1973.

48. Fuchs, S., DeLorenzo, F., and Anfinsen, C.B., Studies on the mechanism of the enzymic catalysis of disulfide interchange in proteins, *J. Biol. Chem.* 242, 398, 1967.

49. Creighton, T.E., Hillson, D.A., and Freedman, R.B., Catalysis by protein-disulphide isomerase of the unfolding and refolding of proteins with disulphide bonds, *J. Mol. Biol.* 142, 43, 1980.

50. Yano, H., Kuroda, S., and Buchanan, B.B., Disulfide proteome in the analysis of protein function and structure, *Proteomics* 2, 1090, 2002.

51. Janolino, V.G., Sliwkowski, M.Y., Swaisgood, H.F., and Horton, H.R., Catalytic effect of sulfhydryl oxidase on the formation of three-dimensional structure in chymotrypsinogen A, *Arch. Biochem. Biophys.* 191, 269, 1978.

52. Thorpe, C. et al., Sulfhydryl oxidases: emerging catalysts of protein disulfide bond formation in eukaryotes, *Arch. Biochem. Biophys.* 405, 1, 2002.

53. Wu, C.K. et al., The crystal structure of augmenter of liver regeneration: a mammalian FAD-dependent sulfydryl oxidase, *Protein Sci.* 12, 1109, 2003.

54. Neumann, H., and Smith, R.L., Cleavage of the disulfide bonds of cystine and oxidized glutathione by phosphorothioate, *Arch. Biochem. Biophys.* 122, 354, 1967.

55. Borman, C.D. et al., Pulse radiolysis studies on galactose oxidase, *Inorg. Chem.* 41, 2158, 2002.

56. Light, A., Hardwick, B.C., Hatfield, L.M., and Sondack, D.L., Modification of a single disulfide bond in trypsinogen and the activation of the carboxymethyl derivative, *J. Biol. Chem.* 244, 6289, 1969.

57. Knights, R.J., and Light, A., Disulfide bond-modified trypsinogen. Role of disulfide 179–203 on the specificity characteristics of bovine trypsin toward synthetic substrates, *J. Biol. Chem.* 251, 222, 1976.

58. Mise, T., and Bahl, O.P., Assignment of disulfide bonds in the α-subunit of human chorionic gonadotropin, *J. Biol. Chem.* 255, 8516, 1980.

59. Zu, Y. et al., Breaking and re-forming the disulfide bond at the high potential respiratory-type Rieske [2Fe-2S] center of *Thermos thermophilus*: characterization of the sulfhydryl states by protein-film voltammetry, *Biochemistry* 41, 14154, 2002.

60. Vanhooren, A. et al., Photoexcitation of tryptophan groups induces reduction of two disulfide bonds in goat α-lactalbumin, *Biochemistry* 41, 11035, 2002.

61. Permyakov, E.A. et al., Ultraviolet illumination-induced reduction of α-lactalbumin disulfide bridges. *Proteins. Struct. Funct. Genet.* 51, 498, 2003.

62. Middendoft, T.R., Aldrich, R.W., and Baylor, P.A., Modification of cyclic nucleotide-gated ion channels by ultraviolet light, *J. Gen. Physiol.* 116, 227, 2000.

63. Miller, B.L. et al., Solid-state photodegradation of bovine somatotropin (bovine growth hormone): evidence for tryptophan-mediated photooxidation of disulfide bonds, *J. Pharm. Sci.* 92, 1698, 2003.

64. Alphey, M.S. et al., Tryporedoxins from *Crithidia fasciculata* and *Trypanosoma brucei*: photoreduction of the redox disulfide using synchrotron radiation and evidence for a conformational switch implicated in function, *J. Biol. Chem.* 278, 25919, 2003.

65. Ravelli, R.B.G., and McSweeney, S.M., The 'fingerprint' that X-rays can leave on structures, *Structure* 8, 315, 2000.

66. Leach, S. J., Meschers, A., and Swanepoel, O.A., The electrolytic reduction of proteins, *Biochemistry* 4, 23, 1965.

67. Overman, L.E. et al., Nucleophilic cleavage of the sulfur-sulfur bond by phosphorus nucleophiles. Kinetic study of the reduction of aryl disulfides with triphenylphosphine and water, *J. Am. Chem. Soc.* 96, 60810, 1975.

68. Rüegg, U.T., Reductive cleavage of *S*-sulfo groups with tributylphosphine, *Methods Ezymol.* 47, 123, 1977.

69. Rüegg, U.T., and Rudinger, J., Reductive cleavage of cystine disulfides with tributylphosphine, *Methods Enzymol.* 47, 111, 1977.

70. Srinivasa, B.R., Sulfyrdryl oxidation of reduced insulin in dilute solution, *Biochem. Int.* 9, 523, 1984.

71. Kirley, T.L., Reduction and fluorescent labeling of cyst(e)ine-containing proteins for subsequent structural analysis, *Anal. Biochem.* 180, 231, 1989.

72. Chin, C.C.Q., and Wold, F., The use of tributylphosphine and 4-(aminosulfonyl)-7-fluoro-2,1,3-benzoxadiazole in the study of protein sulfydryls and disulfides, *Anal. Biochem.* 214, 128, 1993.

73. Herbert, B.D. et al., Improved protein solubility in two-dimensional electrophoresis using tributylphosphine as reducing agent, *Electrophoresis* 19, 845, 1998.

74. Grayson, M., and Farley, C.E., *Chimie Organique du Phosphoros.* Collogues Internationaux du Centre National de la Recherche Scientifique, No. 182, 1969, p. 275.

75. Burns, J.A., Butler, J.C., Moran, J., and Whiteside, G M., Selective reduction of disulfides by tris-(2-carboethoxyethyl)-phosphine, *J. Org. Chem.* 56, 2648, 1991.

76. Gray, W.R., Disulfide structures of highly bridged peptides: a new strategy for analysis, *Protein Sci.* 2, 1732, 1993.

77. Gray, W.R., Echistatin disulfide bridges: selective reduction and linkage assignment, *Protein Sci.* 2, 1749, 1993.

78. Vuong, G.L. et al., Improved sensitivity proteomics by postharvest alkylation and radioactive labeling of proteins, *Electrophoresis* 21, 2594, 2000.

79. Shaw, J. et al., Evaluation of saturation labeling two-dimensional difference gel electrophoretic fluorescent dyes, *Proteomics* 3, 1181, 2003.

80. Shaw, M.M., and Riederer, B.M., Sample preparation for two-dimensional electrophoresis, *Proteomics* 3, 1408, 2003.

81. White, C.E. et al., The fifth epidermal growth factor-like domain of thrombomodulin does not have an epidermal growth factor-like disulfide bonding pattern, *Proc. Natl. Acad. Sci. USA* 93, 10177, 1996.

82. Wu, J., and Watson, J.T., A novel methodology for assignment of disulfide bond pairings in proteins, *Protein Sci.* 6, 391, 1997.

83. Qi, J. et al., Determination of the disulfide structure of sillucin, a highly knotted cysteine-rich peptide by cyanylation/cleavage mass trapping, *Biochemistry* 40, 4531, 2001.

84. Stark, G.R., Cleavage at cysteine after cyanylation, *Methods Enzymol.* 47, 129, 1977.
85. Daniel, R. et al., Mass spectrometric determination of the cleavage sites in *Escherichia coli* dihyroorotase induced by a cysteine-specific agent, *J. Biol. Chem.* 272, 26934, 1997.
86. Tetenbaum, J., and Miller, L.M., A new spectroscopic approach to examining the role of disulfide bonds in the structure and unfolding of soybean trypsin inhibitor, *Biochemistry* 40, 12215, 2001.
87. Singh, R., and Maloney, E.K. Labeling of antibodies by *in situ* modification of thiol groups generated from selenol-catalyzed reduction of native disulfide bonds, *Anal. Biochem.* 304, 147, 2002.
88. Carl, P. et al., Forced unfolding modulated by disulfide bonds in the Ig domain of a cell adhesion molecule. *Proc. Natl. Acad. Sci. USA* 98, 1565, 2001.
89. Donovan, J.W., Spectrophotometric observation of the alkaline hydrolysis of protein disulfide bonds, *Biochem. Biophys. Res. Commn.* 29, 734, 1967.
90. Florence, T.M., Degradation of protein disulphide bonds in dilute alkali, *Biochem. J.* 189, 507, 1980.
91. Canfield, R.E., and Anfinsen, C.B., Chromatography of pepsin and chymotrypsin digests of egg white lysozyme on phosphocellulose, *J. Biol. Chem.* 238, 2684, 1963.
92. Crestfield, A.M., Moore, S., and Stein, W.H., The preparation and enzymatic hydrolysis of reduced and *S*-carboxymethylated proteins, *J. Biol. Chem.* 238, 622, 1963.
93. Stark, G.R., Stein, W.H., and Moore, S., Reaction of the cyanate present in aqueous urea with amino acids and proteins, *J. Biol. Chem.* 235, 214, 1960.
94. Ploug, M., Stoffer, B., and Jensen, A.L., *In situ* alkylation of cysteine residues in a hydrophobic membrane protein immobilized on polyvinylidene diflouride membranes by electroblotting prior to microsequence and amino acid analysis, *Electrophoresis* 13, 148, 1992.

8 The Modification of Methionine

The modification of methionine (2-amino-4-thiomethylbutanoic acid) in a protein is generally accomplished with considerable difficulty. This is possibly a reflection of the fact that, as a relatively hydrophobic residue, methionine is frequently buried in a protein. It is somewhat more challenging to obtain the specific modification of methionine under mild physiological conditions, but advances in the past decade have greatly improved potential in this area. It is possible to obtain highly selective oxidation with some reagents in certain proteins, and the results have been useful. Because the dissociation of a proton from sulfur is unnecessary to generate the nucleophile, relatively specific derivatization by alkylating agents can be accomplished at low pH. Although other residues such as cysteine and histidine are susceptible to alkylation, these residues are protonated and resist modification under acid conditions. Table 8.1 describes reagents used to modify methionine.

The oxidation of methionine (Figure 8.1) has been of increasing interest in the past decade. This interest stems partly from issues associated with the manufacture of recombinant proteins[1-3] and partly from the increase in interest in biological oxidation.[4-8] The reader is directed to an excellent review by Vogt[9] for a discussion of chemical and biological oxidation processes. It is useful to recognize that there is no clear division between chemical and biological processes as, for example, H_2O_2 is produced *in vivo*. The oxidation of methionine proceeds initially to the sulfoxide, which is a reversible process.[7,11]

It is possible to convert methionine sulfoxide to methionine under relatively mild conditions,[15] thus providing for the reversibility of the oxidative reactions described later. A systematic study has shown that of four reducing agents tested, mercaptoacetic acid, 2-mercaptoethanol, dithiothreitol, and N-methylmercaptoacetamide, the last reagent was the most effective. The reactions demonstrated little pH dependence, but did not proceed well at concentrations of acetic acid above 50% (v/v). Complete regeneration of methionine could be accomplished with 0.7 to 2.8 M reagent at 37°C for 21 h. There are also methionine sulfoxide reductases,[12-17] which can catalyze the conversion of methionine sulfoxide to methionine. Conversion of methionine sulfoxide to methionine sulfone is essentially irreversible under common solvent conditions and requires more vigorous reagents such as performic acid.[18]

The oxidation of methionine is used for protein surface mapping.[19] Oxidation is performed in 2 mM sodium ascorbate, 10 mM sodium phosphate, 0.6 µM $NH_4Fe(SO_4)$, 1.3 mM EDTA, pH 6.5. Oxidation is initiated by adding of H_2O_2 to a final concentration of 0.3% and quenched by the addition of an equal volume of 2 M Tris, pH 5.0. The inaccessibility of methionine in many proteins has made some

206   Chemical Reagents for Protein Modification, Third Edition

---

TABLE 8.1
Reagents Used for the Modification of Methionine in Proteins

Reagent	Other Amino Acids Modified	Ref.
Chloramine T	Cysteine, cystine, histidine, tyrosine, tryptophan	1–3
Hydrogen peroxide	Cysteine, cystine, histidine, tryptophan	4
Iodoacetate	Cysteine, histidine, lysine	5
N-Chlorosuccinimide	Tryptophan, cysteine	6
t-Butyl hydroperoxide	—	7

References for Table 8.1

1. Oda, T., and Tokushige, M., Chemical modification of tryptophanase by chloramine T: a possible involvement of the methionine residue in enzyme activity, *J. Biochem.* 104, 178, 1988.
2. Cutruzzola, F., Ascenzi, P., Barra, D., Bolognesi, M., Menegatti, E., Sarti, P., Schenebli, H.-P., Tomova, S., and Amiconi, G., Selective oxidation of Met-192 in bovine alpha-chymotrypsin. Effect on catalytic and inhibitor binding properties, *Biochim. Biophys. Acta* 1161, 201, 1993.
3. Hussain, A.A., Awad, R., Crooks, P.A., and Dittert, L.W., Chloramine T in radiolabeling techniques. I. Kinetics and mechanism of the reaction between chloramine-T and amino acids, *Anal. Biochem.* 214, 495, 1993.
4. Drozdz, R., Naskalski, J.W., and Sznajd, J., Oxidation of amino acids and peptides in reaction with myeloperoxidase, chloride and hydrogen peroxide, *Biochim. Biophys. Acta* 957, 47, 1988.
5. Kleanthous, C., Campbell, D.G., and Coggins, J.R., Active site labeling of the shikimate pathway enzyme, dehydroquinase. Evidence for a common substrate binding site within dehydroquinase and dehydroquinate synthase, *J. Biol. Chem.* 265, 10929, 1990.
6. Padrines, M., Rabaud, M., and Bieth, J.G., Oxidized alpha-1-proteinase inhibitor: a fast-acting inhibitor of human pancreatic elastase, *Biochim. Biophys. Acta* 1118, 174, 1992.
7. Keck, R.G., The use of t-butyl hydroperoxide as a probe for methionine oxidation in proteins, *Anal. Biochem.* 236, 56, 1996.

---

other approaches such as synchrotron X-ray radiation problematic.[20] Methionine-containing peptides and protein samples for mass spectrometry have been oxidized by H_2O_2 directly on MALDI-MS target plates.[21]

Reagents for the selective oxidation of methionine that have attracted recent attention include chloramine T[22,23] (0.1 M phosphate, pH 7.0, or 1.0 M Tris, pH 8.4), sodium periodate[23] (0.1 M sodium acetate, pH 5.0), and hydrogen peroxide.[24] The reaction of methionine with chloramine T can be followed spectrophotometrically.[22] The oxidation of methionine is a possible side reaction of the treatment of proteins with N-bromosuccinimide.[25]

H_2O_2 continues to see considerable use for the modification of methionine in proteins. Sites and co-workers[26] have compared the effect of methionine oxidation with H_2O_2 (25 mM sodium phosphate, 100 mM NaCl, pH 7.0, 1.1% H_2O_2, room temperature, 30 min) and oligonucleotide-directed mutagenesis on the stability of staphylococcal nuclease. Oxidation of two methionine residues (M65 and M95) resulted in a considerable loss of protein stability, whereas modification of the other

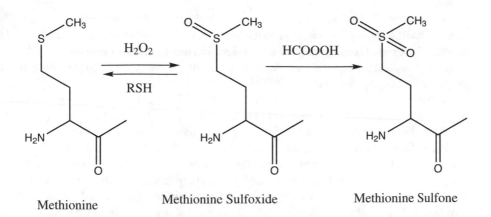

Methionine   Methionine Sulfoxide   Methionine Sulfone

FIGURE 8.1 Oxidation of methionine. The reversible oxidation of methionine to methinone sulfoxide is shown as well as the subsequent further irreversible oxidation of methionine sulfoxide to methionine sulfone.

two residues (M26 and M32) had little effect. Substitution of a leucine residue at position 95 (M95L) resulted in a loss of stability similar to that seen with oxidation, whereas substitution at position 65 (M65L) was less deleterious. A similar experimental approach was taken by Hus, Wang, and co-workers[27] to a N-carbamoyl D-amino acid aminohydrolase from *Agrobacterium radiobacter*. This protein contains nine methionine residues, which were individually replaced with leucine. Two of the mutants had activity similar to that of the wild type, whereas the other seven had reduced activity. The three mutants with solvent-accessible residues retained activity on oxidation with H_2O_2 (0.1 to 1.0 mM H_2O_2 in 200 mM sodium phosphate, pH 7.0, 25°C, 15 min). The other mutants were more resistant to loss of activity on oxidation than the wild type was.

The development of *t*-butyl hydroperoxide by Keck[28] as a selective oxidizing agent for methionine in proteins represents a significant advance. Results obtained with native recombinant interferon and recombinant tissue-type plasminogen activator showed that this reagent was selective for the oxidation of exposed methionine residues in proteins. The reaction were performed in 5 mM succinate, 0.1% Tween 20, pH 5.0 (recombinant interferon), or 0.2 M arginine, 0.1 M sodium phosphate, 0.1% Tween 80, pH 7.0 (recombinant tissue-type plasminogen activator). *t*-Butyl hydroperoxide (0 to 73 mM) was added and the reaction allowed to proceed overnight at room temperature. Tryptic peptides were separated by HPLC and analyzed by mass spectrometry. Two methionine residues were oxidized in recombinant interferon with *t*-butyl hydroperoxide, whereas all five residues were oxidized to a varying extent by H_2O_2 under the same reaction conditions. Three methionine residues were oxidized in native tissue-type plasminogen activator; all five residues were oxidized to varying degrees in the presence of 8.0 M urea. *t*-Butyl hydroperoxide has been successfully used for recombinant human leptin (100 mM sodium borate, pH 9.0, room temperature)[29] and recombinant human granulocyte colony-stimulating factor (25 mM sodium acetate, pH 4.5, 25°C).[30]

Chloramine T has been for the selective oxidation of methionine in human recombinant secretory leukocyte proteinase inhibitor,[31] large conductance calcium-activated potassium channels,[32] actin,[33] and kininogens.[34] Oxidation of methionine residues occurring when chloramine T is used for protein iodination can pose a problem for subsequent analysis.[35,36] There has been infrequent use of sodium periodate for the oxidation of methionine in proteins.[37] The oxidation of methionine in recombinant human interleukin-2 by potassium peroxodisulfate has been reported.[38]

Methionine can be modified with various alkylation agents such as α-halo acetic acids and their derivatives. Gundlach and co-workers [39] examined in some detail the reaction of iodoacetate with methionine. The reaction of iodoacetate with methionine does not appear to be pH dependent and proceeds much slower than the reaction with cysteine under the mildly alkaline conditions used for reduction and carboxymethylation. The resulting sulfonium salt yields homoserine and homoserine lactone when heated at 100°C at pH 6.5. On acid hydrolysis (6 N HCl, 110°C, 22 h), a mixture of methionine and S-carboxymethyl homocysteine together with a small amount of homoserine lactone was obtained. In general, methionine residues only react with the α-halo acids after the disruption of the secondary and tertiary structures of a protein.[40] Selectivity in the modification of methionine in proteins by α-halo acids can be achieved by performing the reaction at acidic pH (pH 3.0 or less). The modification of methionine by ethyleneimine has been reported in a reaction producing a sulfonium salt derivative.[41] The modification of methionine in azurin with bromoacetate has been reported.[42] In the protein, four of six methionine residues were modified at pH 4.0, whereas all methionine residues were reactive at pH 3.2. These modification reactions were performed in 0.1 M sodium formate at ambient temperatures for 24 h with 0.16 M bromoacetate. The modification of methionine in porcine kidney acyl CoA-dehydrogenase occurs with iodoacetate (0.030 M) in 0.1 M phosphate, pH 6.6, at ambient temperature.[43] The identification of methionine as the residue modified by iodoacetate in this protein was supported by the comparison of the chromatogram of the acid hydrolyzate of the modified protein (reacted with ^{14}C-iodacetate) with that of the acid hydrolyzate of synthetic S-([1-^{14}C]carboxymethyl)-methionine.[39] This is necessary because the S-carboxymethyl derivative yielded several different compounds on acid hydrolysis.[39,44]

Naider and Bohak[45] reported that the sulfonium salt derivatives of methionine (e.g., S-carboxymethyl methionine, the reaction product of methionine and iodoacetic acid) can be converted to methionine by reaction with a suitable nucleophile. For example, reaction of S-carboxamidomethyl methionine (in the peptide Gly-Met-Gly) with a sixfold molar excess of mercaptoethanol at pH 8.9 at 30°C resulted in the complete regeneration of methionine after 24 h of reaction. The S-phenacyl derivative of methionine (in the peptide Gly-Met-Gly) was converted to methionine in 1 h under the same reaction conditions. These investigators also showed that chymotrypsin previously treated with phenacyl bromide under conditions that inactivate the enzyme concomitant with the alkylation of Met-192[46] could be reactivated by treatment with 3-mercaptoethanol at pH 7.5 (sodium phosphate). It is of interest that the S-phenacyl methionine in chymotrypsin is converted to methionine at a substantially faster rate than the tripeptide derivative. The authors speculate that the increased reactivity of the chymotrypsin derivative is a reflection of interaction of the phenacyl moiety with the substrate-binding site. Alkylation of methionyl residues in pituitary thyrotropin and lutropin with iodoacetic acid has been reported.[47] Differential reactivity of various methionyl residues was reported on reaction with iodoacetate in 0.2 M formate, pH 3.0, for 18 h at 37°C. Kleanthous and co-workers[48] reported the reversible alkylation of methionine by iodoacetate in dehydroquinase. In this reaction, iodoacetate behaves kinetically as an affinity label, with a K_i of 30 μM and a k_{inact} of 0.014 min^{-1}, pH 7.0 (50 mM potassium phosphate). There is no reaction with iodoacetamide. Two methionine residues are modified during the reaction of dehydroquinase with iodoacetate. In a companion study, Kleanthous and Coggins[49] demonstrated that 2-mercaptoethanol treatment under alkaline conditions (0.5% ammonium bicarbonate, 37°C) could reverse modification at one of the two residues. If the modified protein is denatured, there is no reversal of modification at either residue. The results are interpreted in terms of the proximity of a positive charge (i.e., lysine) in close proximity to one of the two methionyl residues, which provides the basis for the (1) affinity labeling and (2) the 2-mercaptoethanol-mediated reversal of modification.

The ability to reverse the alkylation of methionine under relatively mild conditions as described previously has resulted in the development of a clever affinity approach to the purification of methionine peptides. Several groups[50–52] have reported the isolation of methionine peptide by reaction with bead containing a bromoacetyl function under acidic conditions (e.g., 25% acetic acid) and subsequent reducing agent under alkaline conditions as described previously.

REFERENCES

1. Jensen, J.L. et al., Metal-catalyzed oxidation of bran-derived neurotrophic factor (BDNF): analytical challenge for the identification of modified sites, *Pharm. Res.* 17, 190, 2000.
2. Duenas, E.T. et al., Comparison between light induced and chemically induced oxidation of rhVEGF, *Pharm. Res.* 18, 1455, 2001.
3. Shapiro, R.I. et al., Expression of sonic hedgehog-Fc fusion protein in *Pichia pastoris*. Identification and control of post-translational, chemical, and proteolytic modifications, *Protein Expr. Purif.* 29, 272, 2003.
4. Tien, M. et al., Peroxynitrite-mediated modification of protein at physiological carbon dioxide concentration: pH dependence of carbonyl formation, tyrosine nitration, and methionine oxidation, *Proc. Natl. Acad. Sci. USA* 96, 7809, 1999.
5. Hawkins, C.L., and Davies, M.J., Hypochloride-induced oxidation of proteins in plasma: formation of chloramines and nitrogen-centered radicals and their role in protein fragmentation, *Biochem. J.* 340, 539, 1999.
6. Davies, M.J., Singlet oxygen-mediated damage to proteins and its consequences, *Biochem. Biophys. Res. Commn.* 305, 761, 2003.
7. Imlay, J.A., Pathways of oxidative damage, *Annu. Rev. Microbiol.* 57, 395, 2003.
8. Droge, W., Oxidative stress and aging, *Adv. Exp. Biol. Med.* 543, 191, 2003.
9. Vogt, W., Oxidation of methionyl residues in proteins: tools, targets, and reversal, *Free Radic. Biol. Med.* 18, 93, 1995.

10. Neumann, N.P., Oxidation with hydrogen peroxide, *Methods Enzymol.* 25, 393, 1972.
11. Houghten, R.A., and Li, C.H., Reduction of sulfoxides in peptides and proteins, *Anal. Biochem.* 98, 36, 1979.
12. Brot, N., Weissbach, L., Werth, J., and Weissbach, H., Enzymatic reduction of protein-bound methionine sulfoxide, *Proc. Natl. Acad. Sci. USA* 78, 2155, 1981.
13. Brot, N., and Weissbach, H., Peptide methionine reductase: biochemistry and physiological role, *Biopolymers* 55, 288, 2000.
14. Hoshi, T., and Heinemann, S., Regulation of cell function by methionine oxidation and reduction, *J. Physiol.* 531, 1, 2001.
15. Weissbach, H. et al., Peptide methionine sulfoxide reductase: structure, mechanism of action, and biological function, *Arch. Biochem. Biophys.* 397, 172, 2002.
16. Stadtman, E.R. et al., Cyclic oxidation and reduction of protein methionine residues is an important antioxidant mechanism, *Mol. Cell. Biochem.* 234/235, 3, 2002.
17. Antoine, M., Boschi-Muller, S., and Branlant, G., Kinetic characterization of the chemical steps involved in the catalytic mechanism of methionine sulfoxide reductase A from *Neisseria menignitidis*, *J. Biol. Chem.* 278, 45352, 2003.
18. Hirs, C.H.W., Performic acid oxidation, *Methods Enzymol.* 11, 197, 1967.
19. Sharp, J.S., Becker, J.M., and Hettich, R.L., Protein surface mapping by chemical oxidation: structural analysis by mass spectrometry, *Anal. Biochem.* 313, 216, 2003.
20. Sharp, J.S., Becker, J.M., and Hettich, R.L., Analysis of protein surface accessible residues by photochemical oxidation and mass spectrometry, *Anal. Chem.* 76, 672, 2004.
21. Corless, S., and Cramer, R., On-target oxidation of methionine residues using hydrogen peroxide for composition-restricted matrix-assisted laser desorption/ionization peptide mass-mapping, *Rapid Commn. Mass Spectrom.* 17, 12112, 2003.
22. Trout, G.E., The estimation of microgram amounts of methionine by reaction with chloroamine-T, *Anal. Biochem.* 93, 419, 1979.
23. de la Llosa, P., El Abed., A., and Roy, M., Oxidation of methionine residues in lutropin, *Can. J. Biochem.* 58, 745, 1980.
24. Caldwell, P., Luk, D.C., Weissbach, H., and Brot, N., Oxidation of the methionine residues of *Escherichia coli* ribosomal protein L12 decreases the protein's biological activity, *Proc. Natl. Acad. Sci. USA* 75, 5349, 1978.
25. Spande, T.F., and Witkop, B., Determination of the tryptophan content of proteins with *N*-bromosuccinimide, *Methods Enzymol.* 11, 498, 1967.
26. Kim, Y.H. et al., Comparing the effect on protein stability of methionine oxidation versus mutagenesis: steps toward engineering oxidation resistance in proteins, *Protein Eng.* 14, 343, 2001.
27. Chien, H.-C. et al., Enhancing oxidative resistance of *Agrobacterium radiobacter* *N*-carbamoyl D-amino acid aminohydrolase by engineering solvent-accessible methionine residues, *Biochem. Biophys. Res. Commun.* 297, 282, 2002.
28. Keck, R.G., The use of *t*-butyl hydroperoxide as a probe for methionine oxidation in proteins, *Anal. Biochem.* 236, 56, 1996.
29. Liu, J.L et al., *In vitro* methionine oxidation of recombinant human leptin, *Pharm. Res.* 15, 632, 1998.
30. Lu, H.S. et al., Chemical modification and site-directed mutagenesis of methionine residues in recombinant human granulocyte colony-stimulating factor: effect on stability and biological activity, *Arch. Biochem. Biophys.* 362, 1, 1999.
31. Tomova, S. et al., Selective oxidation of methionyl residues in the recombinant human secretory leukocyte proteinase inhibitor. Effect on inhibitor binding properities, *J. Mol. Recog.* 7, 31, 1994.

32. Tang, X.D. et al., Oxidative regulation of large conductance calcium-activated potassium channels, *J. Gen. Physiol.* 117, 253, 2001.
33. Daile-Donne, I. et al., Methionine oxidation as a major cause of the functional impairment of oxidized actin, *Free Radic. Biol. Med.* 32, 927, 2002.
34. Nieziolek, M. et al., Properties of chemically oxidized kininogens, *Acta Biochim. Pol.* 50, 753, 2003.
35. Bauer, R.J. et al., Alteration of the pharmacokinetics of small proteins by iodination, *Biopharm. Drug. Dispos.* 17, 761, 1996.
36. Kumar, C.C. et al., Chloramine T-induced structural and biochemical changes in echistatin, *FEBS Lett.* 429, 239, 1998.
37. Kennett, M.J. et al., *Acanthamoeba castellanii*: characterization of an adhesion molecule, *Exp. Parisitol.* 92, 161, 1999.
38. Cadée, J.A. et al., Oxidation of recombinant human interleukin-2 by potassium peroxodisulfate, *Pharm. Res.* 18, 1461, 2001.
39. Gundlach, H.G., Moore, S., and Stein, W.H., The reaction of iodoacetate with methionine, *J. Biol. Chem.* 234, 1761, 1959.
40. Gurd, F.R.N., Carboxymethylation, *Methods Enzymol.* 11, 532, 1967.
41. Schroeder, W.A., Shelton, J.R., and Robberson, B., Modification of methionyl residues during aminoethylation, *Biochim. Biophys. Acta* 147, 590, 1967.
42. Marks, R.H.L., and Miller, R.D., Chemical modification of methionine residues in azurin, *Biochem. Biophys. Res. Commn.* 88, 661, 1979.
43. Mizzer, J.P., and Thorpe, C., An essential methionine in pig kidney general acyl-CoA dehydrogenase, *Biochemistry* 19, 5500, 1980.
44. Goren, H.J., Glick, D.M., and Barnard, E.A., Analysis of carboxymethylated residues in proteins by an isotopic method and its application to the bromoacetate-ribonuclease reaction, *Arch. Biochem. Biophys.* 126, 607, 1968.
45. Naider, F., and Bohak, Z., Regeneration of methionyl residues from their sulfonium salts in peptides and proteins, *Biochemistry* 11, 3208, 1972.
46. Schramm, H.J., and Lawson, W.B., Über das activ Zentrum von Chymotrypsin. II. Modifizierung eines Methioninrestes in Chymotrypsin durch einfache Benzolderivate, Hoppe-Seyler's Z, *Physiol. Chem.* 332, 97, 1963.
47. Goverman, J.M., and Pierce, J.G., Differential effects of alkylation of methionine residues on the activities of pituitary thyrotropin and lutropin, *J. Biol. Chem.* 256, 9431, 1981.
48. Kleanthous, C., Campbell, D.G., and Coggins, J.R., Active site labeling of the shikimate pathway enzyme, dehydroquinase. Evidence for a common substrate binding site within dehydroquinase and dehydroquinate synthase, *J. Biol. Chem.* 265, 10929, 1990.
49. Kleanthous, C., and Coggins, J.R., Reversible alkylation of an active site methionine residue in dehydroquinase, *J. Biol. Chem.* 265, 10935, 1990.
50. Weinberger, S.R., Viner, R.J., and Ho, P., Tagless extraction-retentate chromatography: a new global protein digestion strategy for monitoring differential protein expression, *Electrophoresis* 23, 3182, 2002.
51. Grunert, T. et al., Selective solid-phase isolation of methionine-containing peptides and subsequent matrix-assisted laser desorption mass spectrometric detection of methionine- and methionine-sulfoxide-containing tryptic peptides, *Rapid Commn. Mass Spectrom.* 17, 1815, 2003.
52. Shen, M. et al., Isolation and isotope labeling of cysteine- and methionine-containing tryptic peptides, *Mol. Cell. Proteom.* 2, 315, 2003.

residues are observed. Iodination with a 10-fold molar excess of I_2 results in the formation of 3 Mol of diiodotyrosine per mole of cytochrome c.[15] The fourth tyrosyl residue is modified in the presence of 4.0 M urea only. Iodination of tyrosine results in a decrease in the pK_a of the phenolic hydroxyl groups. Iodination with a limiting amount of iodine as described previously results first in the formation of 2 Mol of monoiodotyrosine, then 1 Mol of diiodotyrosine, and 1 Mol of monoiodotyrosine. Tyrosyl residues that can be iodinated are also available for O-acetylation with acetic anhydride [0.1 M potassium phosphate, pH 7.5; acetic anhydride added in two portions over 1 h at 0°C; maintained at pH 7.8 with NaOH (1 M) addition].

The modification of tyrosyl residues in phosphoglucomutase by iodination has been reported.[16] Modification is achieved by reaction in 0.1 M borate, pH 9.5, with 1 mM I_2 (obtained by an appropriate dilution of a stock iodine/iodide solution, 0.05 M I_2 in 0.24 M KI) at 0°C for 10 min. Complete loss of enzymatic activity was observed with these reaction conditions, but the stoichiometry of modification was not established. Nitration of 7 of 20 tyrosyl residues resulted in an 83% loss of catalytic activity. These investigators also studied the reaction of phosphoglucomutase with diazotized sulfanilic acid and N-acetylimidazole.

The previous modifications use the reaction of tyrosyl residues in proteins with iodine/iodide solutions at alkaline pH. Iodination of tyrosyl residues can also be accomplished with iodine monochloride (ICl) at mildly alkaline pH. One such study explores the modification of galactosyltransferase.[17] The modification is accomplished by reaction in 0.2 M sodium borate, pH 8.0. The reaction is initiated by a desired amount of a stock solution of ICl.[18] A stock solution of 0.02 M ICl is prepared by adding 21 ml of 11.8 M HCl (stock concentrated HCl) to ca. 150 ml of H_2O containing 0.555 g KCl, 0.3567 g KIO_3, and 29.23 g NaCl. The solvent is made up to a final volume of 250 ml with H_2O. Free iodine is then extracted with CCl_4, if necessary, and the solution is aerated to remove trace amounts of CCl_4. The resulting solution of ICl is stable for an indefinite period of time under ambient conditions. Reaction proceeds for 1 min at ambient temperature and is terminated by adding a 1:6 volume of 0.5 M $Na_2S_2O_3$ (50 μl for a 0.300-ml reaction mixture). Radiolabeled sodium iodide ($Na^{125}I$) is included to provide a mechanism for establishing the stoichiometry of the reaction. The reaction mixture, after the addition of $Na_2S_2O_3$, is subjected to gel filtration on Bio-Gel P-10 (BioRad Laboratories, Richmond, CA) in 0.1 M Tris, pH 7.4. In experiments designed to assess the relationships between reagent (ICl) concentration and the extent of modification, a maximum of 10 g-atom of iodine was incorporated into galactosyltransferase at a 40-fold molar excess of reagent. Incorporation of iodine is linear up to this excess of reagent and slowly declines at higher concentrations of ICl. Modification of tryptophanyl residues was excluded by direct analysis, and the only iodinated amino acids obtained from the modified protein were monoiodotyrosine and diiodotyrosine. Modification of other residues such as histidine and methionine by oxidation without incorporation of iodine was not excluded.

Iodination can also be accomplished by peroxidase per H_2O_2 per NaI. A recent procedure was described for the modification of tyrosyl residues in insulin.[19] In these studies, 20 mg of porcine insulin in 20 ml 0.4 M sodium phosphate, 6.0 M urea, pH 7.8, was combined with 10 ml $Na^{125}I$ (1 mCi) and 3.6 mg urea, and H_2O_2 (5 μl

0.3 mM solution) and peroxidase (Sigma Chemical Company, 0.2 mg/ml; 5 μl) were added. The preparative reaction was terminated by dilution with an equal volume of 40% (w/v) sucrose. Note that the enzyme-catalyzed iodination proceeds with efficiency in 6.0 M urea. Iodination of tyrosyl residues in peptides and proteins can also be accomplished with chloramine T.[20,21] The solution structure of insulin-like growth factor was investigated by iodination of tyrosyl residues mediated by either chloramine T or lactoperoxidase.[22] Chloramine T was more effective than lactoperoxidase. There continue to be improvements in the chemistry of chloramine T iodination.[23,24]

Radioisotope labeling of proteins with isotopes of iodine has been used extensively to study protein turnover (catabolism) and *in vivo* distribution.[25–28] Caution should be exercised in the interpretation of such results, as most iodination techniques result in heterogeneous products that are trace-labeled with iodine isotopes.[29–31] Site-specific modification of tyrosine with iodine has been used to obtain protein derivatives useful in crystallographic analysis.[32,33]

The development of NAI as a reagent for the selective modification of tyrosyl residues can, in part, be traced to the early observations[34–38] that NAI is an energy-rich compound. The preparation of *O*-acyl derivatives via the action of carboxylic acid anhydrides (i.e., acetic anhydride) has been used for some time, but it is very difficult to obtain selective modification of tyrosine as these reagents readily react with primary amines to form stable *N*-acyl derivatives.[39–41] It is, however, possible to obtain the selective modification of tyrosine with acetic anhydride (Figure 9.1) by reaction at mildly acidic pH (1.0 M acetate, pH 5.8, at 25°C) at ca. 20,000-fold molar excess of acetic anhydride (5.1 × 10⁻² M acetic anhydride, 2.9 × 10⁻⁶ M enzyme).[42] Bernad and colleagues[43] reported on an extensive study comparing the modification of lysyl and tyrosyl residues in lysozyme with dicarboxylic acid anhydrides. In 50 mM HEPES, 1.25 M NaCl, pH 8.2, amino groups (primarily lysine residues) were far more reactive than hydroxyl groups (including tyrosine, serine, and threonine). NAI was first used as a reagent for the modification of tyrosyl residues (Figure 9.2) in bovine pancreatic carboxypeptidase A.[44] This same group of investigators subsequently reported on the use of NAI in the determination of free tyrosyl residues in proteins[45] as opposed to buried residues. This has not necessarily proved to be the case.[46] NAI is commercially available, but it can also be easily synthesized.[37] Our laboratory generally synthesizes the reagent, and always subjects the reagent obtained from a commercial source to recrystallization from benzene after drying with sodium sulfate. Note that, as with many reagents, NAI is hygroscopic and should be stored in a container, preferably a vacuum desiccator, over a suitable desiccant. A stock solution of the reagent is prepared in dry benzene (this stock solution is relatively stable for 2 to 4 weeks at 4°C), and a portion containing the desired amount of reagent is introduced to the reaction vessel. The solvent (benzene) is removed by a stream of dry air or dry nitrogen. The reaction is initiated by adding the protein solution to be modified to the residue of reagent. The reaction is usually performed at pH 7.0 to 7.5. A wide variety of buffers have been used to study the reaction of NAI. A high concentration of nucleophilic species such as Tris should be avoided because of reagent instability.[44] Likewise, although the modification occurs more rapidly at pH values more alkaline than 7.5, reagent

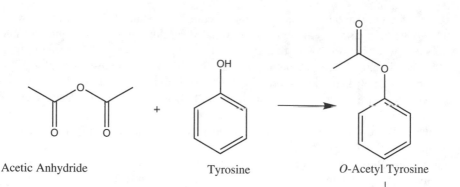

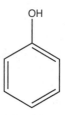

FIGURE 9.1 *O*-Acetylation of tyrosine and subsequent deacetylation with hydroxylamine.

and product (*O*-acetyltyrosine) stability becomes a significant problem at those pH values. *N*-acyl derivatives other than NAI have been prepared.[37,47] N-butyryl-imidazole was demonstrated to be a more potent inhibitor of thrombin than was the acetyl derivative,[48] with modification occurring presumably at a histidine residue in addition to a single tyrosine residue. Table 9.1 presents a partial listing of proteins that have been modified with NAI.

A survey of the literature from 1994 to 2003 shows that NAI continues to be used for the modification of tyrosyl residues in proteins, although not to the extent of TNM. El Kebbaj and Latruffe[49] examined the membrane penetration of NAI. NAI was similar to TNM in its ability to preferentially inactivate D-3-hydroxybutyrate dehydrogenase in inside–out membranes when compared to intact mitochondria. Evaluation of the partition of NAI in water – organic solvent systems showed a 81/19(%) distribution in an H$_2$O/hexane system and a 31/69 distribution in H$_2$O/1-octanol. A number of studies have used both NAI and TNM for the modification of tyrosine residues in the same study. These are described in Table 9.1 (Refs. 18–22, 24). In most situations, if modification with one reagent results in inactivation, modification with the other does as well. There are, however, some studies in which one reagent inhibits and the other does not. The results obtained with HlyC, the

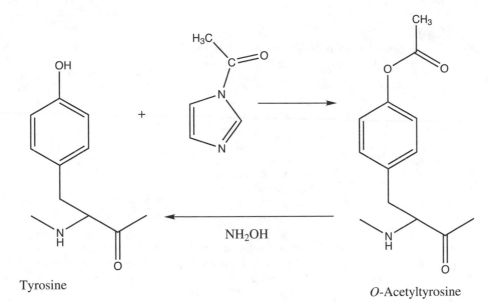

FIGURE 9.2 *O*-Acetylation of tyrosine with *N*-acetylimidazole and reversal with hydroxy-lamine.

internal protein acyltransferase, are described in the discussion of TNM. Graves and co-workers[50] demonstrated that NAI does not acetylate 3-nitrotyrosine. This study also examined the rates of acetylation of tyrosine and 3-fluorotyrosine as a function of pH between 7.5 and 9.5. Formation of the *O*-acetylated derivative was evaluated by HPLC analysis. As NAI is unstable at increasing pH, it was not possible to obtain reliable second-order rate constants; thus, the relative rates of reaction were reported. 3-Fluorotyrosine was acetylated more rapidly than tyrosine at pH 7.5, whereas the opposite was true at pH 9.5. It is suggested that *O*-acetylation of tyrosine by NAI is facilitated by a microenvironment that promotes ionization of the phenolic hydroxyl group. Ji and Bennett[51] showed that peroxynitrite could activate (four to fivefold) reduced rat liver microsomal glutathione *S*-transferase. This activation does not appear to be due to sulfhydryl oxidation and was also observed on reaction with NAI under conditions in which reaction with sulfhydryl groups is unlikely.[47]

There are several approaches to determine the extent of tyrosine modification by NAI. The amount of acetylhydroxamate produced by the reaction of hydroxyl-amine can be determined.[52,53] The procedure described by these investigators involves the addition of 0.25 ml of a hydroxylamine solution (4 M NH$_2$OH·HCl:3.5 M NaOH:0.001 M EDTA, 1:2:1) to 1.0 ml of the acetylated protein sample. After 1 min, 0.5 ml 25% trichloroacetic acid and 0.5 ml 20% FeCl$_2$·6 H$_2$O in 2.5 M HCl is added, and the absorbance of the supernatant fraction is determined at 540 nm. We have found it convenient to use *p*-nitrophenyl acetate as the standard for this reaction. Second, *O*-acetylation of tyrosine produces a decrease in absorption at 278 nm. A

TABLE 9.1
Reaction of Proteins with *N*-Acetylimidazole (NAI)

Protein	Solvent/Temp.	Reagent Excess[a]	O-AcTyr/ Tyr[b]	Ref.
Carboxypeptidase[c]	0.02 M sodium barbital, 2.0 M NaCl, pH 7.5, at 23°C	60	4.3/19[d]	1
Pepsinogen	0.02 M sodium Veronal, 2.0 M NaCl, pH 7.5, at 25°C	60	7/16[e]	2
Pepsin	2.0 M NaCl, pH 5.8, at 25°cf.	60	9/16[g]	2
Trypsin	0.01 M sodium borate, 0.01 M CaCl$_2$, pH 7.6, at 0°C	30	1.7/10[h]	3
Trypsin	0.01 M sodium borate,[i] 0.01 M CaCl$_2$, pH 7.6, at 0°C	465	3.0/10[j]	3
α-Amylase[k]	0.02 M Tris-Cl, pH 7.5, at 25°C	500	3.5/12[l]	4
Subtilisin novo	0.016 M barbital, pH 7.5	100	7/10[m]	5
Subtilisin Carlsberg	0.016 M barbital, pH 7.5	130	8.4/13[m]	5
Hemerythrin	0.05 M sodium borate, 0.05 M Tris, pH 7.5, at 0°C	800	—[n]	6
Thrombin	0.02 M Tris, 0.02 M imidazole, 0.02 M acetate, pH 7.5, at 23°C	300	4.4/12	7
Fructose diphosphatase	0.050 M sodium borate, pH 7.5	—	—	8
Erythrocyte ATPase: stroma	0.010 M Tris, pH 7.4, at 23°C[o]	—	—	9
Erythrocyte ATPase: intact cells	0.010 M Tris, 0.140 M NaCl, pH 7.4, at 23°C[p]	—	—	9
α-Lactalbumin		200	2/5[q]	10, 11
Pancreatic colipase	—	—	—	12, 13
Pancreatic α-amylase	0.01 M phosphate, pH 7.5,[r] 0.1 mM CaCl$_2$ at 25°C	120	5.9/18	14
Sweet potato α-amylase	0.01 M acetate, pH 7.5, at 25°C[r]	120	5.3/17	14
Aspergillus niger glucamylase	0.01 M acetate, pH 7.5, at 25°C	120	11.3/33	14
Emulsin β-D-glucosidase	0.01 M phosphate, pH 6.1, at 25°C[r]	300	—	15
Human placental taurine transporter	10 mM HEPES-Tris, pH 7.4, with 100 mM K$_2$SO$_4$ at 22°C	—	—	16
Renal Na,K ATPase	50 mM sodium borate with 2 mM EDTA, pH 7.5, at 20°C	—	—	17
Zinc metalloenzyme 1				18
Leukotriene A$_4$ hydrolase[s]				
Aspergillus polygalacturonase[s]				19
Coproporhyrinogen oxidase[s]				20
Renal cortical Na/HCO$_3$ cotransporter VI[t]				21
Aminopeptidase A[u]	50 mM HEPES, pH 7.5, 25° C, 30 min			22
α-Crystallin[s]				23
Tryptophan hydroxylase				24
Capase-3 activated DNAse	10 mM MES, pH 7.0			25
Heparinase from Sphingobacterium				26

TABLE 9.1 (continued)
Reaction of Proteins with *N*-Acetylimidazole (NAI)

Protein	Solvent/Temp.	Reagent Excess[a]	O-AcTyr/ Tyr[b]	Ref.
D-Galactose-binding lectin from	50 m *M* Tris, pH 8.0			27
Erythrina specioso				
RNA *N*-glycosidase				28

[a] Moles NAI per mole of protein.

[b] Moles *O*-acetyltyrosine per mole of tyrosine in modified protein.

[c] Bovine pancreatic carboxypeptidase A-Anson.

[d] Changes in catalytic activity reversed by treatment with 0.01 *M* hydroxylamine, pH 7.5, at 23°C. Primary amino groups were not acetylated under those reaction conditions.

[e] Five of ten lysine residues modified.

[f] pH maintained by NaOH from pH-stat.

[g] Lysine not acetylated under these conditions. Reaction with 1.0 *M* hydroxylamine, pH 5.8 (60 min, 37°C) reversed changes in catalytic activity produced on reaction with NAI and presumably deacetylated *O*-acetyl-tyrosyl residues.

[h] Also 1.0 serine and 0.3 lysine.

[i] Also used Tris, TES, HEPES, and barbital buffers without any significant difference in nature of the reaction.

[j] Also 1.7 (probably serine and histidine) and 2.5 lysine residues modified.

[k] From *Bacillus subtilis*.

[l] Approximately two lysine residues modified under these conditions. Only a single tyrosine residue is modified with tetranitromethane. Either reagent (tetranitromethane or NAI) led to a 70 to 80% loss of catalytic activity.

[m] The reaction with *N*-acetylimidazole was performed with subtilisin preparation previously treated with phenylmethanesulfonyl fluoride. The active enzyme catalyzes the rapid hydrolysis of NAI under reaction conditions.

[n] Reaction performed on protein in which lysine residue had been previously blocked by reaction with ethyl acetimidate. *N*-acetylimidazole was added in four 200-fold molar excess portion at 2-h intervals.

[o] Reaction for 1 h at ambient temperature with amount of NAI equivalent (weight/weight basis) to stroma. The reaction mixture was washed with distilled water to remove *N*-acetylimidazole.

[p] Reaction for 1 h at ambient temperature. The quantity of NAI used is not given. It is stated that this reagent should readily pass across the cell membrane, but this conclusion is based on analogy with acetic anhydride.

[q] Extensive modification of amino groups was reported.

[r] NAI added as a solid; pH maintained at 7.5 with pH-stat.

[s] Also used TNM.

[t] Also used TNM and *p*-nitrobenzylsulfonyl fluoride

[u] Site-specific mutagenesis (Y451F or Y451H) decreases K_{cat} but not K_m.

References for Table 9.1

1. Simpson, R.T., Riordan, J.F., and Vallee, B.L., Functional tyrosyl residues in the active center of bovine pancreatic carboxypeptidase A, *Biochemistry* 2, 616, 1963.
2. Perlmann, G.E., Acetylation of pepsin and pepsinogen, *J. Biol. Chem.* 241, 153, 1966.
3. Houston, L.L., and Walsh, K.A., The transient inactivation of trypsin by mild acetylation with *N*-acetylimidazole, *Biochemistry* 9, 156, 1970.

(continued)

TABLE 9.1 (continued)
Reaction of Proteins with N-Acetylimidazole (NAI)

4. Connellan, J.M., and Shaw, D.C., The inactivation of *Bacillus subtilis* α-amylase by N-acetylimidazole and tetranitromethane, Reaction of tyrosyl residues, *J. Biol. Chem.* 245, 2845, 1970.

5. Myers, B., II, and Glazer, A.N., Spectroscopic studies of the exposure of tyrosine residues in proteins with special reference to the subtilisins, *J. Biol. Chem.* 246, 412, 1971.

6. Fan, C.C., and York, J.L., The role of tyrosine in the hemerythrin active site, *Biochem. Biophys. Res. Commn.* 47, 472, 1972.

7. Lundblad, R.L., Harrison, J.H., and Mann, K.G., On the reaction of purified bovine thrombin with N-acetylimidazole, *Biochemistry* 12, 409, 1973.

8. Kirtley, M.E., and Dix, J.C., Effects of acetylimidazole on the hydrolysis of fructose diphosphate and p-nitrophenyl phosphate by liver fructose diphosphatase, *Biochemistry* 13, 4469, 1974.

9. Masiak, S.J., and D'Angelo, G., Effects of N-acetylimidazole on human erythrocyte ATPase activity. Evidence for a tyrosyl residue at the ATP binding site of the (Na+, K+)-dependent ATPase, *Biochim. Biophys. Acta* 382, 83, 1975.

10. Kronman, M.J., Hoffman, W.B., Jeroszko, J., and Sage, G.W., Inter and intramolecular interactions of α-lactalbumin. XI. Comparison of the "exposure" of tyrosyl, tryptophyl and lysyl side chains in the goat and bovine proteins, *Biochim. Biophys. Acta* 285, 124, 1972.

11. Holohan, P., Hoffman, W.B., and Kronman, M.J., Chemical modification of tyrosyl and lysyl residues in goat alpha lactalbumin and the effect on the interaction with the galactosyl transferase, *Biochim. Biophys. Acta* 621, 333, 1980.

12. Erlanson, C., Barrowman, J.A., and Borgström, B., Chemical modifications of pancreatic colipase, *Biochim. Biophys. Acta* 489, 150, 1977.

13. Erlanson-Albertsson, C., The importance of the tyrosine residues in pancreatic colipase for its activity, *FEBS Lett.* 117, 295, 1980.

14. Hoschke, A., Laszlo, E., and Hollo, J., A study of the role of tyrosine groups at the active centre of amylolytic enzymes, *Carbohydr. Res.* 81, 157, 1981.

15. Kiss, L., Korodi, I., and Nanasi, P., Study on the role of tyrosine side-chains at the active center of emulsin β-D-glucosidase, *Biochim. Biophys. Acta* 662, 308, 1981.

16. Kulanthaivel, P., Leibach, F.H., Mahesh, V.B., and Ganapathy, V., Tyrosine residues are essential for the activity of the human placental taurine transporter, *Biochim. Biophys. Acta* 985, 139, 1989.

17. Arguello, J.M., and Kaplan, J.H., N-Acetylimidazole inactivates renal Na, K-ATPase by disrupting ATP binding at the catalytic site, *Biochemistry* 29, 5775, 1990.

18. Mueller, M.J., Samuelsson, B., and Haeggstrom, J.Z., Chemical modification of leukotriene A4 hydrolase. Indications for essential tyrosyl and arginyl residues at the active site, *Biochemistry* 34, 3546, 1995.

19. Stratilova, E. et al., An essential tyrosine residue of *Aspergillus polygalacturonidase*, *FEBS Lett.* 382, 164, 1996.

20. Janes, M.A., Hamilton, M.L., and Lash, T.D., Effect of covalent modification on coproporphyrinogen oxidase from chicken red blood cells, *Prep. Biochem. Biotechnol.* 27, 47, 1997.

21. Bernardo, A.A., Kear, F.T., and Arruda, J.A. The renal cortical Na+/CO3- cotransporter VI: the effect of chemical modification in cotransporter activity, *J. Membr. Biol.* 158, 49, 1997.

TABLE 9.1 (continued)
Reaction of Proteins with _N_-Acetylimidazole (NAI)

22 Vaseux, G. et al., A tyrosine residue essential for catalytic activity in aminopeptidase A, _Biochem. J._ 327, 883, 1997.
23. Bera, S. et al., Chemical modifications and dissociation characterisitics of tyrosine and tryptophan residues in alpha crystalline, _Indian J. Biochem. Biophys._ 34, 419, 1997.
24. Kuhn, D.M., and Geddes, T.J., Peroxynitrite inactivates tryptophan hydroxylase via sulfydryl oxidation. Coincident nitration of enzyme tyrosyl residues has minimal impact on catalytic activity, _J. Biol. Chem._ 274, 29726, 1999.
25. Korn, C. et al., Involvement of conserved histidine, lysine, and tyrosine residues in the mechanism of DNA cleavage by the capase-3-activated DNAse, _Nucl. Acids Res._ 30, 1325, 2003.
26. Yapeng, C. et al., Rapid purification, characterization and substrate specificity of heparinase from a novel species of _Sphingobacterium_, _J. Biochem._ 134, 365, 2003.
27. Konozy, E.H.E. et al., Isolation, purification, and physicochemical characterization of a D-galactose-binding protein lectin from seeds of _Erythrina speciosa_, _Arch. Biochem. Biophys._ 410, 222, 2003.

$\Delta\varepsilon$ of 1160 M^{-1} cm^{-1} has been reported, whereas a subsequent study reported a $\Delta\varepsilon$ of 1210 M^{-1} cm^{-1}.[46] We have had more reliable results with the latter value in our laboratory. We have also found it more accurate to determine changes in absorbance at 278 nm as a function of time, taking into account spectral changes introduced by the addition of reagent to a solvent blank.[54] One of the major advantages of reaction with NAI is the ease of reversal of the reaction. The _O_-acetyl derivative of tyrosine is unstable under mildly alkaline conditions, and presence of a nucleophile such as Tris greatly decreases the stability of the _O_-acetyl derivative. Quantitative deacetylation occurs with hydroxylamine at pH 7.5. As expected, the rate of regeneration of free tyrosine is a function of hydroxylamine concentration. Note that the primary side-reaction products of the reaction of NAI with proteins, ε-_N_-acetyllysine and _N_-acetyl amino-terminal amino acids, are stable to neutral or alkaline hydroxylamine. Assignment of changes in the biological activity of a protein on reaction with NAI to the _O_-acetylation of tyrosine can be verified by the reversibility of such changes in the presence of hydroxylamine.

The possible use of TNM for the modification of tyrosyl residues in proteins was advanced over 50 years ago.[55] However, it was not until some two decades later that the studies of Vallee, Riordan, Sokolovsky, and Harell established the specificity and characteristics of the reaction of TNM (Figure 9.3) with proteins.[56,57] The modification proceeds optimally at mildly alkaline pH. The rate of modification of _N_-acetyltyrosine is twice as rapid at pH 8.0 than at pH 7.0; it is ca. 10 times more rapid at pH 9.5 than at pH 7.0. The reaction of TNM with tyrosine produces 3-nitrotyrosine, nitroformate (trinitromethane anion), and two protons. The spectral properties of nitroformate (ε at 350 nm = 14,000) suggested that monitoring the formation of this species would be a sensitive method for monitoring the time course of the reaction of TNM with tyrosyl residues.[56] Although determining the rate of nitroformate production appears to be effective in studying the reaction of TNM

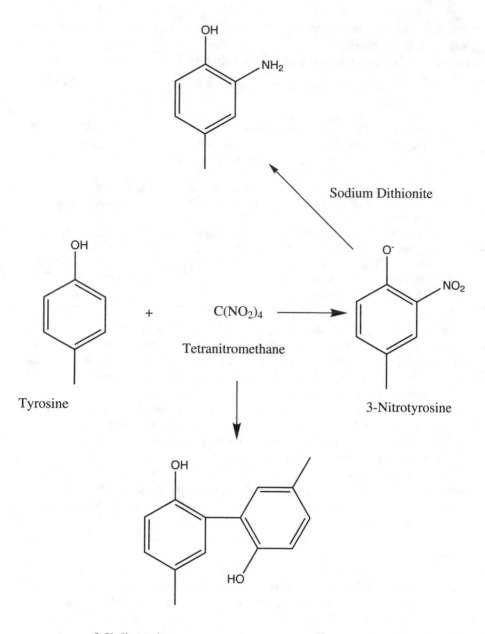

FIGURE 9.3 Reaction of tetranitromethane with tyrosine to form 3-nitrotyrosine and 3,3'-dityrosine.

with model compounds such as *N*-acetyltyrosine, it has not proved useful with proteins.[57,58] Although the reaction of TNM with proteins is reasonably specific for tyrosine, oxidation of sulfhydryl groups has been reported,[57,58] as has reaction with histidine,[57] methionine,[57] and tryptophan.[57,59] The reaction with sulfhydryl groups would seem to be the most common side reaction. Reaction of TNM with cysteine in proteins can result in disulfide bond formation and the formation of oxidation products such as sulfone and sulfenic acid derivatives. It is frequently assumed that the reaction of TNM with sulfhydryl groups can proceed equally well at pH 6.0 and pH 8.0, whereas the reaction with tyrosine occurs at pH 8.0 and not at pH 6.0. As noted later, this is not a good assumption, as there are examples of tyrosine residues nitrated by TNM at pH 6.0. Only rigorous chemical analysis of the products of the reaction can verify the sites of modification.

Although the modification of tryptophanyl residues in proteins by TNM is a somewhat unusual reaction, there are a number of examples of the phenomena.[57,59,60–66] The modified protein is characterized by a broad absorption band with a maximum in the area of 340 to 360 nm. This absorbance, unlike the absorbance of nitrotyrosine at 410 nm, is not pH dependent. The reader is referred to the study of Cuatrecasas and co-workers[59] for additional details.

Reaction with TNM can also result in the covalent cross-linkage of tyrosyl residues resulting in inter- and intramolecular association of peptide chains.[67,68] The cross-linking of tyrosyl residues in proteins via reaction with TNM is an example of zero-length or contact-site cross-linking[69,70] In studies of the mechanism of TNM nitration of phenols, Bruice and co-workers[71] observed more cross-linkage (formation of Plummerer's ketone) than in nitration with the reaction of TNM with p-cresol at neutral pH. The magnitude of this problem is dependent on variables such as protein concentration and solvent conditions (i.e., pH). With respect to this latter consideration, it is noted that acidification of reaction mixtures tends to favor the cross-linkage reaction.[70] As expected, the extent of cross-linkage observed varies with the protein studied. For example, reaction of pancreatic deoxyribonuclease with TNM results in extensive formation of dimer.[72] A related reaction is the free-radical-induced cross-linking between tyrosyl residues and thymine, which provides a basis for the formation of nucleic acid – protein conjugates occurring as a result of ionizing radiation.[73] Treatment of apoovotransferrin with periodate (50 mM HEPES, pH 7.4, with 5 mM sodium periodate) resulted in protein cross-linking via 3,3′-dityrosine.[74]

The extent of modification of tyrosyl residues by TNM in proteins can be assessed by either spectrophotometric means or amino acid analysis.[58,75] At alkaline pH (pH ≥ 8), 3-nitrotyrosine has an absorption maximum at 428 nm with $\varepsilon = 4100$ M^{-1} cm^{-1}; the absorption maximum of tyrosine at 275 nm increases from $\varepsilon = 1360$ to 4000 M^{-1} cm^{-1}. At acid pH (pH ≤ 6), the absorption maximum is shifted from 428 to 360 nm, with an isosbestic point at 381 nm ($\varepsilon = 2200$ M^{-1} cm^{-1}). We have found it convenient to determine the A_{428} in 0.1 M NaOH. Amino acid analysis after acid hydrolysis has also proved to be a convenient method of assessing the extent of 3-nitrotyrosine formation. 3-Nitrotyrosine is stable to acid hydrolysis (6 N HCl, 105°C, 24 h). This approach has the added advantage that other modifications of tyrosine such as free-radical-mediated cross-linkage can be either excluded or quantitatively determined. If nitration to form 3-nitrotyrosine is the only modification of tyrosyl residues in a protein occurring on reaction with TNM, the sum of 3-nitrotyrosine and tyrosine should be equivalent to the amount of tyrosine in the unmodified protein. The study of the reaction of peroxynitrite with proteins has the effect of increasing the number of approaches to the analysis of 3-nitrotyrosine in proteins.[76] Mass spectrometry is proving to be of increasing value in the analysis of nitrotyrosine in proteins.[77–79] Mass spectrometry permitted the identification of a dinitrotyrosine secondary to reaction with TNM. Antibodies to 3-nitrotyrosine in proteins have been developed and are useful not only for the analysis of purified proteins[80–82] but also for the enrichment of nitrotyrosine-containing proteins,[82] *in situ* localization in cells,[83] and identification on two-dimensional gel electrophoretograms.[84] Reduction with sodium dithionite improves the specificity for the detection of nitrotyrosine-containing proteins on Western blots; nitrotyrosine-positive bands are eliminated on reduction, leaving false-positive spots.[84,85]

There are several consequences of nitration of a tyrosyl residue. The most obvious is the placing of a somewhat bulky substituent (the nitro group) *ortho* to the phenolic hydroxyl function. The properties of the substituent nitro group "push" electrons into the benzene ring (inductive effect), lowering the pK_a of the phenolic hydroxyl from ca. 10.3 to 7.3. This means that the phenolic hydroxyl of the nitrated tyrosyl residue will be in a partially ionized state at physiological pH. The nitro function can be reduced to the corresponding amine under relatively mild conditions ($Na_2S_2O_4$, 0.05 M Tris, pH 8.0).[86] The conversion of 3-nitrotyrosine to 3-amino-tyrosine is associated with the loss of the absorption maximum at 428 nm and the change in the pK_a of the phenolic hydroxyl group from ca. 7.0 to 10.0. On occasion, reduction of the nitro function in this manner reverses the modification of function observed on nitration. The resultant amine function can be subsequently modified.[87] Fox and colleagues[88] modified the single tyrosine residue in *Escherichia coli* acyl carrier protein with TNM, subsequently reduced the 3-nitrotyrosyl residue to 3-amino-tyrosine with sodium dithionite, and modified the 3-aminotyrosine with dansyl chloride at pH 5.0 (50 mM sodium acetate) to obtain the dansyl acyl carrier protein. The dansyl acyl carrier protein was subsequently used for fluorescence anisotropy studies of enzyme–substrate complex formation in stearoyl-ACP desaturase.[89]

In addition to changing the properties of a given tyrosyl residue, nitration also introduces a spectral probe (3-nitrotyrosine) that can be used to detect conformational change in the protein. The concept of using reagents to introduce probes with unique spectral and fluorescent properties has been introduced in Chapter 1. 3-Nitro-tyrosine has an absorption maximum at 428 nm at alkaline pH. Riordan and co-workers[90] first used this spectral property with studies on nitrated carboxypeptidase A to study changes in the microenvironment around the modified residue. The addition of β-phenylpropionate, a competitive inhibitor of carboxypeptidase and nitrated carboxypeptidase, decreased the absorbance of mononitrocarboxypeptidase at 428 nm. This change is consistent with an increase in the hydrophobic quality of the microenvironment surrounding the modified tyrosyl residue.

In their early description of the reaction of TNM with tyrosine, Riordan and co-workers[58] suggested that the reactivity of tyrosyl residues in proteins with TNM was a measure of exposure to solvent. This concept of free and buried residues was introduced earlier by the same group in their studies on the reaction of tyrosine with NAI in proteins.[45] In this model, free tyrosyl residues are considered to be in direct contact with the solvent and have pK_a values between 9.5 and 10.5, whereas buried tyrosyl residues are relatively inaccessible to solvent and have pK_a values above

10.5. Myers and Glazer[46] challenged the general applicability of this correlation between reactivity and solvent accessibility in studies on subtilisin. They argued that tyrosyl residues in apolar locations are preferentially modified by TNM or N-acetyl-imidazole. The observations of Skov and co-workers[91] on the modification of cytochrome c with TNM support the argument presented by Myers and Glazer. Two of the four tyrosyl residues in equine cytochrome c are converted to 3-nitrotyrosine on reaction with TNM at pH 8.0 (0.05 M Tris). The two residues modified, Tyr-47 and Tyr-67, are located in the interior, whereas the two residues on the surface of the protein, Tyr-74 and Tyr-97, are not modified. As noted by Myers and Glazer,[46] local environment might affect both residue reactivity with TNM as well as spectral studies on the derivatives. TNM is not considered to be water soluble,[92] but is soluble in polar solvents such as ethanol and apolar solvents such as ethyl ether. This latter fact is clearly stated in a review of TNM by Riordan and Vallee in 1972.[58] As noted by these authors, TNM can be washed with water to remove impurities and high concentrations of reagent should be avoided to obviate the possibility of phase separation. In his description of the synthesis of TNM from acetic anhydride and nitric acid in 1955, Liang[93] reported that TNM separates from the aqueous phase during the process. However, studies on the chemistry of TNM at concentrations of 50 μM in aqueous solutions have been performed without difficulty.[94] Most experiments using TNM involve the introduction of the reagent as a solution in ethanol or less frequently with methanol. The final concentration of ethanol is frequently 5 to 10% of the final reaction volume, which likely enhances the solubility of TNM in the reaction mixture. I am not aware of any studies that specifically address the effect of ethanol (or other organic water-miscible solvents) on the solubility of TNM in an aqueous solution.

This suggests that the reactivity of a given tyrosyl residue with TNM is not easily predicted on the basis of physical location in the protein. More likely, reactivity is a function of local environmental factors. The reaction mechanism as suggested by Bruice and co-workers[71] postulates the formation of an intermediate phenoxide–TNM charge transfer complex involving electron transfer from the aromatic ring. This would imply that the reaction of TNM with a tyrosine residue occurs with the ionized species. These studies were performed with TNM and p-methylphenol (4-methylphenol; p-cresol), and the reaction was followed by the disappearance of TNM as measured by the formation of trinitromethane anion (nitroformate) at 350 nm. The products of the reaction were trinitromethane anion, nitrite, and 4-methyl, 3-nitrophenol. The rate of reaction ($k_2 = 5.1 \times 10^4$ M^{-1} min^{-1} was slower with phenol ($k_2 = 2.0 \times 10^3$ M^{-1} min^{-1}) and much slower with 4-chlorophenol (5.7 $\times$ 10^2 M^{-1} min^{-1}) and 4-cyanophenol (0.26 M^{-1} min^{-1}). There is no obvious correlation between these rates and the dissociation constant for the phenolic hydroxyl group; there is a correlation with reaction rate and the Hammett constant σ$^-$. Analysis of the products of the reaction of TNM and 4-methylphenol demonstrated the presence of 4-methyl, 3-nitrophenol in a 23% yield and a 30% yield of 1,2,10,11-tetrahydro-6,11-dimethyl-2-oxodibenzfuran (Pummerer's ketone), a product derived from the free-radical-mediated cross-linking of the parent 4-methylphenol. Given the average pK_a value for tyrosine of 10.13 (see Table 1.1), this would imply that the average tyrosine residues would be mostly unreactive within the pH range of 6.0 to 8.0, the suggested optimal range for the modification of tyrosine in proteins. This is supported by the work of Bruice and co-workers,[71] who demonstrated that the rate of reaction of TNM with the phenolic form must be at least four orders of magnitude lesser than that observed with the phenolate form. At pH values above 8, the rate of reaction of TNM with tyrosine would appear to increase consistent with the importance of the ionization of the phenolic function; however, the side reactions with methionine, tryptophan, and histidine increase. Early studies on the modification of tyrosine residues with TNM in proteins suggest that the reactive tyrosyl residues in carboxy-peptidase and staphylococcal nuclease have abnormally low pK_a values.[58] There are some problems with this concept. Studies on the reaction with 4-methylphenol and TNM yielded a ca. 20% yield of 3-nitro-4-methylphenol, which could be only modestly increased by excess TNM, whereas reaction in proteins could yield 100% of the modification of a given tyrosyl residue.[95] Early studies[56,57] suggested that the formation of trinitromethane anion (nitroformate) could be used to stoichiometrically follow the reaction of TNM with tyrosyl residues in proteins. This has not proved to be the situation as, in general, the release of trinitromethane ion greatly exceeds the extent of modification of tyrosine and continues beyond the nitration reaction in proteins. Although it is not clear why this occurs, Fendler and Liecht[94] have shown that certain detergent micelles catalyze the reaction of TNM with hydroxide ion. Micellar hexadecyltrimethylammonium bromide and polyoxyethylene (15) non-ylphenol (Igepal C-730) enhance the rate constant for the reaction, whereas micellar sodium dodecyl sulfate has no effect. There is a rich literature[96,97] on the photochemical reactions of aromatic compounds with TNM, which might be applicable to interpret the reaction of TNM with proteins. At least one study[98] has examined the effect of light on the reaction of TNM with proteins. The study reported the effect of light on the reaction of TNM with bacteriorhodopsin with light, in which light influenced the intrinsic properties of the protein and also influenced the reaction with a carbodiimide.[99] This reaction is discussed in greater detail later. Other studies on the chemistry of TNM that might be instructive include a study by Capellos and co-workers[100] which showed that, on excitation with an eximer laser, TNM formed the NO_2 radical and the TNM anion in either aerated or deaerated polar solvents (e.g., methanol), which the trinitromethyl radical was suggested to form in a nonpolar solvent (e.g., hexane). TNM also reacts with unsaturated compounds[101] and has been used as a colorimetric method[102] to determine unsaturation. The study on the difference of reactions in polar and nonpolar solvents might be useful in interpreting environmental effects on the reaction of TNM with individual residues in proteins. Given the complexity of the interpretation of the previous studies and extension to the reaction of tyrosine in proteins, it is clear that the reaction of TNM with proteins can be complex and care should be taken in the interpretation of experimental results.[103]

Several studies on the reaction of tyrosine with tyrosine in proteins in which there is information about the microenvironment surrounding the modified residues are discussed next. Strosberg[104] and co-workers reported on the reaction of TNM with hen egg-white lysozyme. With a 12-fold molar excess of TNM in 50 mM Tris-HCl, 1 M NaCl, pH 8.0, at 20°C for 1 h, the derivative obtained contained 1 Mol of nitrotyrosine per mole protein as determined by amino acid analysis. A derivative containing 2 Mol of nitrotyrosine per mole protein is obtained with a 47-fold molar excess of TNM, whereas a derivative containing 3 Mol of nitrotyrosine per mole of protein is obtained with a 100-fold molar excess of TNM in 50 mM Tris-HCl, 1 M NaCl, 8.0 M urea, pH 8.0, at 50°C for 1 h. Analysis of the various products showed that with a 12-fold molar excess of reagent, two mononitrated derivatives are obtained: the major one (69%) is modified at Tyr-23 and the minor product (21%) at Tyr-20. With a 47-fold molar excess of TNM, two dinitrated products are obtained: the major product (71%) is modified at both Tyr-23 and Tyr-20 and the minor product (15%) modified at Tyr-23 and Tyr-53. The derivative modified with a 100-fold molar excess of reagent in the presence of 8.0 M urea was homogenous and was modified at Tyr-23, Tyr-20, and Tyr-53. A later study[105] examined the characteristics of hen egg-white lysozyme modified at Tyr-23 or at both Try-23 and Tyr-20. Both derivatives were obtained by reaction in 50 mM Tris-HCl, pH 8.0, at 37°C for 1 h. The Tyr-23/Tyr-20 derivative was obtained at a 15-fold molar excess of TNM, whereas the Tyr-23 was selectively obtained at a twofold molar excess of reagent. The Tyr-23 derivative could also be obtained by electrochemical nitration.[106] Izzo and co-workers[107] used Raman and NMR spectroscopy to study nitrated tyrosine residues in hen egg-white lysozyme. At a 10-fold molar excess of TNM to protein in 0.05 M Tris, 1.0 M NaCl, pH 8.0, two tyrosine residues are modified. The pK_a values of the modified residues were 6.76 and 6.52, as determined by Raman resonance spectroscopy. Other studies suggested that the first residue was in an aqueous environment, whereas the characteristics of the second residue were consistent with a hydrophobic environment with a hydrogen-bonded site. The presence of a hydrogen bond might indicate a decreased pK_a for the phenolic hydroxyl group, suggesting the potential for increased susceptibility to nitration with TNM. Although both residues are likely exposed to the solvent, one is a surface residue and the other in the interior of the protein.

The reaction of TNM with bacteriorhodopsin has been mentioned previously. Bacteriorhodopsin is a bacterial membrane protein that is activated by light to generate an electrochemical gradient. Absorption of light by the protein initiates a cycle of reactions, which results in proton translocation across the membrane.[108] Scherrer and Stoeckenius[98] showed that two tyrosyl residues, Tyr-26 and Try-63, could be modified with TNM. The modification of tyrosine residues in bacteriorhodopsin was sensitive to light (yellow light, Corning 3-69 filter, 250-W projector lamp). At pH 9.0 (50 mM Tris-HCl) in the dark at 23°C with a 10-fold molar excess of TNM, Tyr-26 is modified. Tyr-64 is selectively modified when the reaction is performed in the light at pH 6.0 (40 mM sodium phosphate) at 0°C. In a subsequent study,[109] selective ionization of Tyr-64 (pK_a ca. 9.0) was observed to occur with illumination. Photochemically induced dynamic nuclear polarization NMR spectroscopy of bacteriorhodopsin suggested that only one tyrosine residue is exposed to the solvent.[110]

Crambin is a small water-insoluble protein that contains two tyrosine residues. The reaction with TNM is performed in 50% EtOH/H$_2$O at pH 9.0 (0.0005 M Tris) and monitored by NMR spectroscopy.[111] Tyr-29 is modified with a 50-fold molar excess of the reagent in 10 min at 298°K (25°C). Addition of a second 50-fold molar excess of TNM and continuation of the reaction resulted in the additional modification of Tyr-44. A reverse susceptibility is observed with iodination (I_2, pH 8.5, with 0.0005 M Tris-HCl in 50% EtOH/H_2O). The reaction of tyrosine residues with TNM is consistent with solvent-exposed residues being more susceptible to reaction, but the iodination pattern is not. It was concluded that the susceptibility to modification by either TNM or I_2 is controlled by the microenvironment of the residue, independent of solvent exposure.

The modification of tyrosine residues in thermolysin with TNM was performed in 40 mM Tris, pH 8.0, containing 10 mM $CaCl_2$ at 25°C for 1 h.[112] Thermolysin contains 28 tyrosine residues and ca. 9 residues were modified under these conditions as determined by absorbance at 381 nm. When the reaction was allowed to proceed for 15 h under the same conditions, ca. 16 residues were modified. Analysis of the reaction rate allowed the identification of three classes of reactive tyrosyl residues that reacted at different rates with TNM. The apparent second-order rate constants for the three classes are 3.32, 0.52, and 0.18 M^{-1} min^{-1} compared to a second-order rate constant for the reaction of TNM with N-acetyltyrosine ethyl ester of 1.99 M^{-1} min^{-1} under the same reaction conditions. In the same study, spectrophotometric (295 nm) titration of thermolysin by pH-jump demonstrated that 16 tyrosine residues were readily ionized, whereas 12 additional tyrosine residues required an apparent conformational change in the protein for ionization after adjustment to pH 12. These investigators divided the readily ionized tyrosine residues into three classes with pK_a values of 10.2, 11.4, and 11.8. They concluded that the microenvironment of the more slowly reacting tyrosine residues is either negatively charged or hydrophobic.

El Kabbaj and Latruffe[49] studied the ability of TNM to penetrate the inner mitochondrial membrane by evaluating the inactivation of D-3-hydroxybutyrate dehydrogenase by TNM in its normal location and in inside–out membranes. TNM readily inactivated the enzyme on inside–out membranes, but was much less effective with intact mitochondrial membranes. The investigators also reported on the distribution of TNM between aqueous and organic solvents. The % partition was 17/83 in water/hexane and 11/89 in water/n-octanol. The investigators suggested that the behavior of TNM is consistent with both partial amphiphilic and hydrophobic properties.

This limited discussion suggests that it is not possible to accurately predict the environment of a tyrosyl residue modified by TNM in a protein. Care should be exercised in the interpretation of results when such observations are used to describe the topography of a protein.[113–117] Table 9.2 presents selected examples of the application of TNM for the site-specific modification of proteins.

Christen and Riordan[118] reported an interesting effect of the substrate on the reaction of an enzyme with TNM. Generally, protection of an enzyme from loss of activity secondary to chemical modification with the concomitant lack of modification of an amino acid residue is observed in the presence of substrate. Christen and Riordan, however, observed that aspartate aminotransferase was readily inactivated by TNM only in the presence of both substrates, glutamate and α-ketoglutarate (syncatalytic modification). This inactivation was associated with the modification of an additional tyrosyl residue. Kirschner and Schachman[119] reported some interesting spectral studies on the catalytic subunit of aspartate transcarbamylase modified by TNM (0.1 M potassium phosphate, pH 6.7). Addition of both substrates, carbamyl phosphate and succinate, resulted in a decrease in absorbance at 430 nm. This is

TABLE 9.2

Use of Tetranitromethane to Modify Tyrosyl Residues in Proteins

Protein Concentration	Solvent/Temp.	Molar Excess TNM	Residues Modified	Ref.
Carboxypeptidase A (10 mg/ml)	0.05 M Tris, 2 M NaCl at 20°C	4	1.2/18	1
Staphylococcal nuclease (2 mg/ml)	0.05 M Tris, pH 8.1, at 23°C	2	1.1/7	2
		4	1.6/7	
		8	3.4/7	
		12	3.7/7	
		16	4.2/7	
		20	4.7/7	
		30	5.0/7	
		60	4.8/7	
	4 M guanidine	60	6.4/7	
Horse heart cytochrome c (1 mM)	0.05 M Tris, pH 8.0	16	2/4	3
Aspartate aminotransferase (5 mg/ml)	0.05 M Tris, pH 7.5, at 22°C	30	—	4
Thrombin (0.06 mg/ml)	0.03 M sodium phosphate, pH 8.0, at 24°C	1000	4.9/12	5
2-Keto-4-hydroxyglutarate aldolase (1.2 mg/ml)	0.05 M Tris, pH 8.0			6
Porcine carboxypeptidase-β (5–7 mg/ml)	0.05 M Tris, pH 8.0, at 23°C	8	1.2/21	7
Bovine pituitary growth hormone (4 mg/ml)	0.05 M Tris, pH 8.0, at 0°C	30	2.7/6	8
Ovine pituitary growth hormone (4 mg/ml)	0.05 M Tris, pH 8.0, at 0°C	30	3.0/6	8
Aspartate transcarbamylase (catalytic subunit, 4 mg/ml)	0.1 M potassium phosphate, pH 6.7, at 23°C	2/3/8		9
Mucor miehei protease (1 mg/ml)	0.1 M phosphate, pH 8.0, at 15°C	20	2.8/21	10
Turnip yellow mosaic virus capsids (1.75 mg/ml)	0.05 M Tris, pH 8.0, at 22°C	50	3/3	11
α_1-Antiprotease inhibitor (1 mg/ml)	0.05 M Tris, pH 8.0, at 25°C	120	3/7	12
	5 M guanidine at 25°C	60	7/7	12
α_1-Acid glycoprotein	0.1 M Tris, pH 8.0, at 23°C	10	2.7/12	13
Carboxypeptidase A crystals (5 mg/ml)	0.05 M Tris, pH 8.0, at 20°C		1/9	14
Bovine growth hormone (2 mg/ml)	0.03 M Ringer phosphate at 25°C	12	3/7	15
Equine growth hormone (2 mg/ml)	0.03 M Ringer phosphate at 25°C	12	3/7	15
Lactose repressor protein	0.1 M Tris, 0.1 M mannose, pH 7.8, at 23°C	800	2.4/8	16

(continued)

TABLE 9.2 (continued)
Use of Tetranitromethane to Modify Tyrosyl Residues in Proteins

Protein Concentration	Solvent/Temp.	Molar Excess TNM	Residues Modified	Ref.
Aspartate transcarbamylase (8.3 mg/ml)	0.1 M Tris acetate at 25°C	750	2.2/10	17
Aspartate transcarbamylase (5 mg/ml)	0.1 M Tris, pH 8.0 at 25°C		2.2/10	18
Human serum albumin	pH 8.0	80	9/18 1.2/18	19, 20
Prolactin (1 mg/ml)	0.05 M Tris, pH 8.0, at 23°C	175	1.9/7	21
Porcine pancreatic phospholipase (1 mg/ml)	0.05 M Tris, 0.1 M NaCl, 0.01 M CaCl$_2$, pH 8.0, at 30°C	10	—	22
Equine pancreatic phospholipase (1 mg/ml)	0.05 M Tris, 0.1 M NaCl, 0.01 M CaCl$_2$, pH 8.0, at 30°C	10	—	22
Bovine pancreatic phospholipase (1 mg/ml)	0.05 M Tris, 0.1 M NaCl, 0.01 M CaCl$_2$, pH 8.0 at 30°C	10	—	22
Troponin C (1 mg/ml)	0.05 M Tris, 0.002 M EGTA at 23°C	8	3/3	23
Mouse myeloma protein (5×10^{-5} M)	0.01 M Tris, pH 8.2, at 23°C	10		24
Escherichia coli elongation factor G (4–6 mg/ml)	0.1 M Tris, 0.01 M KCl, 5% glycerol, 0.2 mM EDTA, pH 8.0, at 25°C	250	4/20	25
Elapid venom cardiotoxins (7 mg/ml)	0.1 M Tris, pH 7.0, at 25°C or 0.05 M Tris, pH 8.0, at 25°C	—	—	26
Lactose repressor (0.1–1.0 mg/ml)	0.1 M Tris, pH 8.0 or 0.24 M potassium phosphate, 5% glucose, pH 8.0, at 23°C	50	—	27
L-Lactate monooxygenase (1.8 μM)	0.05 M Tris, pH 8.0, 7.5 at 30°C			28
Tryptophanase apoenzyme (0.1 μM)	0.05 M triethanolamine, pH 8.0 at 30°C	—	—	29
A-Lactamase (1.3 mg/ml)	0.05 M Tris, pH 8.0 at 25°C	5.20		30
Bacillus subtilis neutral protease	50 mM Tris-Cl, pH 8.0, with 5 mM CaCl$_2$ at 25°C	60	2	31
Fructose-1,6-bisphosphatase	50 mM Tris-Cl, pH 8.0	50	4	32
A-1-Anti-protease inhibitor	50 mM Tris-Cl, pH 8.0, at 22°C	105	3/6	33
Bovine thrombin	50 mM sodium phosphate, pH 8.0, with 100 mM NaCl or 50 mM Tris, pH 8.0 with 100 mM NaCl	50–200	1/12	34

TABLE 9.2 (continued)
Use of Tetranitromethane to Modify Tyrosyl Residues in Proteins

Protein Concentration	Solvent/Temp.	Molar Excess TNM	Residues Modified	Ref.
Mouse monoclonal antibodies	Tris-buffered saline, pH 7.3	50-2000	3-14.4	35
Lipase/acyltransferase from *Aeromonas hydrophilia*	50 m*M* Tris, 100 m*M* NaCl, pH 8.0, 23°C	53	2	36
HlyC (protein acyltransferase)	50 m*M* Tris, pH 8.0, or 100 m*M* sodium phosphate, pH 6.0	—	—	37
Retinoic acid receptor α	20 m*M* Tris, pH 8.0, 1 m*M* PMSF, 20 μg leupeptin/ml, 20 μg aprotinin/ml	—	—	38
HSNP-C' (helix-stabilizing nucleoid protein	0.1 *M* sodium phosphate, pH 7.0, 0.1 m*M* EDTA	—	—	39
Human angiotensin II	50 m*M* Tris, pH 8.0	50	—	40
Bovine serum albumin	50 m*M* Tris, pH 8.0	50		40
Glycine receptor in phospholipid vesicles	25 m*M* sodium phosphate, pH 7.4	10-100	6	41

References for Table 9.2

1. Riordan, J.F., Sokolovsky, M., and Vallee, B.L., The functional tyrosyl residues of carboxypeptidase A. Nitration with tetranitromethane, *Biochemistry* 6, 3609, 1967.
2. Cuatrecasas, P., Fuchs, S., and Anfinsen, C.B., The tyrosyl residues at the active site of staphylococcal nuclease. Modifications by tetranitromethane, *J. Biol. Chem.* 243, 4787, 1968.
3. Skov, K., Hofmann, T., and Williams, G.R., The nitration of cytochrome c, *Can. J. Biochem.* 47, 750, 1969.
4. Christen, P., and Riordan, J.F., Syncatalytic modification of a functional tyrosyl residue is aspartate aminotransferase, *Biochemistry* 9, 3025, 1970.
5. Lundblad, R.L., and Harrison, J.H., The differential effect of tetranitromethane on the proteinase and esterase activity of bovine thrombin, *Biochem. Biophys. Res. Commn.*, 45, 1344, 1971.
6. Lane, R.S., and Dekker, E.E., Oxidation of sulfhydryl groups of bovine liver 2-keto-4-hydroxyglutarate aldolase by tetranitromethane, *Biochemistry* 11, 3295, 1972.
7. Sokolovsky, M., Porcine carboxypeptidase B. Nitration of the functional tyrosyl residue with tetranitromethane, *Eur. J. Biochem.* 25, 267, 1972.
8. Glaser, C.B., Bewley, T.A., and Li, C.H., Reaction of bovine and ovine pituitary growth hormones with tetranitromethane, *Biochemistry* 12, 3379, 1973.
9. Kirschner, M.W., and Schachman, H.K., Conformational studies on the nitrated catalytic subunit of aspartate transcarbamylase, *Biochemistry* 12, 2987, 1973.
10. Rickert, W.S., and McBride-Warren, P.A., Structural and functional determinants of *Mucor miehei* protease. IV. Nitration and spectrophotometric titration of tyrosine residues, *Biochim. Biophys. Acta* 371, 368, 1974.
11. Re, G.G., and Kaper, J.M., Chemical accessibility of tyrosyl and lysyl residues in turnip yellow mosaic virus capsids, *Biochemistry* 14, 4492, 1975.
12. Busby, T.F., and Gan, J.C., The reaction of tetranitromethane with human plasma a₁-antitrypsin, *Int. J. Biochem.* 6, 835, 1975.

(continued)

TABLE 9.2 (continued)
Use of Tetranitromethane to Modify Tyrosyl Residues in Proteins

13. Kute, T., and Westphal, U., Steroid-protein interactions. XXXIV. Chemical modification of α_1 acid glycoprotein for characterization of the progesterone binding site, *Biochim. Biophys. Acta* 420, 195, 1976

14. Muszynska, G., and Riordan, J.F., Chemical modification of carboxypeptidase A crystals. Nitration of tyrosine 248, *Biochemistry* 15, 46, 1976.

15. Daurat-Larroque, S.T., Portuguez, M.E.M., and Santome, J.A., Reaction of bovine and equine growth hormones with tetranitromethane, *Int. J. Pept. Protein Res.* 9, 119, 1977.

16. Alexander, M.E., Burgum, A.A., Noall, R.A., Shaw, M.D., and Matthews, K.S., Modification of tyrosine residues of the lactose repressor protein, *Biochim. Biophys. Acta* 493, 367, 1977.

17. Landfear, S.M., Lipscomb, W.N., and Evans, D.R., Functional modifications of aspartate transcarbamylase induced by nitration with tetranitromethane, *J. Biol. Chem.* 253, 3988, 1978.

18. Lauritzen, A.M., Landfear, S.M., and Lipscomb, W.N., Inactivation of the catalytic subunit of aspartate transcarbamylase by nitration with tetranitromethane, *J. Biol. Chem.* 255, 602, 1980.

19. Malan, P.G., and Edelhoch, H., Nitration of human serum albumin and bovine and human goiter thyroglobulins with tetranitromethane, *Biochemistry* 9, 3205, 1970.

20. Moravek, L., Saber, M.A., and Meloun, B., Steric accessibility of tyrosine residues in human serum albumin, *Collect. Czech. Chem. Commn.* 44, 1657, 1979.

21. Andersen, T.T., Zamierowski, M.M., and Ebner, K.E., Effect of nitration on prolactin activities, *Arch. Biochem. Biophys.* 192, 112, 1979.

22. Meyer, H., Verhoef, H., Hendriks, F.F.A., Slotboom, A.J., and de Haas, G.H., Comparative studies of tyrosine modification in pancreatic phospholipases. I. Reaction of tetranitromethane with pig, horse, and ox phospholipases A_2 and their zymogens, *Biochemistry* 18, 3582, 1979.

23. McCubbin, W.D., Hincke, M.T., and Kay, C.M., The utility of the nitrotyrosine chromophore as a spectroscopic probe in troponin C and modulator protein, *Can. J. Biochem.* 57, 15, 1979.

24. Gavish, M., Neriah, Y.B., Zakut, R., Givol, D., Dwek, R.A., and Jackson, W.R.C., On the role of Tyr 34_L in the antibody combining site of the dinitrophenyl binding protein 315, *Mol. Immunol.* 16, 957, 1979.

25. Alakhov, Y.B., Zalite, I.K., and Kashparov, I.A., Tyrosine residues in the C-terminal domain of the elongation factor G are essential for its interaction with the ribosome, *Eur. J. Biochem.* 105, 531, 1980.

26. Carlsson, F.H.H., The preparation of 3-nitrotyrosyl derivatives of three elapid venom cardiotoxins, *Biochim. Biophys. Acta* 624, 460, 1980.

27. Hsieh, W.-T., and Matthews, K.S., Tetranitromethane modification of the tyrosine residues of the lactose repressor, *J. Biol. Chem.* 256, 4856, 1981.

28. Durfor, C.N., and Cromartie, T.H., Inactivation of L-lactate monooxygenase by nitration with tetranitromethane, *Arch. Biochem. Biophys.* 210, 710, 1981.

29. Nihira, T., Toraya, T., and Fukui, S., Modification of tryptophanase with tetranitromethane, *Eur. J. Biochem.* 119, 273, 1981.

30. Wolozin, B.L., Myerowitz, R., and Pratt, R.F., Specific chemical modification of the readily nitrated tyrosine of the R_{TEM}b-lactamase and of *Bacillus cereus* β-lactamase. I. The role of this tyrosine in β-lactamase catalysis, *Biochim. Biophys. Acta* 701, 153, 1982.

31. Kobayashi, R., Kanatani, A., Yoshimoto, T., and Tsuru, D., Chemical modification of neutral protease from *Bacillus subtilis* var. *amylosacchariticus* with tetranitromethane: assignment of tyrosyl residues nitrated, *J. Biochem. (Tokyo)*, 106, 1110, 1989.

32. Liu, F., and Fromm, H.J., Investigation of the relationship between tyrosyl residues and the adenosine 5'-monophosphate binding site of rabbit liver fructose-1,6-bisphosphatase as studied by chemical modification and nuclear magnetic resonance spectroscopy, *J. Biol. Chem.* 264, 18320, 1989.

(continued)

TABLE 9.2 (continued)
Use of Tetranitromethane to Modify Tyrosyl Residues in Proteins

33. Mierzwa, S., and Chan, S.K., Chemical modification of human alpha-1-proteinase inhibitor by tetranitromethane. Structure-function relationship, *Biochem. J.* 246, 37, 1987.
34. Lundblad, R.L., Noyes, C.M., Featherstone, G.L., Harrison, J.H., and Jenzano, J.W., The reaction of bovine alpha-thrombin with tetranitromethane. Characterization of the modified protein, *J. Biol. Chem.* 263, 3729, 1988.
35. Tawfik, D.S. et al., pH on-off switching of antibody-binding by site-specific chemical modification of tyrosine, *Protein Eng.* 7, 431, 1994.
36. Robertson, D.L. et al., Influence of active site and tyrosine modification on the secretion and activity of the *Aeromonas hydrophilia* lipase/acyltransferase, *J. Biol. Chem.* 269, 2146, 1994.
37. Trent, S.M., Warsham, L.M.S., and Ernst-Fonberg, M.L., HlyC, the internal protein acyltransferase that activates hemolysin toxin: the role of conserved tyrosine and arginine residues in enzymatic activity as probed by chemical modification and site-directed mutagenesis, *Biochemistry* 38, 8831, 1999.
38. Mailfait, S. et al., Critical role of tyrosine 277 in the ligand-binding and transactivating properties of retinoic acid receptor α, *Biochemistry* 39, 2183, 2000.
39. Celestina, F., and Suryanarayana, T., Biochemical characterization and helix stabilizing properties of HSNP-C' from the thermoacidophilic archareon *Sulfolobus acidocaldarius*, *Biochem. Biophys. Res. Commn.* 267, 614, 2000.
40. Petersson, A.-S. et al., Investigation of tyrosine nitration in proteins by mass spectrometry, *J. Mass Spectrom.* 36, 616, 2001.
41. Leite, J.F., and Cascio, M., Probing the topology of the glycine receptor by chemical modification coupled to mass spectrometry, *Biochemistry* 41, 6140, 2002.

consistent with the increase in the pK_a of the phenolic hydroxyl (6.25 to 6.62). These investigators found that for the 3-nitrotyrosyl group in the modified aspartate transcarbamylase, $\varepsilon_{430} = 430 \times 10^{-3}\ M^{-1}\ cm^{-1}$ and $\varepsilon_{390} = 2.8 \times 10^{-3}\ M^{-1}\ cm^{-1}$ (390 nm the isosbestic point).

Modification of the binding site of antibodies with TNM resulted in pH-dependent binding of antibody in a physiological pH range.[120] A mouse monoclonal antibody (antidinitrophenyl) was modified with TNM (200-fold molar excess, pH 8.3, 1.5 h, 4°C) and the binding of the modified antibody to antigen (dinitrophenylbovine serum albumin) evaluated by ELISA techniques and compared with that of a control antibody preparation. Binding was slightly reduced at pH 6.0, reduced by ca. 90% at pH 8.0, and completely absent at pH 9.0. This is probably a reflection of the decreased pK_a of the phenolic hydroxyl group in 3-nitrotyrosine. The tight binding of biotin to avidin is reduced by modification of a tyrosine residue in avidin by TNM.[121] The modified avidin can be used for the affinity chromatography of biotinylated proteins under relatively mild conditions. 3-Nitrotyrosine is a more effective fluorescence quencher than is tyrosine, and this property has been used by several groups for fluorescence resonance energy transfer (FRET) in the solution structural analysis of proteins. Rischel and Poulsen[122] modified a specific tyrosine residue in apomyoglobin. These investigators measured the energy transfer from tryptophanyl side chains to this modified tyrosine residue to determine end-to-end distance during protein folding. Two tryptophanyl residues are located at Positions 7 and 14 and tyrosyl residues at 103 and 146. Tyr-146 is specifically modified by

TNM. Two tyrosine residues were modified in a lipase/acyltransferase from *Aeromonas hydrophilia* by TNM, resulting in an 80% loss of enzyme activity;[123] replacing one tyrosine with phenylalanine (Y230F) by site-specific mutagenesis caused the same loss of activity. Ernst-Fonberg and co-workers[124] studied the reaction of HlyC, the internal protein acyltransferase that activates hemolysin, with TNM and NAI. At pH 8.0, reaction with TNM resulted in a 70% loss of activity, with the modification of 2.3 Mol of tyrosine per mole protein, whereas only 20% was lost at pH 6.0, suggesting that oxidation of a sulfhydryl group was not responsible for the loss of activity. Comparison of the reaction of TNM with proteins at pH 6.0 and 8.0 is frequently used to evaluate the possible oxidation of sulfhydryl groups. It is assumed that the modification of cysteine is equivalent at pH 6.0 and 8.0, whereas nitration of tyrosine occurs at pH 8.0 and not at pH 6.0. This is usually the situation, but there are examples of nitration of tyrosine at pH 6.0. The modification of HlyC by NAI did not result in the loss of activity with the modification of 1.1 Mol of tyrosine per mole of protein. It is not known whether this residue (or residues) is the same as that modified by TNM. These investigators showed that the acetylated protein could be subsequently inactivated by TNM. This might reflect the modification of a different tyrosine residue (or residues), but it is also likely that the *O*-acetyl groups are not stable under the conditions of reaction with TNM (50 mM Tris, pH 8.0). HlyC is a member of a family of membrane-active toxins (RTX toxins). Comparison of amino acid sequence in members of the RTX family suggested that two tyrosyl residues were conserved. Site-specific mutagenesis showed that Y50G and Y150G had reduced enzyme activity whereas Y70F and Y150F had wild-type activity. With optimal folding conditions, Y70G has 12% of the native acyltransferase activity and Y150G had 44% of the native enzyme activity. An essential tyrosine residue in nitroalkane oxidase was identified by reaction with TNM.[125] This study is of considerable interest as saturation kinetics were observed for TNM, suggesting the formation of a reversible enzyme–TNM complex (K_d = 12.9 mM) before the nitration reaction. The modified tyrosine residue is likely the functional group with a pK_a of 9.5 important for catalysis. Modification of the retinoic acid receptor α with TNM at pH 8.0 resulted in the modification of three tyrosine residues and the loss of retinoic acid binding.[126] One of these three residues, Y277, was protected from modification by the presence of retinoic acid; substitution of this residue by alanine (Y277A) also resulted in the loss of retinoic acid binding. Hage and co-workers[127] modified tyrosine residues in human serum albumin with TNM. Selective modification of Tyr-411 was obtained at a fourfold molar excess of TNM (50 mM Tris, pH 8.0).[128] The modified albumin was used to evaluate drug binding in affinity capillary electrophoresis. The ability of 3-nitrotyrosine to serve as a fluorescence resonance energy transfer receptor was cited previously.[122] Squier and co-workers used this property to study structural linkages between the transmembrane and cytosolic domains of phospholamban.[129] A single cysteine residue was inserted into phospholamban by site-specific mutagenesis and subsequently modified with *N*-(1-pyrenyl)maleimide. A single tyrosine residue was modified with TNM (sixfold molar excess of TNM, 100 mM Tris, 100 mM NaCl, pH 8.0, containing polyoxyethylene 9 lauryl ether, 25°C, 1 h). The pyrenyl maleimide served as the FRET donor following excitation at 333 nm.

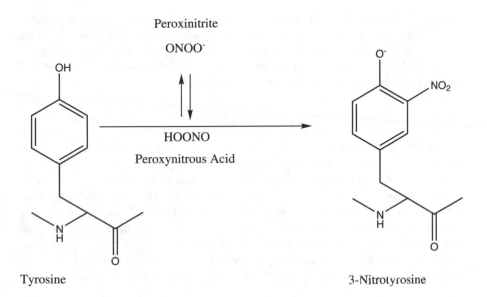

Peroxinitrite

ONOO⁻

HOONO

Peroxynitrous Acid

Tyrosine   3-Nitrotyrosine

FIGURE 9.4 Nitration of tyrosine by peroxynitrite to form 3-nitrotyrosine.

A discussion of tyrosine nitration in proteins would be incomplete without a brief consideration of peroxynitrite-mediated nitration (Figure 9.4). Although per-oxynitrite is not as convenient a reagent as TNM, there are facile methods for the synthesis and storage of the reagent.[130–132] There has been extensive work in this area, and there are far more current studies on peroxynitrite than on TNM. However, the work on peroxynitrite has been directed mostly toward physiological implications rather than site-specific chemical modification of proteins.[76,133,134] For this reason, there is only limited consideration of this specific aspect of tyrosine nitration. One article[135] deserves particular consideration as it related to the environmental aspects of tyrosine nitration. This study used 23-residue transmembrane peptides with single tyrosine residues at Positions 4, 8, and 12. The peptides were inserted into a multi-lamellar liposome composed of 1,2-dilauroyl-sn-glycero-3-phosphatidyl choline. Fluorescence spectra of the peptides incorporated into the membrane showed increasing derivatives with the highest value and the Tyr-4 with the lowest. When peroxynitrite is generated *in situ*, nitration of the Tyr-12 derivatives is more than that of the Tyr-8 derivative, which in turn is more than that of the Tyr-4 peptide. The authors note that peroxynitrite is in equilibrium with peroxynitrous acid ($pK_a = 6.8$). It is suggested that peroxynitrous acid diffuses into the membrane, where it undergoes hemolytic decomposition to form nitric oxide radical, which then reacts with tyrosine to form nitrotyrosine.

Tyrosyl residues in proteins can also be modified by reaction with cyanuric fluoride.[136,137] The reaction proceeds at alkaline pH (9.1) via modification of the phenolic hydroxyl group, with a change in the spectral properties of tyrosine. The phenolic hydroxyl groups must be ionized (phenoxide ion) for reaction with cyanuric fluoride. The modification of tyrosyl residues in elastase[138] and yeast hexokinase[139] with cyanuric fluoride has been reported. Modification of tyrosyl residues can occur as a side reaction with other residue-specific reagents such as 7-chloro-4-nitrobenzo-2-oxa-1,3-diazole (7-chloro-4-nitrobenzofurazan; NBD-Cl; Nbf-Cl).[140,141] NBD-Cl reacts primarily with amino groups and sulfhydryl groups in proteins.[142–144] The reaction product obtained with tyrosine, unlike that obtained with either amino groups or sulfhydryl groups, is not fluorescent and has an absorption maximum at 385 nm compared with 475 nm for amino derivatives and 425 nm for sulfhydryl derivatives.[140] The reaction of NBD-Cl with sulfenic acid has been reported.[143] The transfer of the nitrobenzofurazan moiety from the phenolic hydroxyl group of tyrosine to a nitrogen nucleophile in a protein has been observed.[141] Modification of the phenolic hydroxyl group with 2,4-dinitrofluorobenzene has also been reported.[145] A novel reaction of PMSF with a tyrosyl residue in an archaeon super-oxide dismutase has been reported,[146] and Means and Wu[147] reported the modification of a tyrosine residue in human serum albumin with diisopropylphosphorofluoridate.

Diazonium salts readily couple with proteins (Figure 9.5) to form colored derivatives with interesting spectral properties.[148–151] Reaction with diazonium salts is

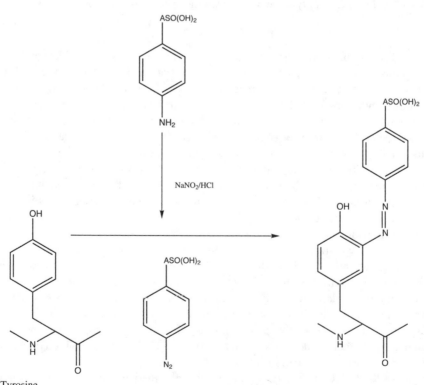

FIGURE 9.5 Reaction of tyrosine with diazotized arsanilic acid.

accomplished at alkaline pH (pH 8 to 9, bicarbonate/carbonate or borate buffer). It is relatively difficult to obtain specific residue class modification with the aromatic diazonium salts, but tyrosine, lysine, and histidine are rapidly modified.[152,153] The reaction of chymotrypsinogen A with diazotized arsanilic acid has been investigated.[154] Diazotization of arsanilic acid is accomplished by treatment of *p*-arsanilic acid with nitrous acid (0.55 mM sodium nitrite in 0.15 M HCl at 0°C). After adjusting the pH to 5.5 with NaOH, the reagent is diluted to a final concentration of 0.02 M. Reaction with chymotrypsinogen is accomplished in 0.5 M sodium bicarbonate buffer, pH 8.5, with a 20-fold excess of the reagent at 0°C. The reaction is terminated by adding a sufficient quantity of aqueous phenol (0.1 M) to react with excess reagent. The extent of the formation of monoazotyrosyl and monoazohistidyl derivatives is determined by spectral analysis.[152,153] The extent of reagent incorporation is determined by atomic absorption analysis for arsenic. Tyrosine (~1.0 Mol/Mol) and lysine (~4 Mol/Mol) were the only amino acid residues modified to any significant extent under these reaction conditions. The arsaniloazo functional group provides a spectral probe that can be used to study conformational change in proteins. In this particular study, there was a substantial change in the circular dichroism spectrum during the activation of the modified chymotrypsinogen preparation by trypsin.

The reaction of α-chymotrypsin with three diazonium salt derivatives (analogs) of *N*-acetyl-D-phenylalanine methyl ester[155] has been studied. The corresponding aromatic amine was converted to the diazonium salt by the action of nitrous acid (sodium nitrite per 0.6 M HCl at 0°C) and, after neutralization (NaOH) and dilution with 0.2 M sodium borate, pH 8.4, was used immediately for the modification of α-chymotrypsin (diazonium salt at a 10-fold molar excess) in 0.2 M sodium borate, pH 8.4, at ambient temperature for 1 h. The reaction was terminated by gel filtration (G-25 Sephadex) in 0.001 M HCl. Amino acid analysis showed that only tyrosine is modified under these reaction conditions. Subsequent analysis showed that Tyr-146 is modified by each of the three reagents. The peptide with the modified tyrosine residue (possessing a yellow color) adsorbs to the gel filtration matrix (G-10 equilibrated with 0.001 M HCl) and is eluted with 50% acetic acid. This phenomenon is somewhat similar to that observed with tryptophan-containing peptides modified with 2-hydroxy-5-nitrobenzyl bromide.[156] Pancreatic ribonuclease has been modified by a diazonium salt derivative of uridine 2'(3')5'-diphosphate.[157] Modification occurs at a specific tyrosine residue (Tyr-73). Modification of ribonuclease with 5'-(4-diazophenyl phosphoryl)-uridine-2'(3')-phosphate was accomplished by *in situ* generation of the diazonium salt from the corresponding amine by NaNO$_2$/HCl in the cold. The pH was then adjusted to pH 8.4 (NaOH); the solution was added to ribonuclease in 0.1 M borate, pH 8.4, and the reaction was allowed to proceed for 1 h at ambient temperature. The reaction was terminated by gel filtration (G-25 Sephadex) in 0.1 M acetic acid. The extent of modification was determined by spectral and amino acid analysis. Tyrosine was the only amino acid residue modified. Although it is relatively easy to assess the loss of tyrosyl residues, diazotization can be precisely determined only after reduction to the corresponding amine with sodium sulfite. These investigators also examined the reaction of ribonuclease with *p*-diazophenylphosphate under the same conditions of solvent and temperature.

Reaction with this reagent was far less specific, with losses of lysine, histidine, and tyrosine (3 Mol/Mol ribonuclease). The reaction of bovine carboxypeptidase A with various diazonium salts has been explored in greater detail than reactions of the previously discussed proteins. Vallee and co-workers[158,159] reported the reaction of bovine carboxypeptidase A crystals with diazotized p-arsanilic acid (conditions not specified) and obtained specific modification of Tyr-248. The peptide containing the modified tyrosine residue was purified by using antibodies directed against the arsaniloazotyrosyl group. The antibodies were obtained from, rabbits using arsaniloazovalbumin and arsaniloazobovine β-globulin as antigens. The reaction of bovine carboxypeptidase A with diazotized 5-amino-1H-tetrazole has been reported.[160] Diazotized 5-amino-1H-tetrazole also specifically reacts with Tyr-248 in bovine carboxypeptidase A (in 0.67 M potassium bicarbonate/carbonate, 1.0 M NaCl, pH 8.8). A sevenfold molar excess of reagent was used, and the reaction was terminated after 30 min by adding Tris buffer. The extent of modification of tyrosine to tetrazolylazotyrosine is determined by absorbance at 483 nm (Figure 9.2; $\varepsilon = 8.7 \times 10^3 \, M^{-1} \, cm^{-1}$). Modification of Tyr-248 in carboxypeptidase A by this reagent permits the subsequent modification of Tyr-198 by TNM.

p-Nitrobenzenesulfonyl fluoride (NBSF) is another environmentally sensitive reagent that can be used for the modification of tyrosyl residues in proteins. Liao and co-workers[161] developed this reagent for the selective modification of tyrosyl residues in pancreatic DNase. The modification reaction with NBSF can be performed in solvents (i.e., 0.1 M Tris-Cl, pH 8.0; 0.1 M N-ethylmorpholine acetate, pH 8.0) typically used for the modification of tyrosine residues by other reagents such as TNM or N-acetylimidazole. The rate of reagent hydrolysis is substantial and increases with increasing pH. In a recent study, NBSF has been used to characterize tyrosyl residues in NAD(P)H:quinone reductase.[162] Analysis of the product of the reaction showed that NBSF-modified tyrosyl residues were located in hydrophobic regions of the protein. In another study,[163] the reaction of NBSF with tyrosyl residues in the human placental taurine transporter was compared with modification observed with TNM, N-acetylimidazole, and NBD-Cl. Tetranitromethane and NBD-Cl were the most effective reagents. NBSF was an order of magnitude less effective, whereas NAI was 500 times less potent.

REFERENCES

1. Donovan, J.W., Changes in ultraviolet absorption produced by alteration of protein conformation, *J. Biol. Chem.* 244, 1961, 1969.
2. Markland, F.S., Phenolic hydroxyl ionization in two subtilisins, *J. Biol. Chem.* 244, 694, 1969.
3. Laws, W.R., and Shore, J.D., Spectral evidence for tyrosine ionization linked to a conformational change in liver alcohol dehydrogenase ternary complex, *J. Biol. Chem.* 254, 2582, 1979.
4. Kuramitso, S. et al., Ionization of the catalytic groups and tyrosyl residues in human lysozyme, *J. Biochem.* 87, 771, 1980.

5. Kobayashi, J., Hagashijima, T., and Miyazawa, T., Nuclear magnetic resonance analyses of side chain conformations of histidine and aromatic acid derivatives, *Int. J. Pept. Protein Res.* 24, 40, 1984.

6. Poklar, N., Vesnaver, G., and Laponje, S., Studies by UV spectroscopy of thermal denaturation of beta-lactoglobulin in urea and alkylurea solutions, *Biophys.Chem.* 47, 143, 1993.

7. Autelitano, F. et al., Covalent labeling of muscarinic acetylcholine receptors by tritiated aryldiazonium photoprobes, *Biochem. Pharmacol.* 53, 501, 1997.

8. Lee, P.H. et al., An efficient binding chemistry for glass polynucleotide microarrays, *Bioconjug. Chem.* 13, 97, 2002.

9. Tiller, J.C. et al., Stabilization of activity of oxidoreductases by their immobilization onto special functionalized glass and novel aminocellulose film using different coupling reagents, *Biomacromolecules* 3, 1021, 2002.

10. Gotoh, Y. et al., Preparation of lactose-silk fibroin conjugates and their application as a scaffold for hepatocyte attachment, *Biomaterials* 25, 1131, 2004.

11. Roholt, O.A., and Pressman, D., Iodination-isolation of peptides from the active site, *Methods Enzymol.* 25, 438, 1972.

12. Tsomides, T.J., and Eisen, H.N., Stoichiometric labeling of peptides by iodination on tyrosyl or histidyl residues, *Anal. Biochem.* 210, 129, 1993.

13. Rosenfeld, R. et al., Sites of iodination in recombinant human brain-derived neurotrophic factor and its effect on neurotrophic activity, *Protein Sci.* 2, 1664, 1993.

14. Huntley, T.E., and Strittmatter, P., The reactivity of the tyrosyl residues of cytochrome b_5, *J. Biol. Chem.* 247, 4648, 1972.

15. McGowan, E.B., and Stellwagen, E., Reactivity of individual tyrosyl residues of horse heart ferricytochrome c toward iodination, *Biochemistry* 9, 3047, 1970.

16. Layne, P.P., and Najjar, V.A., Evidence for a tyrosine residue at the active site of phosphoglucomutase and its interaction with vanadate, *Proc. Natl. Acad. Sci. USA* 76, 5010, 1979.

17. Silva, J.S., and Ebner, K.E., Protection by substrates and α-lactalbumin against inactivation of galactosyltransferase by iodine monochloride, *J. Biol. Chem.* 255, 11262, 1980.

18. Izzo, J.L., Bale, W.F., Izzo, M.J., and Ronconc, A., High specific activity labeling of insulin with ^{131}I, *J. Biol. Chem.* 239, 3743, 1964.

19. Linde, S. et al., Monoiodoinsulin labelled in tyrosine residue 16 or 26 of the insulin B-chain. Preparation and characterization of some binding properties, *Hoppe-Seyler's Z. Physiol. Chem.* 362, 573, 1981.

20. Hunter, W.M., and Greenwood, F.C., Preparation of iodine-131 labelled human growth hormone of high specific activity, *Nature* 194, 495, 1962.

21. Heber, D. et al., Improved iodination of peptides for radioimmunoassay and membrane radioreceptor assay, *Clin. Chem.* 24, 796, 1978.

22. Maly, P., and Lüthi, C., The binding sites of insulin-like growth factor I (IGF I) to type I IGF receptor and to a monoclonal antibody. Mapping by chemical modification of tyrosine residues, *J. Biol. Chem.* 263, 7068, 1988.

23. Hussain, A.A. et al., Chloramine T in radiolabeling techniques. II. A nondestructive method for radiolabeling biomolecules by halogenation, *Anal. Biochem.* 224, 221, 1995

24. Tashtoush, B.M. et al., Chloramine T in radiolabeling techniques. IV. Penta-*O*-acetyl-*N*-chloro-*N*-methylglucamine as an oxidizing agent, *Anal. Biochem.* 288, 16, 2001.

25. Nikula, T.K. et al., Impact of the high tyrosine fraction in complementarity determining regions: measured and predicted effects of radioiodination on IgG immunoreactivity, *Mol. Immunol.* 32, 865, 1995.

26, Bauer, R.J. et al., Alteration of the pharmacokinetics of small proteins by iodination. *Biopharm. Drug Dispos.* 17, 761, 1996.

27. Mathew, S. et al., Characterization of the interaction between α_2-macroglobulin and fibroblast growth factor-2: the role of hydrophobic interactions, *Biochem. J.* 374, 123, 2003.

28. Smith, C.L., and Peterson, C.L., Couple tandem affinity purification and quantitative tyrosine iodination to determine subunit stoichiometry of protein complexes, *Methods* 31, 104, 2003.

29. Linde, S., Hansen, B., and Lemmark, A., Preparation of stable radioiodiniated polypeptide hormones and proteins using polyacrylamide gel electrophoresis, *Methods Enzymol.* 92, 309, 1983.

30. Kamatso, Y., and Hayashi, H., Revaluating the effects of tyrosine iodination of recombinant hirudin on its thrombin inhibitor kinetics, *Thromb. Res.* 87, 3343, 1997.

31. Braschi, S. et al., Role of the kidney in regulating the metabolism of HDL in rabbits: evidence that iodination alters the catabolism of apolipoprotein A-1 by the kidney, *Biochemistry* 39, 5441, 2000.

32. Ghosh, D. et al., Determination of a protein structure by iodination: the structure of iodinated acetylxylan esterase, *Acta Crystallogr. D Biol. Crystallogr.* 55, 779, 1999.

33. Leinala, E.K., Davies, P.L., and Jia, Z., Elevated temperature and tyrosine iodination aid in the crystallization and structure determination of an antifreeze protein, *Acta Crystallogr. D Biol. Crystallogr.* 58, 1081, 2002.

34. Wieland, T., and Schneider, G., N-Acylimidazoles as acyl derivatives of high energy, *Ann. Chem. Justus Liebigs* 580, 159, 1953.

35. Stadtman, E.R., and White, F.H., Jr., The enzymic synthesis of N-acetylimidazole, *J. Am. Chem. Soc.* 75, 2022, 1953.

36. Stadtman, E.R., On the energy-rich nature of acetyl imidazole, an enzymatically active compound. In *A Symposium on the Mechanism of Enzyme Action* (McElroy, W.D., and Glass, B., Eds.). Johns Hopkins Press, Baltimore, MD, 1954, p. 581.

37. Fife, T.H., Steric effects in the hydrolysis of N-acylimidazoles and ester of *p*-nitrophenol, *J. Am. Chem. Soc.* 87, 4597, 1965.

38. Lee, J.P., Bembi, R., and Fife, T.H., Steric effects in the hydrolysis reactions of N-acylimidazoles. Effect of aryl substitution in the leaving group, *J. Org. Chem.* 62, 2872, 1997.

39. Riordan, J.F., and Vallee, B.L., Acetylation, *Methods Enzymol.* 11, 565, 1967.

40. Riordan, J.F., and Vallee, B.L., O-Acetylation, *Methods Enzymol.* 25, 570, 1972.

41. Karibian, D., Jones, C., Gertler, A., Dorrington, K.J., and Hofmann, T., On the reaction of acetic and maleic anhydrides with elastase. Evidence for a role of the NH$_2$-terminal valine, *Biochemistry* 13, 2891, 1974.

42. Ohnishi, M., Suganuma, T., and Hiromi, K., The role of a tyrosine residue of bacterial liquefying α-amylase in the enzymatic hydrolysis of linear substrates as studied by chemical modification with acetic anhydride, *J. Biochem. (Tokyo)* 76, 7, 1974.

43. Bernad, A., Nieto, M.A., Vioque, A., and Palacian, E., Modification of the amino groups and hydroxyl groups of lysozyme with carboxylic acid anhydrides: a comparative study, *Biochim. Biophys. Acta* 873, 350, 1986.

44. Simpson, R.T., Riordan, J.F., and Vallee, B.L., Functional tyrosyl residues in the active center of bovine pancreatic carboxypeptidase A, *Biochemistry* 2, 616, 1963.

45. Riordan, J.F., Wacker, W.E.C., and Vallee, B.L., N-Acetylimidazole: a reagent for determination of "free" tyrosyl residues of proteins, *Biochemistry* 4, 1758, 1965.

46. Myers, B., II, and Glazer, A.N., Spectroscopic studies of the exposure of tyrosine residues in proteins with special reference to the subtilisins, *J. Biol. Chem.* 246, 412, 1971.

47. Cronan, J.E., Jr., and Klages, A.L., Chemical synthesis of acyl thioesters of acyl carrier protein with native structure, *Proc. Natl. Acad. Sci. USA* 78, 5440, 1981.

48. Lundblad, R.L., The reaction of bovine thrombin with N-butyrylimidazole. Two different reactions resulting in the inhibition of catalytic activity, *Biochemistry* 14, 1033, 1975.

48a. Lundblad, R.L., Observations of the hydrolysis of p-nitrophenyl acylates by purified bovine thrombin, *Thromb. Diath. Haemorrh.* 30, 248, 1973.

49. El Kabbaj, M.S., and Latruffe, N., Chemical reagents of polypeptide side chains. Relationship between solubility properties and ability to cross the inner mitochondrial membranes, *Cell. Mol. Biol.* 40, 41, 1994.

50. Martin, B.L. et al., Chemical influences on the specificity of tyrosine phosphorylation, *J. Biol. Chem.* 265, 7108, 1990.

51. Ji, Y., and Bennett, B.M., Activation of microsomal glutathione S-transferase by peroxynitrite, *Mol. Pharmacol.* 63, 136, 2003.

52. Tildon, J.T., and Ogilvie, J.W., The esterase activity of bovine mercaptalbumin. The reaction of the protein with p-nitrophenyl acetate, *J. Biol. Chem.* 247, 1265, 1972.

53. Balls, A.K., and Wood, H.N., Acetyl chymotrypsin and its reaction with ethanol, *J. Biol. Chem.* 219, 245, 1955.

54. Riordan, J.F., and Vallee, B.L., O-Acetyltyrosine, *Methods Enzymol.* 25, 500, 1972.

55. Herriott, R.M., Reactions of native proteins with chemical reagents, *Adv. Protein Chem.* 3, 169, 1947.

56. Riordan, J.F., Sokolovsky, M., and Vallee, B.L., Tetranitromethane. A reagent for the nitration of tyrosine and tyrosyl residues in proteins, *J. Am. Chem. Soc.* 88, 4104, 1966.

57. Sokolovsky, M., Harell, D., and Riordan, J.F., Reaction of tetranitromethane with sulfhydryl groups in proteins, *Biochemistry* 8, 4740, 1969.

58. Riordan, J.F., and Vallee, B.L., Nitration with tetranitromethane, *Methods Enzymol.* 25, 515, 1972.

59. Cuatrecasas, P., Fuchs, S., and Anfinsen, C.B., The tyrosyl residues at the active site of staphylococcal nuclease. Modifications by tetranitromethane, *J. Biol. Chem.* 243, 4787, 1968.

60. Spande, T.F., Fontana, A., and Witkop, B., An unusual reaction of skatole with tetranitromethane, *J. Am. Chem. Soc.* 91, 6169, 1969.

61. Riggle, W.L., Long, J.A., and Borders, C.L., Jr., Reaction of turkey egg-white lysozyme with tetranitromethane. Modification of tyrosine and tryptophan, *Can. J. Biochem.* 51, 1433, 1973.

62. Sokolovsky, M., Fuchs, M., and Riordan, J.F., Reaction of tetranitromethane with tryptophan and related compounds, *FEBS Lett.* 7, 167, 1970.

63. Teuwissen, G., Masson, P.L., Osinski, P., and Heremans, J.F., Metal-combining properties of human lactoferrin. The effect of nitration of lactoferrin with tetranitromethane, *Eur. J. Biochem.* 35, 366, 1973.

64. Katsura, T., Lam, E., Packer, L., and Seltzer, S., Light dependent modification of bacteriorhodopsin by tetranitromethane. Interaction of a tyrosine and a tryptophan residue with bound retinal, *Biochem. Int.* 5, 445, 1982.

65. Atassi, M.Z., and Habeeb, A.F.S.A., Enzymic and immunochemical properties of lysozyme. I. Derivatives modified at tyrosine. Influence of nature of modification on activity, *Biochemistry* 8, 1385, 1969.
66. Haddad, I.Y. et al., Nitration of a surfactant protein A results in decreased ability to aggregate lipids, *Am. J. Physiol.* 270, L281, 1996.
67. Doyle, R.J., Bello, J., and Roholt, O.A., Probable protein crosslinking with tetranitromethane, *Biochim. Biophys. Acta* 160, 274, 1970.
68. Boesel, R.W., and Carpenter, F.H., Crosslinking during the nitration of bovine insulin with tetranitromethane, *Biochem. Biophys. Res. Commn.* 38, 678, 1970.
69. Kunkel, G.R., Mehrabian, M., and Martinson, H.G., Contact-site cross-linking agents, *Mol. Cell. Biochem.* 34, 3, 1981.
70. Nadeau, O.W., Traxler, K.W., and Carlson, G.M., Zero-length crosslinking of the beta subunit of phosphorylase kinase to the N-terminal half of its regulatory alpha subunit, *Biochem. Biophys. Res. Commn.* 251, 637, 1998.
71. Bruice, T.C., Gregory, M.J., and Walters, S.L., Reactions of tetranitromethane. I. Kinetics and mechanism of nitration of phenols by tetranitromethane. *J. Am. Chem. Soc.* 90, 1612, 1968.
72. Hugli, T.E., and Stein, W.H., Involvement of a tyrosine residue in the activity of bovine pancreatic deoxyribonuclease A, *J. Biol. Chem.* 246, 7191, 1971.
73. Margolis, S.A., Coxon, B., Gajewski, E., and Dizdaroglu, M., Structure of a hydroxyl radical induced cross-link of thymine and tyrosine, *Biochemistry* 27, 6353, 1988.
74. Hsuan, J.J., The cross-linking of tyrosine residues in apo-ovotransferrin by treatment with periodate anions, *Biochem. J.* 247, 467, 1987.
75. Crow, J.P., and Ishiropoulos, H., Detection and quantitation of nitrotyrosine residues in proteins: *in vivo* marker of peroxynitrite, *Methods Enzymol.* 269, 185, 1996.
76. Greenacre, S.A.B., and Ischeriopoulos, H., Tyrosine nitration: localisation, quantification, consequences for protein function and signal transduction, *Free Rad. Res.* 34, 541, 2001.
77. Schmidt, P. et al., Specific nitration at tyrosine 430 revealed by high resolution mass spectrometry as basis for redox regulation of bovine prostacyclin synthase, *J. Biol. Chem.* 278, 12813, 2003.
78. Peterson, A.-S. et al., Investigation of tyrosine nitration in proteins by mass spectrometry, *J. Mass Spectrom.* 36, 616, 2001.
79. Willard, B.B. et al., Site-specific quantitation of protein nitration using liquid chromatography/tandem mass spectrometry, *Anal. Chem.* 75, 2370, 2003.
80. Daiber, A. et al., A new pitfall in detecting biological end products of nitric oxide: nitration, nitros(yl)ation and nitrite/nitrate artifacts during freezing, *Nitric Oxide* 9, 44, 2003.
81. Irie, Y. et al., Histone H1.2 is a substrate for dinitrase, an activity that reduces nitrotyrosine immunoreactivity in proteins, *Proc. Natl. Acad. Sci. USA* 100, 5634, 2003.
82. Nikov, G. et al., Analysis of nitrated proteins by nitrotyrosine-specific affinity probes and mass spectrometry, *Anal. Biochem.* 320, 214, 2003.
83. Ogino, K. et al., Immunohistochemical artifact for nitrotyrosine in eosinophils or eosinophil containing tissue, *Free Radic. Res.* 36, 1163, 2002.
84. Miyagi, M. et al., Evidence that light modulates protein nitration in rat retina, *Mol. Cell. Proteom.* 1, 293, 2003.
85. Zu, Y. et al., Antibodies that recognize nitrotyrosine, *Methods Enzymol.* 269, 201, 1996.

86. Sokolovsky, M., Riordan, J.F., and Vallee, B.L., Conversion of 3-nitrotyrosine to 3-aminotyrosine in peptides and proteins, *Biochem. Biophys. Res. Commn.* 27, 20, 1967.
87. Riordan, J.F., Sokolovsky, M., and Vallee, B.L., Environmentally sensitive tyrosyl residues. Nitration with tetranitromethane, *Biochemistry* 6, 358, 1967.
88. Haas, J.A., Frederick, M.A., and Fox, B.G., Chemical and post-translational modifications of *Escherichia coli* acyl carrier protein for preparation of dansyl carrier protein, *Protein Exp. Purif.* 20, 274, 2000.
89. Haas, J.A., and Fox, B.G., Fluorescence anisotropy studies of enzyme-substrate complex formation in stearoyl-ACP-desaturase, *Biochemistry* 41, 14472, 2002.
90. Riordan, J.F., Sokolovsky, M., and Vallee, B.L., The functional tyrosyl residues of carboxypeptidase A. Nitration with tetranitromethane, *Biochemistry* 6, 3609, 1967.
91. Skov, K., Hofmann, T., and Williams, G.R., The nitration of cytochrome c, *Can. J. Biochem.* 47, 750, 1969.
92. Lide, D.R. (Ed.), *Handbook of Chemistry and Physics*, 82nd ed. CRC Press, Boca Raton, FL, pp. 3–207, 2002.
93. Liang, P., Tetranitromethane, *Org. Synth. Coll.* 3, 803, 1955.
94 Fendler, J.H., and Liechti, R.R., Miceller catalysis of the reaction of hydroxide ion with tetranitromethane, *J. Chem. Soc. Perkins* II, 1041, 1972.
95. Walters, S.L., and Bruice, T.C., Reactions of tetranitromethane. II. Kinetics and products for the reactions of tetranitromethane with inorganic ions and alcohols, *J. Am. Chem. Soc.* 93, 2269, 1971.
96. Eberson, L., and Hartshorn, M.P., The formation and reactions of adducts from the photochemical reactions of aromatic compounds with tetranitromethane and other X-NO$_2$ reagents, *Aust. J. Chem.* 51, 1061, 1998.
97. Rasmusseon, M. et al., Ultrafast formation of trinitromethanide (C(NO$_2$)$_3^-$) by photoinduced dissociative electron transfer and subsequent ion pair coupling reaction in acetonitrile and dichloromethane, *J. Phys. Chem. B* 105, 2027, 2001.
98. Scherrer, P., and Stoeckenius, W., Selective nitration of tyrosine-26 and -64 in bacteriorhodopsin with tetranitromethane, *Biochemistry* 23, 6195, 1984.
99. Renthal, R. et al., Light activates the reaction of bacteriorhopsin aspartic acid-115 with dicyclohexylcarbodiimide, *Biochemistry* 24, 4275, 1985.
100. Capellos, C. et al., Transient species and product formation from electronically excited tetranitromethane, *J. Chem. Soc. Faraday Trans.* 2, 82, 2195, 1986.
101. Eberson, L., Hartshorn, M.P., and Persson, O., The reactions of some dienes with tetranitromethane, *Acta Chem. Scand.* 52, 450, 1998.
102. Fieser, M., and Fieser, L., *Reagents for Organic Synthesis*, Vol 1. Wiley Interscience, New York, 1967, p. 1147.
103. Jewett, S.W., and Bruice, T.C., Reactions of tetranitromethane. Mechanisms of the reaction of tetranitromethane with pseudo acids, *Biochemistry* 11, 3338, 1972.
104. Strosberg, A.D., Van Hoeck, B., and Kanarek, L., Immunochemical studies on hen's egg white lysozyme. Effect of selective nitration of the three tyrosine residues, *Eur. J. Biochem.* 19, 36, 1971.
105. Richards, P.G., Walton, D.J., and Heptinstall, J., The effects of nitration on the structure and function of hen egg-white lysozyme, *Biochem. J.* 315, 473, 1996.
106. Richards, P.G., Coles, B., Heptinstall, J., and Walton, D.J., Electrochemical modification of lysozyme: anodic reaction of tyrosine residues, *Enzyme Microb. Technol.* 16, 795, 1994.

107. Izzo, G.E., Jordan, F., and Mendelsohn, R., Resonance raman and 500-MHz ^{1}H NMR studies of tyrosine modification in hen egg white lysozyme, *J. Amer. Chem. Soc.* 104, 3178, 1982.

108. Brown, L.S., Reconciling crystallography and mutagenesis: a synthetic approach to the creation of a comprehensive model for proton pumping by bacteriorhopsin, *Biochim. Biophys. Acta* 1460, 49, 2000.

109. Scherrer, P., and Stoeckenius, W., Effects of tyrosine-26 and tyrosine-64 nitration on the photoreactions of bacteriorhodopsin, *Biochemistry* 24, 7733, 1985.

110. Mayo, K.H. et al., Mobility and solvent exposure of aromatic residues in bacterio-rhodopsin investigated by ^{1}H-NMR and photo-CIDNP-NMR spectroscopy, *FEBS Lett.* 235, 163, 1988.

111. Lecomte, J.T.J., and Llinás, M., Characterization of the aromatic proton magnetic resonance spectrum of crambin, *Biochemistry* 23, 4799, 1984.

112. Lee, S.-B., Inouye, K., and Tomura, B., The states of tyrosyl residues in thrermolysin as examined by nitration and pH-dependent ionization, *J. Biochem.* 121, 231, 1997.

113. Ploug, M. et al., Chemical modification of the urokinase-type plasminogen activator and its receptor using tetranitromethane. Evidence for the involvement of specific tyrosine residues in both molecules during receptor-ligand interaction, *Biochemistry* 34, 12524, 1995.

114. Zappacosta, F. et al., Surface topology of minibody by selective chemical modifications and mass spectrometry, *Protein Sci.* 6, 1901, 1997.

115. Leite, J.F., and Cascio, M., Probing the topology of the glycine receptor by chemical modification couple to mass spectrometry, *Biochemistry* 41, 6140, 2002.

116. Cascio, M., Glycine receptors: lessons on topology and structural effects of the lipid bilayer, *Biopolymers* 66, 359, 2002.

117. D'Ambrosio, C. et al., Probing the dimeric structure of porcine aminoacylase 1 by mass spectrometric and modeling procedures, *Biochemistry* 42, 4430, 2003.

118. Christen, P., and Riordan, J.F., Syncatalytic modification of a functional tyrosyl residue in aspartate aminotransferase, *Biochemistry* 9, 3025, 1970.

119. Kirschner, M.W., and Schachman, H.K., Conformational studies on the nitrated catalytic subunit of aspartate transcarbamylase, *Biochemistry* 12, 2987, 1973.

120. Tawfik, D.S. et al., pH on-off switching of antibody–hapten binding by site-specific chemical modification of proteins, *Protein Eng.* 7, 431, 1994.

121. Morag, E., Bayer, E.A., and Wilchek, M., Reversibility of biotin-binding by selective modification of tyrosine in avidin, *Biochem J.* 316, 193, 1996.

122. Rischel, C., and Poulsen, F.M., Modification of a specific tyrosine enables tracing of the end-to-end distance during apomyoglobin folding, *FEBS Lett.* 374, 105, 1995.

123. Robertson, D.L. et al., Influence of active site and tyrosine modification on the secretion and activity of the *Aeromonas hydrophilia* lipase/acyltransferase, *J Biol. Chem.* 269, 2146, 1994.

124. Trent, M.S., Washam, L.M.S., and Ernst-Fonberg, M.L., HlyC, the internal protein acyltransferase that activates hemolysis toxin: the role of conserved tyrosine and arginine residues in enzymatic activity as probed by chemical modification and site-directed mutagenesis, *Biochemistry* 38, 8831, 1999.

125. Gadda, G., Banerjee, A., and Fitzpatrick, P.F., Identification of an essential tyrosine residue in nitroalkane oxidase by modification with tetranitromethane, *Biochemistry* 39, 1162, 2000.

126. Mailfait, S. et al., Critical role of tyrosine 277 in the ligand-binding and transactivating properties of retinoic acid receptor α, *Biochemistry* 39, 2183, 2000.

127. Kim, H.S., Austin, J., and Hage, D.S., Identification of drug-binding sites on human serum albumin using affinity capillary electrophoresis and chemically modified proteins as buffer additives, *Electrophoresis* 23, 956, 2002.

128. Fehske, K.J., Müller, W.E., and Wollert, U., A highly reactive tyrosine residue as part of the indole binding and benzodiazepine binding site of human serum albumin, *Biochim. Biophys. Acta* 577, 346, 1979.

129. Li, J., Bigelow, D.J., and Squier, T.C., Phosphorylation by cAMP-dependent protein kinase modulates the structural coupling between the transmembrane and cytosolic domains of phospholamban, *Biochemistry* 42, 10674, 2003.

130. Beckman, J.S. et al., Oxidative chemistry of peroxynitrite, *Methods Enzymol.* 233, 229, 1994.

131. Pryor, W.A. et al., A practical method for preparing peroxynitrite solutions of low ionic strength and free of hydrogen peroxide, *Free Rad. Biol. Med.* 18, 75, 1995.

132. Uppu, R.M. et al., Selecting the most appropriate synthesis of peroxynitrite, *Methods Enzymol.* 269, 285, 1996.

133. Ischiropoulos, H., Biological tyrosine nitration: a pathophysiological function of nitric oxide and reactive oxygen species, *Arch. Biochem. Biophys.* 356, 1, 1998.

134. Ischiropoulos, H., Biological selectivity and functional aspects of protein tyrosine nitration, *Biochem. Biophys. Res. Commn.* 305, 776, 2003.

135. Zhang, H. et al., Transmigration nitration of hydrophobic tyrosyl peptides. Localization, characterization, mechanisms of nitration, and biological implications, *J. Biol. Chem.* 278, 8969, 2003.

136. Kurihara, K., Horinishi, H., and Shibata, K., Reaction of cyanuric halides with proteins. I. Bound tyrosine residues of insulin and lysozyme as identified with cyanuric fluoride, *Biochim. Biophys. Acta* 74, 678, 1963.

137. Gorbunoff, M.J., Cyanuration, *Methods Enzymol.* 25, 506, 1972.

138. Gorbunoff, M.J., and Timasheff, S.N., The role of tyrosines in elastase, *Arch. Biochem. Biophys.* 152, 413, 1972.

139. Coffe, G., and Pudles, J., Chemical reactivity of the tyrosyl residues in yeast hexokinase. Properties of the nitroenzyme, *Biochim. Biophys. Acta* 484, 322, 1977.

140. Ferguson, S.J., Lloyd, W.J., Lyons, M.H., and Radda, G.K., The mitochondrial ATPase. Evidence for a single essential tyrosine residue, *Eur. J. Biochem.* 54, 117, 1975.

141. Ferguson, S.J., Lloyd, W.J., and Radda, G.K., The mitochondrial ATPase. Selective modification of a nitrogen residue in the β-subunit, *Eur. J. Biochem.* 54, 127, 1975.

142. Sesaki, H., Wong, E.F., and Siu, C.H., The cell adhesion molecule DdCAD-1 in *Dicytostelium* is targeted to the cell surface by a nonclassical transport pathway involving contractile vacuoles, *J. Cell. Biol.* 138, 939, 1997.

143. Denu, J.M., and Tanner, K.G., Specific and reversible inactivation of protein tyrosine phophatases by hydrogen peroxide: evidence for a sulfenic acid intermediate and implications for redox regulation, *Biochemistry* 37, 5633, 1998.

144. Nieslanik, B.S., and Atkins, W.M., The catalytic Tyr-9 of glutathione S-transferase A1-1 controls the dynamics of the C terminus, *J. Biol. Chem.* 275, 17447, 2000.

145. Andrews, W.W., and Allison, W.S., 1-Fluoro-2,4-dinitrobenzene modifies a tyrosine residue when it inactivates the bovine mitochondrial F_1-ATPase, *Biochem. Biophys. Res. Commn.* 99, 813, 1981.

146. De Vendittis, E. et al., Phenylmethylsulfonyl fluoride inactivates an archael superoxide dismutase by chemical modification of a specific tyrosine residue. Cloning, sequencing and expression of the gene coding for *Sulfolobus solfataricus* superoxide dismutase, *Eur. J. Biochem.* 268, 1794, 2001.

147. Means, G.E., and Wu, H.L., The reactive tyrosine residue of human serum albumin: characterization of its reaction with diisopropylfluorophosphate, *Arch. Biochem. Biophys.* 194, 526, 1979.

148. de Almeida Olivera, M.G. et al., Tyrosine 151 is part of the substrate activation binding site of bovine trypsin. Identification by covalent labeling with *p*-diazonium-benzamidine and kinetic characterization of Tyr-151-(*p*-benzamidino)-azo-β-trypsin, *J. Biol. Chem.* 268, 26893, 1993.

149. Landsteiner, K., *The Specificity of Serological Reactions.* Harvard University Press, Cambridge, 1945.

150. Fraenkel-Conrat, H., Bean, R.S., and Lineweaver, H., Essential groups for the interaction of ovomucoid (egg white trypsin inhibitor) and trypsin, and for tryptic activity, *J. Biol. Chem.* 177, 385, 1949.

151. Riordan, J.F., and Vallee, B.L., Diazonium salts as specific reagents and probes of protein conformation, *Methods Enzymol.* 25, 521, 1972.

152. Tabachnick, M., and Sobotka, H., Azoproteins. I. Spectrophotometric studies of amino acid azo derivatives, *J. Biol. Chem.* 234, 1726, 1959.

153. Tabachnick, M., and Sobotka, H., Azoproteins. II. A spectrophotometric study of the coupling of diazotized arsanilic acid with proteins, *J. Biol. Chem.* 235, 1051, 1960.

154. Fairclough, G.F., Jr., and Vallee, B.L., Arsanilazochymotrypsinogen. The extrinsic Cotton effects of an arsanilazotyrosyl chromophore as a conformation probe of zymogen activation, *Biochemistry* 10, 2470, 1971.

155. Gorecki, M., Wilchek, M., and Blumberg, S., Modulation of the catalytic properties of α-chymotrypsin by chemical modification at Tyr 146, *Biochim. Biophys. Acta* 535, 90, 1978.

156. Robinson, G.W., Reaction of a specific tryptophan residue in streptococcal proteinase with 2-hydroxy-5-nitrobenzyl bromide, *J. Biol. Chem.* 245, 4832, 1970.

157. Gorecki, M., and Wilchek, M., Modification of a specific tyrosine residue of ribonuclease A with a diazonium inhibitor analog, *Biochim. Biophys. Acta* 532, 81, 1978.

158. Johansen, J.T., Livingston, D.M., and Vallee, B.L., Chemical modification of carboxypeptidase A crystals. Azo coupling with tyrosine-248, *Biochemistry* 11, 2584, 1972.

159. Harrison, L.W., and Vallee, B.L., Kinetics of substrate and product interactions with arsanilazotyrosine-248 carboxypeptidase A, *Biochemistry* 17, 4359, 1978.

160. Cueni, L., and Riordan, J.F., Functional tyrosyl residues of carboxypeptidase A. The effect of protein structure on the reactivity of tyrosine-198, *Biochemistry* 17, 1834, 1978.

161. Liao, T.-H., Ting, R.S., and Young, J.E., Reactivity of tyrosine in bovine pancreatic deoxyribonuclease with *p*-nitrobenzenesulfonyl fluoride, *J. Biol. Chem.* 257, 5637, 1982.

162. Haniu, M., Yuan, H., Chen, S., Iyanagi, T., Lee, T.D., and Shively, J.E., Structure-function relationship of NAD(P)H:quinone reductase: characterization of NH_2-terminal blocking group and essential tyrosine and lysine residues, *Biochemistry* 27, 6877, 1988.

163. Kulanthaivel, P., Leibach, F.H., Mahesh, V.B., and Ganapathy, V., Tyrosine residues are essential for the activity of the human placental taurine transporter, *Biochim. Biophys. Acta* 985, 139, 1989.

10 The Chemical Modification of Tryptophan

The specific chemical modification of tryptophan in proteins is one of the more challenging problems in protein chemistry. First, solvent conditions for providing specificity of modification are, in general, somewhat harsh. This, however, is not an absolute and it possible to obtain satisfactory results under mild reaction conditions with some proteins. Second, there is considerable possibility of either the concomitant or the separate modification of other amino acid residues. Third, analysis for the determination of the exact extent of modification requires a rigorous approach combining spectral analysis and amino acid analysis[1-3] after hydrolysis in a solvent that will not destroy tryptophan. Friedman has written an excellent review[4] of the measurement of tryptophan by amino acid analysis, with a discussion of the specific reaction of ninhydrin with tryptophan.

Treatment of tryptophan with hydrogen peroxide results in oxidation of the indole ring.[5-8] Usually, the reaction is performed at alkaline pH (1.0 M sodium bicarbonate, pH 8.4) with the H_2O_2/dioxane mixture prepared as described by Hachimori and co-workers.[5] The loss of tryptophan is monitored by the change in absorbance at 280 nm.[5,7,8] The difference in the molar extinction coefficient between tryptophan and the fully oxidized derivative is 3490 $M^{-1}cm^{-1}$. Underberg and colleagues[9] reviewed the methods for the qualitative and quantitative analysis of tryptophan oxidation in peptides and proteins, including UV spectroscopy, fluorescence, and HPLC analysis. HPLC analysis of tryptophan oxidation products has been described.[10] Details are provided for the separation of kynureine, 5-hydroxytryptophan, tryptophan, and dioxindolealanine on a C18 column. The reader is directed to an excellent study by Mach and co-workers[11] for extinction coefficients of tryptophan, tyrosine, and cystine in proteins. Fluorescence has also been used to measure tryptophan modification in proteins.[12] Reshetanyak and colleagues[13] recently reviewed the fluorescence properties of tryptophan in proteins.

The reaction of N-bromosuccinimide (NBS) with tryptophan (Figure 10.1) continues to be a useful method for the site-specific chemical modification of this residue in proteins.[14-25] NBS is only rarely used at present for the quantitative analysis of tryptophan in proteins. For the quantitative determination of the tryptophan content in proteins, the basic approach is to add NBS to the protein (pH 4.0 to 5.0, usually acetate buffers) until there is no further decrease in absorbance at 280 nm. The change in the molar extinction coefficient of tryptophan on conversion to the oxindole derivative is taken to be 4×10^3 M^{-1} cm^{-1}.[26] It has been our experience that the reaction must be performed either in 8.0 M urea (pH adjusted to 4.0) or with the

247

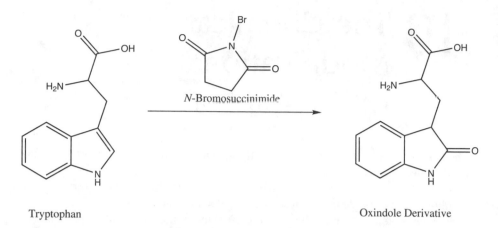

Tryptophan   Oxindole Derivative

FIGURE 10.1 Reaction of N-bromosuccinimide with tryptophan.

reduced, carboxymethylated derivative.[27] The spectra must be obtained as soon as possible after adding NBS, because unless the excess reagent and low-molecular-weight products of the reaction are rapidly removed, there can be reversal of the decrease in absorbance.[28] This is not a trivial consideration, because in at least one study[20] there is a real difference in the extent of modification as determined by spectroscopy and amino acid analysis. The rigorous evaluation[28] of the reaction of NBS with model tryptophanyl and tyrosyl compounds reported from Keitaro Hiromi's laboratory provides considerable insight into the problems to be encountered with the study of intact proteins. At ratios of NBS to N-acetyltryptophan ethyl ester higher than 2, there is an apparent reversal of the decrease in absorbance at 280 nm. The maximal decrease in absorbance occurs at a ratio of NBS to tryptophan of 2. If data are obtained by stopped-flow spectroscopy, the molar excess of NBS does not have an effect on the maximum decrease observed, but when the spectrum is obtained 5 min after the initiation of the reaction, there is a decrease in the observed magnitude of change in absorbance at 280 nm. Evaluation of spectral changes in a protein is further complicated by the reaction of NBS with tyrosine. These observations do not appear to be considered by current investigators and probably deserve consideration in experimental design.

The use of spectroscopy after reaction with NBS for the analysis of tryptophan content in proteins has been largely supplanted by ion-exchange analysis following modified hydrolytic procedures.[1,2,4,29] However, the acid ninhydrin method developed by Gaitonde and Davey[30] for the determination of tryptophan in proteins and peptides deserves comment. The original technique was modified by Sodak and others[31] and more recently by Molnár-Perl and Pintér-Szakáks.[32] Although not used extensively for purified proteins, it has received considerable use in food chemistry.[33,34] Nagaraja and co-workers[35] described a very sensitive system for the determination of tryptophan, based on reaction with diazotized p-phenylenediamine in sulfuric acid forming a product absorbing at 520 nm. Absorbance is linear with tryptophan concentration in the range of 0.25 to 11 μg/ml.

NBS continues to be used for the site-specific modification of tryptophan in proteins. In general, the modification reaction is performed in 0.1 M sodium acetate, pH 4 to 5. The NBS should be recrystallized from water before use. The presence of halides such as chloride or bromide in the solvent must be avoided, because the addition of NBS will oxidize these ions to the elemental form with disastrous and irreproducible effect on the proteins under study. In general, a twofold molar excess of NBS per mole of tryptophan is necessary to achieve modification. Daniel and Trowbridge[36] found that (at pH 4.0) the reaction of NBS with acetyl-L-tryptophan ethyl ester required 1.5 Mol of NBS per mole of the acetyl-L-tryptophan ethyl ester, whereas trypsinogen required 2.0 to 2.3 Mol NBS per mole of tryptophan oxidized and trypsin required 1.5 to 2.0 Mol NBS per mole of tryptophan oxidized. At pH 4.0, only tryptophan in dihydrofolate reductase reacts with NBS, whereas at pH 6.0 a sulfhydryl group apparently is preferentially oxidized by the reagent before the reaction of tryptophan.[37] Poulos and Price reported on the reaction of a tryptophanyl residue in bovine pancreatic DNase and NBS.[38] Prior reaction of the DNase with another tryptophan reagent, 2-hydroxy-5-nitrobenzyl bromide, modified a residue different from the one modified by NBS. These investigators used spectral analysis to determine the extent of tryptophan modification. Subsequent studies from another laboratory[15] on the modification of DNase with NBS suggested that apparently 2 Mol of tryptophan are modified per mole of enzyme at 100% inactivation with a sixfold molar excess of NBS in 0.01 M CaCl$_2$ at pH 4.0. By amino acid analysis (after hydrolysis in 6 N HCl containing mercaptoacetic acid, phenol, and 3-(2-aminoethyl) indole for 24 h at 110°C), these investigators showed that all three tryptophanyl residues are modified under these experimental conditions. The study on the modification of tryptophan in galactose oxidase[39] is worth comment in that these investigators report the amino acid composition of the modified protein after hydrolysis in 3 N p-toluenesulfonic acid. There was excellent agreement between the extent of tryptophan modification as judged by direct amino acid analysis and the value observed by spectral analyses.

Several other facets of the use of NBS for the modification of tryptophanyl residue in proteins should be considered. The use of the reagent at mildly acidic pH has been mentioned previously. Increasing the pH not only increases the possibility of modification at amino acids other than tryptophan, but there is also a decrease in the modification of tryptophan. This is shown by the studies[40] on the modification of a glucoamylase from *Aspergillus saitoi*. There is a modest decrease in modification as the pH is increased from 4.0 to 6.0, with a dramatic decrease at pH 7.0. At pH values close to neutrality, there is the increased possibility of modification of amino residues other than tryptophan. The pH dependence of the NBS tryptophan modification might well reflect local conformational effect rather than intrinsic chemistry. In studies[41] on the reaction on *Escherichia coli* lac repressor protein with NBS at pH 7.8 (1.0 M Tris), cysteine was modified as readily as tryptophan, with lesser modification of methionine and tyrosine. The great majority of NBS modifications of proteins are performed at pH values less than 5.0 to avoid modification of other functional groups, but success can be achieved under less acidic conditions. Kumar and co-workers[42] modified tryptophanyl residues in transcarboxylase in 0.25 M

potassium phosphate, pH 6.5, containing 0.1 M dithiothreitol and 0.1 M phenyl-methylsulfonyl fluoride. They also demonstrated that the loss of activity on modification was not a reflection of gross conformational change by examination of the quenching of intrinsic fluorescence and changes in the susceptibility to tryptic cleavage. In general, modification should occur at a four to sixfold molar excess (with respect to total tryptophan) of NBS. Under most reaction conditions, modification of tryptophanyl residues with NBS is quite rapid. Time-dependent reactions have, however, been observed, such as that reported for xylanase.[43] Reaction with 2-hydroxy-5-nitrobenzyl bromide was also time dependent under these reaction conditions.

Bray and Clarke[44] used NBS to evaluate the role of tryptophan in *Schizophyllum coomune* xylanase. It was possible to eliminate the importance of tryptophan in substrate binding, whereas modification with tetranitromethane indicated a role for tyrosine in the binding of xylooligosaccharides. A fourth-derivative spectroscopic method was used to quantitate modified tyrosine and tryptophan residues.[19] Takahashi and co-workers[45,46] used the reaction with NBS to identify two tryptophan residues in the leader peptidase from *Escherichia coli* important for catalytic activity. Ray and co-workers[47] used modification with NBS to identify tryptophan as the exclusive aromatic donor in fluorescence transfer to NDAd in a UPD-glucose-4-epimerase from *Kluyveromyces fragilis*.

Xue and co-workers[48] studied the reaction of NBS with tryptophanyl-tRNA synthetase (25 mM sodium acetate, pH 4.5, 22°C). This is an extremely interesting study, as a large molar excess of NBS to enzyme (1500-fold) is required for complete inactivation. An even larger molar excess of BPNS-skatole[3-bromo-3-methyl-3-(2-nitromercapto)-3H-indole] is required for inactivation. Tryptophanyl-tRNA synthetase has a single tryptophan (Trp-92), which is buried in the interface between the subunits of this dimeric enzyme.[49] Addition of ATP to the modification reaction greatly increases the rate of inactivation, reflecting a structural change in the protein.

Ash and co-workers[50] studied the reaction of rat liver arginase with NBS. The reaction was performed in 100 mM potassium acetate, pH 5.0, and terminated by dilution with pH 9.0 CHES buffer. Enzyme inactivation occurs under these conditions. However, oligonucleotide-directed mutagenesis (W122F, W164F) suggested that a functional tryptophan was not required for enzyme activation. Further study of mutant proteins demonstrated that treatment of a histidine mutant (H141N) with NBS did not result the loss of enzyme activity. It is suggested that modification of this histidine residue by NBS is responsible for the loss of activity. These investigators also observed that peptide bond cleavage was observed with a molar excess (NBS to protein) of 35 or higher. Clottes and Vial[51] used the reaction with NBS (10 mM sodium acetate, pH 5.5, 10 min, room temperature; reaction stopped by adding a twofold molar excess of free tryptophan) to study the individual contribution of the four tryptophan residues to the intrinsic fluorescence of MM-creatinine kinase. The quantum yield of each of the tryptophan residues was different, reflecting their individual microenvironments. Josse and co-workers[52] used NBS to study the role of tryptophan in human serum paraoxonase catalysis. Whereas reaction with NBS resulted the loss of catalytic activity, reaction with 2-hydroxy-5-nitrobenzyl bromide had no effect.

Soulages and Arrese[53] prepared five different single-tryptophan mutants of apophorin III to study the lipid-binding domains. Fluorescence quenching was measured for each of the five mutant proteins on interaction with 1-palmitoyl-2-oleoyl-*sn*-glycerol-3-phosphocholine in discoid lipoprotein micelles. NBS (50 mM Tris, pH 7.8; fivefold molar excess to protein) abolished the fluorescence of all derivatives, indicating that none of the inserted tryptophan residues resided in a transmembrane state.

Deshpande and colleague[54] studied the modification of an alkaline protease (*Conidiobolus* sp.) by NBS (100 mM sodium acetate, pH 5.0; 25°C, 20 min, 0- to 13.3-fold molar excess) and determined that three tryptophan residues were available for modification under these conditions. UV-circular dichroism studies demonstrated the absence of gross structural modification as a consequence of modification. The intrinsic fluorescence was decreased as result of modification under these conditions. There was no change in the fluorescence profile for modified protein (excitation, 295 nm; emission, 340 nm). Reaction with NBS under different solvent conditions (100 mM potassium phosphate, pH 7.5) demonstrated that loss of catalytic activity was associated with the modification of one mole of tryptophan per mole of enzyme, with a second-order rate constant of $3.45 \times 10^3\ M^{-1}\ min^{-1}$. Other studies suggested that a histidine residue provided an electropositive environment for the modified tryptophan residue. The enzyme activity inactivation studies were performed at pH 7.5 compared with the titration studies that were performed at pH 5.0. Although raising the pH might result in the increased modification of other amino acid residues, it is also possible that there is an increase in reaction specificity as result of the more alkaline pH. Yamaguchi and colleagues[55] studied the reaction of NBS with an α-amylase inhibitor from white kidney bean (*Phaseolus vulgaris*). There are two observations of great importance to investigators who use NBS to modify proteins. First, it is an excellent example of the modification of tyrosine by NBS under conditions usually considered specific for tryptophan (pH 5.0). Second, the modification of tryptophan is linear, with pH decreasing as pH increases. The effect of pH on the modification of tyrosine is a little more complex, with no modification observed at pH 5.0. Modification of tyrosine increases with both increasing and decreasing pH. Table 10.1 summarizes the use of NBS in the study of proteins.

The reaction of NBS with proteins can also result in the cleavage of peptide bonds at tryptophan, tyrosine, and histidine.[56] Thus, the careful investigator will also evaluate the integrity of the polypeptide chains of the protein of interest.[50] Whereas peptide bond cleavage is usually an unwanted side reaction, Feldhoff and Peters[57] have devised a procedure that has enhanced specificity for tryptophan. Their procedure uses 8.0 M urea, 2.0 M acetic acid as the solvent with a 20-fold molar excess of NBS. Their approach offers at least two advantages: (1) the protein is denatured so that all residues are equally available, and (2) the NBS reacts with urea to yield *N*-bromourea, a less severe oxidizing agent, which should have increased specificity for tryptophanyl residues. The use of *N*-chlorosuccinimide for peptide bond cleavage of tryptophanyl residues is considered a superior approach.[58] A direct comparison of NBS and *N*-chlorosuccinimide in the modification of the single tryptophanyl residue in the *Clostridium perfringes* α-toxin has been made.[59] Modification with NBS (50 mM sodium acetate, pH 5.0) resulted in the total loss of tryptophan and

TABLE 10.1
Some Examples of the Use of N-Bromosuccinimide (NBS) for the Modification of Proteins

Protein	Reaction Conditions	Molar Excess[a]	Ref.
Trypsinogen	pH 7.0	1–4	1
Trypsin	pH 4.0	1–4	1
Dihydrofolate reductase	0.1 M sodium phosphate, pH 6.0	15	2
	0.1 M sodium acetate, pH 4.0	12	2
Bovine pancreatic DNase	0.1 M sodium acetate, pH 4.0, containing 0.033 M CaCl$_2$	6	3
Bovine pancreatic DNase	pH 4.0, 0.010 M CaCl$_2$	1–6	4
Dihydrofolate reductase	0.05 M potassium phosphate, pH 6.5	20	5
Pyrocatechas (*Brevibacterium fuscum*)	0.1 M phosphate, pH 7.0	—	6
Relaxin	0.2 M sodium acetate, pH 4.7	—	7
Rhodopsin	0.1 M Tris acetate, pH 7.4, containing 1% emulphogen	50	8
		100	8
Pig kidney amino acylase	0.1 M sodium acetate, pH 5.0, 1.0 M urea	50	9
Galactose oxidase	0.005 M sodium acetate, pH 4.15	7	10
Bovine thrombin	0.1 M sodium acetate, pH 4.0	1	11
		2	11
Papain	50 mM sodium acetate, pH 4.75	6	12
Lac repressor protein	1.0 M Tris HCl, pH 7.8	8	13
α-Mannosidase (*Phaseolus vulgaris*)	1.0 M sodium acetate, pH 4.0	35	14
α-Amylase (*Bacillus subtilis*)	0.1 M sodium phosphate, pH 7.0	8, 50	15
Dihydrofolate reductase	0.015 M bis Tris HCl, pH 6.5, 0.5 M KCl	4	16
Xylanase	50 mM sodium acetate, pH 4.5	—	17
Transcarboxylase	250 mM potassium phosphate, pH 6.5	—	18
Cellulase	50 mM NaOAc, pH 5.0	30	19
Winged bean	0.1 M sodium citrate, pH 6.0	10	20
Clostridium glucoamylase	20 mM sodium acetate, pH 4.5, 25°C		21
Human serum vitamin D-binding protein	20 mM sodium acetate, pH 4.0, 25°C	0–10	22
XynC xylanase from *Fibrobacter succinogenes* S85	50 mM sodium acetate, pH 5.0, 22°C	0–10	23
Ferridoxin:NADP$^+$ oxidoreductase	250 mM potassium phosphate, pH 7.7, dark, 4°C		24
Low-molecular weight Xylanase from *Chainia*	50 mM sodium acetate, pH 5.0, 25°C, 15 min	0–100	25
Xylose reductase from *Neurospora crassa*	50 mM sodium acetate, pH 6.0, 25°C		26
Tick anticoagulant peptide			27
Alkaline phosphatase (*Syclla serrata*)			28
Alpha-galactosidase (*Trichoderma reesei*)			29

TABLE 10.1 (continued)
Some Examples of the Use of *N*-Bromosuccinimide (NBS) for the Modification of Proteins

Protein	Reaction Conditions	Molar Excess[a]	Ref.
Alpha-amylase inhibitor (*Phaseolus vulgaris*)	100 m*M* sodium acetate, 50 m*M* NaCl, 1.0 m*M* CaCl$_2$, pH 5.0		30
Glutaryl 7-amino-cephalosporanic acid acylase	0.1 *M* potassium phosphate, pH 7.0, 25°C		31
Xylanase C (*Fibrobacter succinogenes* S85)	25 m*M* sodium acetate, pH 5.0, room temperature	1–15	32
Alpha-hemolysin (*Escherichia coli*)	20 m*M* Tris-HCl, pH 6.5	0.25–2	33
Alkaline protease inhibitor	50 m*M* potassium phosphate, pH 7.5	4.6–16.2	34
Hypoxanthine-guanine-xanthine phosphoribosyl transferase	50 m*M* sodium acetate, 6 m*M* MgCl$_2$, pH 6.0		35
D-Galactose-binding leaf lectin (*Erythrina indica*)	50 m*M* acetate, pH 6.0, 30°C, 60 min		36
Human cytidine deaminase	0.1 *M* citrate, pH 6.0	0–5	37
Adenosine deaminase	10 m*M* phosphate, pH 7.4	0–50	38
Coagulase (*Staphylococcus intermedius*)			39
Xylanse (*Acrophialophora nainiana*)			40
Patatin (*Solanum tuberosum* L.)			41

References for Table 10.1

1. Daniel, V.W., III, and Trowbridge, C.G., The effect of *N*-bromosuccinimide upon trypsinogen activation and trypsin catalysis, *Arch. Biochem. Biophys.* 134, 506, 1969.
2. Freisheim, J.H., and Huennekens, F.M., Effect of *N*-bromosuccinimide on dihydrofolate reductase, *Biochemistry* 8, 2271, 1969.
3. Poulos, T.L., and Price, P.A., The identification of a tryptophan residue essential to the catalytic activity of bovine pancreatic deoxyribonuclease, *J. Biol. Chem.* 246, 4041, 1971.
4. Sartin, J.L., Hugli, T.E., and Liao, T.-H., Reactivity of the tryptophan residues in bovine pancreatic deoxyribonuclease with *N*-bromosuccinimide, *J. Biol. Chem.* 255, 8633, 1980.
5. Warwick, P.E., D'Souza, L., and Freisheim, J.H., Role of tryptophan in dihydrofolate reductase, *Biochemistry* 11, 3775, 1972.
6. Nagami, K., The participation of a tryptophan residue in the binding of ferric iron to pyrocatechase, *Biochem. Biophys. Res. Commn.* 51, 364, 1973.
7. Schwabe, C., and Braddon, S.A., Evidence for the essential tryptophan residue at the active site of relaxin, *Biochem. Biophys. Res. Commn.* 68, 1126, 1976.
8. Cooper, A., and Hogan, M.E., Reactivity of tryptophans in rhodopsin, *Biochem. Biophys. Res. Commn.* 68, 178, 1976.
9. Kördel, W., and Schneider, F., Chemical modification of two tryptophan residues abolishes the catalytic activity of aminoacylase, *Hoppe Seyler's Z. Physiol. Chem.* 357, 1109, 1976.
10. Kosman, D.J., Ettinger, M.J., Bereman, R.D., and Giordano, R.S., Role of tryptophan in the spectral and catalytic properties of the copper enzyme, galactose oxidase, *Biochemistry* 16, 1597, 1977.
11. Uhteg, L.C., and Lundblad, R.L., The modification of tryptophan in bovine thrombin, *Biochim. Biophys. Acta* 491, 551, 1977.

(continued)

TABLE 10.1 (continued)
Some Examples of the Use of N-Bromosuccinimide (NBS)
for the Modification of Proteins

12. Glick, B.R., and Brubacher, L.S., The chemical and kinetic consequences of the modification of papain by N-bromosuccinimide, *Can. J. Biochem.* 55, 424, 1977.
13. O'Gorman, R.B., and Matthews, K.S., N-bromosuccinimide modification of lac repressor protein, *J. Biol. Chem.* 252, 3565, 1977.
14. Paus, E., The chemical modification of tryptophan residues of α-mannosidase from *Phaseolus vulgaris*, *Biochim. Biophys. Acta* 533, 446, 1978.
15. Fujimori, H., Ohnishi, M., and Hiromi, K., Tryptophan residues of saccharifying α-amylase from *Bacillus subtilis*. A kinetic discrimination of states of tryptophan residues using N-bromosuccinimide, *J. Biochem.* 83, 1503, 1978.
16. Thomson, J.W., Roberts, G.C.K., and Burgen, A.S.V., The effects of modification with N-bromosuccinimide on the binding of ligands to dihydrofolate reductase, *Biochem. J.* 187, 501, 1980.
17. Keskar, S.S., Srinivasan, M.C., and Deshpande, V.V., Chemical modification of a xylanase from a thermotolerant *Streptomyces*. Evidence for essential tryptophan and cysteine residues at the active site, *Biochem. J.* 261, 49, 1989.
18. Kumar, G.K., Beegen, H., and Wood, H.G., Involvement of tryptophans at the catalytic and subunit-binding domains of transcarboxylase, *Biochemistry* 27, 5972, 1988.
19. Clarke, A.J., Essential tryptophan residues in the function of cellulase from *Schizophyllum commune*, *Biochim. Biophys. Acta* 912, 424, 1987.
20. Higuchi, M., Inoue, K., and Iwai, K., A tryptophan residue is essential to the sugar-binding site of winged bean basis lectin, *Biochim. Biophys. Acta* 829, 51, 1985.
21. Ohnishi, H. et al., Functional roles of trp337 and glu632 in *Clostridium* glucoamylase, as determined by chemical modification, mutagenesis, and the stopped-flow methods, *J. Biol. Chem.* 269, 3503, 1994.
22. Swamy, N., Brisson, M., and Ray, R., Trp-145 is essential for binding of 25-hydroxyvitamin D to human serum vitamin D-binding protein, *J. Biol. Chem.* 270, 2636, 1995.
23. Zhu, H. et al., Enzymatic specificities and modes of action of the two catalytic domains of the XynC xylanase from *Fibrobacter succinogenes* S85, *J. Bacteriol.* 176, 3885, 1994.
24. Hirasawa, M. et al., The effects of N-bromosuccinimide on ferredoxin:NADP$^+$ oxidoreductase, *Arch. Biochem. Biophys.* 320, 280, 1995.
25. Bandivacdekor, K.R., and Deshpande, V.V., Structure-function relationship of xylanase: fluorometric analysis of the tryptophan environment, *Biochem. J.* 315, 583, 1996.
26. Ravat, V.B., and Rao, M.B., Purification, kinetic characterization and involvement of tryptophan residue at the NADP binding site of xylose reductase from *Neurospora crassa*, *Biochim. Biophys. Acta* 1293, 222, 1996.
27. Sillen, A., Vor, K., and Engelborgh, Y., Fluorescence study of the conformational properties of a recombinant tick anticoagulant peptide (*Ornithodorus moubata*) using multifrequency phase fluorometry, *Photochem. Photobiol.* 64, 785, 1996.
28. Zheng, W.Z. et al., An essential tryptophan residue of green crab (*Syclla serrata*) alkaline phosphatase, *Biochem. Mol. Biol. Int.* 41, 951, 1997.
29. Kachurin, A.M. et al., Tryptophan residues in alpha-galactosidase from *Trichoderma reesei*, *Biochemistry (Mosc.)* 63, 1183, 1998.
30. Takahashi, T. et al., Identification of essential amino acid residues of an alpha-amylase inhibitor from *Phaseolus vulgaris* white kidney beans, *J. Biochem.* 126, 838, 1999.
31. Lee, Y.S. et al., Involvement of arginine and tryptophan residues in the catalytic activity of glutaryl 7-aminocephalosporonic acid acylase from *Pseudomonas* sp. strain GK16, *Biochim. Biophys. Acta* 1523, 123, 2000.
32. McAllister, K.A., Marrone, L., and Clarke, A.J., The role of tryptophan residues in substrate binding to the catalytic domains A and B of xylanase C from *Fibrobacter succinogenes* S85, *Biochim. Biophys. Acta* 1480, 342, 2000.

TABLE 10.1 (continued)
Some Examples of the Use of N-Bromosuccinimide (NBS)
for the Modification of Proteins

33. Verza, G., and Bakás, L., Location of tryptophan residues in free and membrane bound *Escherichia coli* α-hemolysin and their role on the lytic membrane properties, *Biochim. Biophys. Acta* 1464, 27, 2000.

34. Vernekar, J.V. et al., Novel bifunctional alkaline protease inhibitor: protease inhibitory activity as the biochemical basis of antifungal activity, *Biochem. Biophys. Res. Commn.* 285, 1018, 2001.

35. Managala, N., Basus, V.J., and Wang, C.C., Role of the flexible loop of hypoxanthine-guanine-xanthine phosphoribosyltransferase from *Tritrichomonas foetus* in enzyme catalysis, *Biochemistry* 40, 4303, 2001.

36 Konozy, E.H.E. et al., Purification, some properties of a D-galactose-binding leaf lectin from *Erythrina indica* and further characterization of seed lectin, *Biochimie* 84, 1035, 2002.

37. Vincenzeeti, S. et al., Possible role of two phenylalanine residues in the active site of human cytidine deaminase, *Protein Eng.* 13, 791, 2000.

38. Mordanyan, S. et al., Tryptophan environment in adenosine deaminase. I. Enzyme modification with N bromosuccinimide in the presence of adenosine and EHNA analogues, *Biochim. Biophys. Acta* 1546, 185, 2001.

39. Komori, Y. et al., Characterization of coagulase from *Staphyloccus intermedius*, *J. Nat. Toxins* 10, 111, 2001.

40. Cardosa, O.A., and Filho, E.Y., Purification and characterization of a novel cellulase-free xylanase from *Acrophialophora nainiana*, *FEMS Microbiol. Lett.* 223, 309, 2203.

41. Liu, Y.W. et al., Patatin, the tuber storage protein of potato (*Solanum tubersum* L.), exhibits antioxidant activity *in vitro*, *J. Agric. Food Chem.* 51, 4389, 2003.

marked reduction in tyrosine (71% decrease) and methionine (79% decrease). Reaction with *N*-chlorosuccinimide (50 m*M* Tris, pH 8.5) resulted in the total loss of both tryptophan and methionine, but no significant change in tyrosine. Reaction with chloroamine T (50 m*M* Tris, pH 8.5) resulted in the loss of methionine only. Activity was lost only with the modification of tryptophan. Peptide bond cleavage was not observed under these reaction conditions. The literature on the use of NBS for the cleavage of peptide bonds suggests a preference for tyrosyl peptide bonds under mild acid conditions (50% acetic acid).[60–63] Table 10.2 presents some examples of peptide bond cleavage by NBS.

The conversion of tryptophanyl residues to 1-formyltryptophanyl residues (Figure 10.2) has been reported. The reaction conditions are somewhat harsh, but the procedure is reversible. As such it should prove quite useful for small peptides and has been applied to several proteins. Coletti-Previero and co-workers[64] successfully applied this procedure to bovine pancreatic trypsin. Trypsin was dissolved in formic acid saturated with HCl at a concentration of 2.5 mg/ml at 20°C. The formylation reaction is associated with an increase in absorbance at 298 nm.[65] Therefore, it is possible to follow the reaction spectrophotometrically. The reaction is judged complete when there is no further increase in absorbance at 298 nm. The reaction with trypsin was complete after an incubation period of 1 h. The solvent was partially removed *in vacuo* over KOH pellets, followed by lyophilization. The formyltryptophan derivative is unstable at alkaline pH. At pH 9.5 (pH-stat), conversion back to tryptophan is complete after a 200-min incubation at 20°C. Holmgren successfully

TABLE 10.2
Cleavage of Peptide Bonds by NBS

Protein	Reaction Conditions	NBS Molar Excess	Ref.
Histone H1[a]	50% Acetic acid		1
Human HMW kininogen[a]	50% Acetic acid		2
Histone H1[a]	50% Acetic acid		3
Proline-rich bactenecins	50% Acetic acid, 10 h, room temperature	4	4
Glucose carrier from human erythrocytes			5, 6
Bovine serum albumin	50 mM citrate, pH 6.0, with 1% SDS and 8 M urea	50–500 μg/ml	7
Various peptides			8
Rat liver arginase	100 mM potassium acetate, pH 5.0	35	9
Histone H1.b allelic variants from quail erythrocytes			10

[a] Cleavage at tyrosine residues.

References for Table 10.2

1. Kmiecik, D. et al., Primary structure of two variants of a sperm-specific histone H1 from the annelid *Platynereis dumerilii*, *Eur. J. Biochem.* 150, 259, 1985.
2. Kellerman, J. et al., Completion of the primary structure of human high molecular mass kininogen. The amino acid sequence of the entire heavy chain and evidence for its evolution by gene triplication, *Eur. J. Biochem.* 154, 471, 1986.
3. Hoyer-Fender, S., and Grossbach, V., Histone H1 heterogeneity in the midge, *Chironomus thummi*. Structural comparison of the H1 variants in an organism where their intrachromosomal localization is possible, *Eur. J. Biochem.* 176, 139, 1988.
4. Frank, R.W. et al., Amino-acid sequence of two proline-rich bactenecins: antimicrobial peptides of bovine neutrophils, *J. Biol. Chem.* 265, 18871, 1990.
5. May, J.M., Buchs, A., and Cartersu, C., Localization of a reactive exofacial sulfhydryl on the glucose carrier of human erythrocytes, *Biochemistry* 29, 10393, 1990.
6. Holman, G.D., and Rees, W.D., Photolabelling of the hexose transporter at external and internal sites: fragmentation patterns and evidence for a conformational change, *Biochim. Biophys. Acta* 897, 395, 1987.
7. Takeda, K., and Wada, A., Secondary structural changes of *N*-bromosuccinimide-cleaved bovine serum albumin in solutions of sodium dodecyl sulfate, urea, and guanidine hydrochloride, *J. Colloid Interface Sci.* 144, 45, 1991.
8. Grieve, P.A., Jones, A., and Alewood, P.F., Analytical methods for differentiated minor sequence variations in related peptides, *J. Chromatogr.* 646, 175, 1993.
9. Daghigh, F. et al., Chemical modification and inactivation of rat liver arginase by *N*-bromosuccinimide: reaction at His141, *Arch. Biochem. Biophys.* 327, 107, 1996.
10. Palyga, J., and Neelin, J.M., Isolation and preliminary characterization of histone H1.b allelic variants from quail erythrocytes, *Genome* 41, 709, 1998.

applied this procedure to thioredoxin.[66] A more recent study on *N*-formylation of tryptophanyl residues in proteins involved the study of epitopes in horse heart cytochrome c.[67] The single tryptophanyl residue was formylated with formic acid

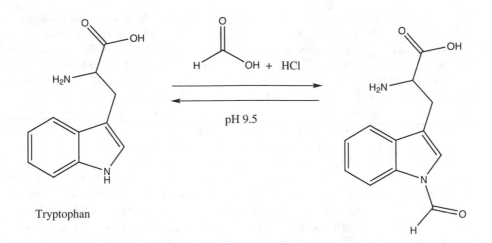

FIGURE 10.2 Formylation of tryptophan with formic acid.

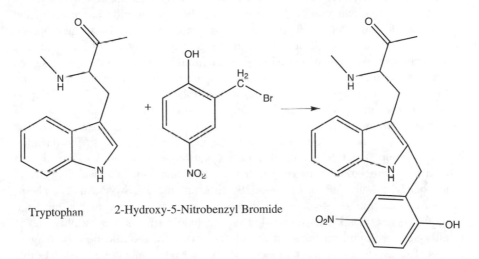

FIGURE 10.3 Reaction of tryptophan with 2-hydroxy-5-nitrobenzyl bromide.

saturated with HCl. The modified protein had markedly reduced affinity for a mono-clonal antibody resulting from local conformational change. It is not apparent that this technique has seen much use in the past decade.

One of the most useful modification procedures for tryptophanyl residues in proteins involves the use of 2-hydroxy-5-nitrobenzyl bromide (Figure 10.3) and its various derivatives. 2-Hydroxy-5-nitrobenzyl bromide, frequently referred to as Koshland's reagent, was introduced by Koshland and co-workers.[68,69] Barman and Koshland[70] reported the use of 2-hydroxy-5-nitrobenzyl bromide for the quantitative determination of tryptophanyl residues in proteins. Although this approach to the quantitative determination of tryptophanyl residues in proteins has been largely replaced by the development of new methods for the hydrolysis of proteins, it is still useful in certain instances. For this procedure, the sample is incubated for 16 to 20 h at 37°C in 1.0 ml 10 M urea (the urea should be recrystallized (EtOH/H$_2$O) before use), pH 2.7 (pH adjusted with concentrated HCl). This solution is cooled to ambient temperature, and ca. 5.0 mg of 2-hydroxy-5-nitrobenzyl bromide (in 0.1 ml acetone) is added followed by vigorous stirring. (We have found the Pierce Reacti-Vials® very useful for this purpose.) Occasionally, a precipitate of 2-hydroxy-5-nitrobenzyl alcohol (the hydrolytic product of 2-hydroxy-5-nitrobenzyl bromide) forms, which can be removed by centrifugation. The labeled protein is obtained free of reagent by gel filtration. This step is generally performed under acidic conditions (e.g., 0.18 M acetic acid, 10% acetic acid, or 10% formic acid). Depending on the protein under study, it might be necessary to perform this step in 10 M urea (pH 2.7) to maintain the solubility of the modified protein. A portion of the modified protein is taken to pH >12 with NaOH. The extent of incorporation is determined at 410 nm, using an extinction coefficient of 18,000 M^{-1} cm^{-1}. It is necessary to determine the concentration of protein by a technique other than absorbance at 280 nm because of the modification of tryptophan. We have found it convenient to use either amino acid analysis after acid hydrolysis or the ninhydrin reaction[71] after alkaline hydrolysis.[72]

The most frequent use of 2-hydroxy-5-nitrobenzyl bromide has been in the specific modification of tryptophan in peptides and proteins. Under appropriate reaction conditions (pH 4.0 or below), the reagent is highly specific for reaction with tryptophan. We have, on occasion, seen the modification of methionine residues under these conditions. This reagent also has the advantage of being a reporter group in the sense that the spectrum of the hydroxynitrobenzyl derivative is sensitive to changes in the microenvironment. This decrease observed in absorbance at 410 nm associated with an increase in absorbance at 320 nm on adding dioxane is similar to that seen with acidification and reflects the increase in the pK_a of the phenolic hydroxyl group. Titration curves of oxidized and reduced laccase[73] modified with 2-hydroxy-5-nitrobenzyl bromide suggested that the residues modified with 2-hydroxy-5-nitrobenzyl bromide are in an essentially aqueous microenvironment. The chemistry of the reaction of 2-hydroxy-5-nitrobenzyl bromide with tryptophan has been studied in some detail.[74] Disubstitution on the indole ring is a possibility and is usually seen as a sudden break in the plot of extent of modification vs. reagent excess. The reader is directed to a recent study by Kodíek and colleagues.[75] In this elegant study, MALDI-TOF mass spectrometry was used on products derived from the modification of tryptophan in a model peptide (GEGKGWGEGK) with 2-hydroxy-5-nitrobenzyl bromide. Five products were obtained, which reflected various degrees of substitution at the single tryptophan residue. The effect of 2-hydroxy-5-nitrobenzyl bromide was evaluated over a 1- to 200-fold molar excess range. The degree of modification increased as the concentration of reagent increased. A disubstituted product was observed at an equimolar excess of reagent; at a 200-fold molar excess, the most abundant product was a trisubstituted derivative and five different products were detected. The extent of modification increased with increasing pH, with no major change in product distribution.

We have found the following procedure to be useful. The protein or peptide to be modified is taken in 0.1 to 0.2 M sodium acetate buffer, pH 4 to 5. Reaction with other nucleophilic centers on the protein will become more of a problem as one approaches neutral pH. A 100-fold molar excess of 2-hydroxy-5-nitrobenzyl bromide (dissolved in a suitable water-miscible organic solvent such as acetone or dimethyl sulfoxide) is added in the dark. After 5 min (this time period was arbitrarily selected; the reaction can be considered to be essentially instantaneous to either modify tryptophan or undergo hydrolysis), the reaction mixture is taken by gel filtration into a solvent suitable for subsequent analysis. The extent of modification is determined under basic conditions as described previously for the use of this reagent in the quantitative determination of tryptophan.

The use of 2-hydroxy-5-nitrobenzyl bromide does present problems in that the reagent is extremely sensitive to hydrolysis and is not very soluble under aqueous conditions. These difficulties are avoided and the characteristics of the reaction preserved by using the dimethyl sulfonium salt obtained from the reaction of 2-hydroxy-5-nitrobenzyl bromide with dimethyl sulfide.[76] This compound is easily synthesized or can be obtained from various commercial sources. This water-soluble sulfonium salt derivative has recently been used to modify a tryptophanyl residue in rabbit skeletal myosin subfragment-1.[77] Peptide-containing modified tryptophanyl residues were purified by immunoaffinity chromatography, using rabbit antibody to bovine serum albumin previously modified with dimethyl-(2-hydroxy-5-nitrobenzyl) sulfonium bromide.

Horton and Koshland[78] also developed a clever approach for modification of hydrolytic enzymes. If 2-hydroxy-5-nitrobenzyl bromide is substituted at the phenolic hydroxyl, it is essentially unreactive, as originally shown for the methoxy derivative. Horton and Young[79] prepared 2-acetoxy-5-nitrobenzyl bromide. This derivative, like the methoxy derivative, is essentially unreactive. There is considerable structural identity between 2-acetoxy-5-nitrobenzyl bromide and *p*-nitrophenyl acetate, which is a nonspecific substrate for chymotrypsin. α-Chymotrypsin removes the acetyl group from 2-acetoxy-5-nitrobenzyl bromide, thus generating 2-hydroxy-5-nitrobenzyl bromide at the active site (Figure 10.4), which then either rapidly reacts with a neighboring nucleophile or undergoes hydrolysis. Uhteg and Lundblad[80] used both the acetoxy and butyroxy derivatives in the study of thrombin. A similar approach was used in the study of papain with 2-chloromethyl-4-nitrophenyl *N*-carbobenzoxy-glycinate.[81] It was subsequently shown that this modification occurs at a specific tryptophan residue in papain.[82] There has been one report of peptide bond cleavage secondary to reaction with 2-hydroxy-5-nitrobenzyl bromide.[83] Peptide bond cleavage was observed with a 50-fold molar excess of reagent at pH 8.5 (20 mM sodium phosphate, 25°C). Hage and co-workers[84] modified human serum albumin with 2-hydroxy-5-nitrobenzyl bromide for use in the affinity capillary electrophoresis of warfarin, ibuprofen, suprofen, and flurbiprofen. Table 10.3 gives some studies that have used 2-hydroxy-5-nitrobenzyl bromide.

Nitrophenylsulfenyl derivatives are reagents with reaction characteristics similar to those of 2-hydroxy-5-nitrobenzyl bromide[85] (Figure 10.5). The reaction product resulting from the sulfonylation of lysozyme[86] with 2-nitrophenylsulfenyl chloride (40-fold molar excess) pH 3.5 (0.1 M sodium acetate), has spectral characteristics that can be used to determine the extent of reagent incorporation ($\varepsilon_{365\ nm} = 4 \times 10^3\ M^{-1}\ cm^{-1}$). These reagents show considerable specificity for the modification of tryptophan at pH $\leq$ 4.0. Possible side reactions with other nucleophiles such as amino

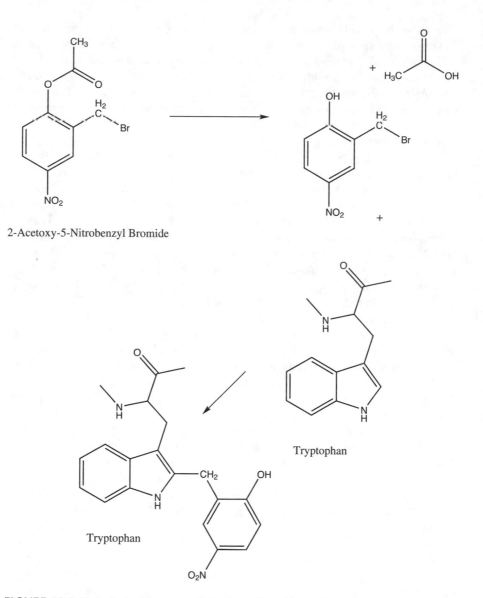

2-Acetoxy-5-Nitrobenzyl Bromide

Tryptophan

Tryptophan

FIGURE 10.4 Hydrolysis of 2-acetoxy-5-nitrobenzylbromide to release 2-hydroxy-5-nitroben-zyl bromide for subsequent reaction with tryptophan.

groups need to be considered. In the case of human chorionic somatomammotropin and human pituitary growth hormone,[87] reaction with *o*-nitrophenylsulfenyl chloride (2-nitrophenylsulfenyl chloride) was achieved in 50% acetic acid but not in 0.1 sodium acetate, pH 4.0. Wilchek and Miron[88] reported on the reaction of 2,4-dinitrophenylsulfenyl chloride with tryptophan in peptides and protein and subsequent conversion of the modified tryptophan to 2-thiotryptophan by reaction with 2-mercaptoethanol at pH 8.0. The thiolysis of the modified tryptophan is responsible

TABLE 10.3
Some Examples of the Use of 2-Hydroxy-5-Nitrobenzyl Bromide to Modify Proteins

Protein	Solvent	Molar Excess	Ref.
Pepsin	0.1 M NaCl	300	1
Streptococcal proteinase	0.46 M sodium phosphate, pH 3.1	200	2
Pancreatic deoxyribonuclease	0.050 M CaCl$_2$	100	3
Carbonic anhydrase	0.1 M phosphate, pH 6.8	100	4
Trypsin	0.1 M NaCl, 0.02 M CaCl$_2$, pH 4.2 (pH-stat)	ca. 100	5
Human chorionic somatomammotropin	0.05 M glycine, pH 2.8	—	6
Naja naja neurotoxin	0.2 M acetic acid	40	7
Glyceraldehyde-3-phosphate dehydrogenase	pH 6.75	30	8
α-Mannosidase (*Phaseolus vulgaris*)	0.1 M sodium acetate, pH 3.7	100	9
Thrombin	0.2 M acetate, pH 4.0	100	10
Laccase	pH 6.95[i]	50	11
	pH 4.00[i]	50	11
	pH 3.30[i]	110	11
Human serum albumin	10 M urea, pH 4.4	1000	12
Xylanase	50 mM NaOAc, pH 5.0	—	13
Winged beam	0.1 M sodium citrate, pH 3.1	100	14
		200	
		400	
Rat ovarian LH/hCG receptor	20 mM Tris-HCl, pH 6.0, 30 min, 24°C	15 mM	15
Mitochondrial F$_1$-ATPase ε-subunit	0.25 M sucrose, 1 mM EDTA, 0.2 M Tris-HCl, pH 6.7		16,17
Extracellular (1 → 6)-glucanase from *Acremonium persicinum*	0.5 M sodium acetate, pH 5.0, with 20 μg/ml		18,19
Pisum phytochrome A			20

References for Table 10.3

1. Dopheide, T.A.A., and Jones, W.M., Studies on the tryptophan residues in porcine pepsin, *J. Biol. Chem.* 243, 3906, 1968.
2. Robinson, G.W., Reaction of a specific tryptophan residue in streptococcal proteinase with 2-hydroxy-5-nitrobenzyl bromide, *J. Biol. Chem.* 245, 4832, 1970.
3. Poulos, T.L., and Price, P.A., The identification of a tryptophan residue essential to the catalytic activity of bovine pancreatic deoxyribonuclease, *J. Biol. Chem.* 246, 4041, 1971.
4. Lindskog, S., and Nilsson, A., The location of tryptophanyl groups in human and bovine carbonic anhydrases. Ultraviolet difference spectra and chemical modification, *Biochim. Biophys. Acta* 295, 117, 1973.
5. Imhoff, J.M., Keil-Dlouha, V., and Keil, B., Functional changes in bovine α- and β-trypsins caused by the substitution of tryptophan-199, *Biochimie* 55, 521, 1973.
6. Neri, P., Arezzini, C., Botti, R., Cocola, F., and Tarli, P., Modification of the tryptophanyl residue and its effect on the immunological and biological activity of human chorionic somatomammotropin, *Biochim. Biophys. Acta* 322, 88, 1973.

(continued)

TABLE 10.3 (continued)
Some Examples of the Use of 2-Hydroxy-5-Nitrobenzyl Bromide
to Modify Proteins

7. Karlsson, E., Eaker, D., and Drevin, H., Modification of the invariant tryptophan residue of two *Naja naja* neurotoxins, *Biochim. Biophys. Acta* 328, 510, 1973.
8. Heilman, H.D., and Pfleiderer, G., On the role of tryptophan residues in the mechanism of action of glyceraldehyde-3-phosphate dehydrogenase as tested by specific modification, *Biochim. Biophys. Acta* 384, 331, 1975.
9. Paus, E., The chemical modification of tryptophan residues of α-mannosidase from *Phaseolus vulgaris*, *Biochim. Biophys. Acta* 533, 446, 1978.
10. Uhteg., L.C., and Lundblad, R.L., The modification of tryptophan in bovine thrombin, *Biochim. Biophys. Acta* 491, 551, 1977.
11. Clemmer, J.D., Carr, J., Knaff, D.B., and Holwerda, R.A., Modification of laccase tryptophan residues with 2-hydroxy-5-nitrobenzyl bromide, *FEBS Lett.* 91, 346, 1978.
12. Fehske, K.J., Müller, W.F., and Wollert, U., The modification of the lone tryptophan residue in human serum albumin by 2-hydroxy-5-nitrobenzyl bromide. Characterization of the modified protein and the binding of L-tryptophan and benzodiazepines to the tryptophan-modified albumin, *Hoppe Seyler's Z. Physiol. Chem.* 359, 709, 1978.
13. Keskar, S.S., Srinivasan, M.C., and Deshpande, V.V., Chemical modification of a xylanase from a thermotolerant *Streptomyces*. Evidence for essential tryptophan and cysteine residues at the active site, *Biochem. J.* 261, 49, 1989.
14. Higuchi, M., Inoue, K., and Iwai, K., A tryptophan residue is essential to the sugar-binding site of winged bean basis lectin, *Biochim. Biophys. Acta* 829, 51, 1985.
15. Kolena, J. et al., Involvement of tryptophan in the structural alterations of the rat ovarian LH/hGC receptor, *Exp. Clin. Endocrinol. Diabetes* 105, 304, 1997.
16. Solaini, G. et al., Modification of the mitochondrial F1-ATPase ε-subunit, enhancement of the ATPase activity of the IF1-F1 complex and the IF1-binding dependence of the conformation of the ε subunit, *Biochem. J.* 327, 443, 1997.
17. Baracca, A., Interaction and effects of 2-hydroxy-5-nitrobenzyl bromide on the bovine heart mitochondrial F1-ATPase, *Int. J. Biochem.* 25, 1269, 1993.
18. Pitson, S.M. et al., Purification and characterization of an extracellular $(1 \rightarrow 6)$--glucanase from the filamentous fungus, *Aeromonium persicium*, *Biochem. J.* 316, 841, 1996.
19. Moore, A.E., and Stone, B.A., A -1,3-glucan hydrolase from *Nicotiana glutinosa*. II Specificity, action pattern and inhibitor studies, *Biochim. Biophys. Acta* 258, 248, 1972.
20. Wells, J.A. et al., A conformational change associated with the phototransformation of *Pisum* phytochrome A as probed by fluorescence quenching, *Biochemistry* 33, 708, 1994.

for changes in the spectral properties of the derivative. The characteristics of the modified tryptophan have resulted in the development of a facile purification scheme for peptides containing the modified tryptophan residues.[89,90] Mollier and co-workers[91] examined the reaction of *o*-nitrophenylsulfenyl chloride (2-nitrophenyl-sulfenyl chloride) with notexin (a phospholipase obtained from *Notechis scutalatus scutalatus* venom, which contains two tryptophanyl residues). Reactions with 2-nitrophenylsulfenyl chloride (twofold molar excess) in 50% (v/v) acetic acid resulted in two derivative proteins on HPLC analysis. One derivative contained two modified tryptophanyl residues (20 and 110), whereas the other derivative was modified only at Position 20. Several recent applications of this modification are of importance for investigators working in proteomics. Nishimura and workers[92] developed a heavy (^{13}C) form of 2-nitrobenzenesulfonyl chloride for the differential

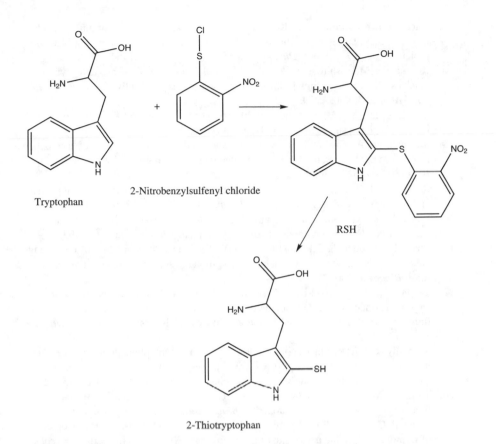

Tryptophan

2-Nitrobenzylsulfenyl chloride

RSH

2-Thiotryptophan

FIGURE 10.5 Reaction of tryptophan with 2 nitrophenylsulfenyl chloride.

labeling of tryptophan residues in protein mixtures. Application of the isotope-coded affinity tag strategy[93] to tryptophanyl residues has significant advantage in that tryptophan is one of the least abundant residues in proteins. This chemical approach has also been used for the selective isolation of tryptophan peptides in which the sufonyl chloride function is on a solid matrix. Reaction is accomplished with tryptophan peptides and elution accomplished by mild reduction.

REFERENCES

1. Liu, T.-Y., and Chang, Y.H., Hydrolysis of proteins with p-toluenesulfonic acid. Determination of tryptophan, *J. Biol. Chem.* 246, 2842, 1971.
2. Simpson, R.J., Neuberger, M.R., and Liu, T.-Y., Complete amino acid analysis of proteins from a single hydrolysate, *J. Biol. Chem.* 251, 1936, 1976.
3. Molnár, I., Tryptophan analysis of peptides and proteins, mainly by liquid chromatography, *J. Chromatogr.* 763, 1, 1997.
4. Friedman, M., Applications of the ninhydrin reaction for analysis of amino acids, peptides, and proteins to agricultural and biomedical sciences, *J. Agric. Food Chem.* 52, 385, 2004.

5. Hachimori, Y., Horinishi, H., Kurihara, K., and Shibata, K., States of amino residues in proteins. V. Different reactivities with H_2O_2 of tryptophan residues in lysozyme, proteinases and zymogens, *Biochim. Biophys. Acta* 93, 346, 1964.
6. Kotoku, I., Matsushima, A., Bando, M., and Inada, Y., Tyrosine and tryptophan residues and amino groups in thrombin related to enzymic activities, *Biochim. Biophys. Acta* 214, 490, 1970.
7. Sanda, A., and Irie, M., Chemical modification of tryptophan residues in ribonuclease from a *Rhizopus* sp., *J. Biochem.* 87, 1079, 1980.
8. Matsushima, A., Takiuchi, H., Saito, Y., and Inada, Y., Significance of tryptophan residues in the D-domain of the fibrin molecule in fibrin polymer formation, *Biochim. Biophys. Acta* 625, 230, 1980.
9. Reubsaet, J.L.E. et al., Analytical techniques used to study the degradation of proteins and peptides: chemical instability, *J. Pharm. Biomed. Anal.* 17, 955, 1998.
10. Simat, T., Meyer, K., and Steinhart, H., Syntheses and analysis of oxidation and carbonyl condensation compounds of tryptophan, *J. Chromatogr. A* 661, 93, 1994.
11. Mach, H., Middaugh, C.R., and Lewis, R.V., Statistical determination of the average values of the extinction coefficients of tryptophan and tyrosine in native proteins, *Anal. Biochem.* 200, 74, 1992.
12. Takita, T., et al., Lysyl-tRNA synthetase from *Bacillus stearothermophilus*: the tryp312 residue is shielded in a non-polar environment and is responsible for the fluorescence changes observed in the amino acid activation reaction, *J. Mol. Biol.* 325, 677, 2003.
13. Reshetnyak, Y.K. et al., Decomposition of protein tryptophan fluorescence spectra into log-normal components. III. Correlation between fluorescence and microenvironmental parameters of individual tryptophan residues, *Biophys. J.* 81, 1735, 2001.
14. Turk, T., Macek, P., and Gubensek, F., The role of tryptophan in structural and functional properties of equinatoxin II, *Biochim. Biophys. Acta* 1119, 1, 1992.
15. Hsieh, P.-C., Shenoy, B.C., Haase, F.C., Jentoft, J.E., and Phillips, N.F.B., Involvement of tryptophan(s) at the active site of polyphosphate/ATP glucokinase from *Mycobacterium tuberculosis*, *Biochemistry* 32, 6243, 1993.
16. Kaneko, S.-I., Ichiba, T., Hirano, N., and Hachimori, A., Modification of a single tryptophan of the inorganic pyrophosphatase from thermophilic bacterium PS-3: possible involvement in its substrate binding, *Biochim. Biophys. Acta* 1077, 281, 1991.
17. Stevens, W.K., and Nesheim, M.E., Structural changes in the protease domain of prothrombin upon activation as assessed by *N*-bromosuccinimide modification of tryptophan residues in prethrombin-2 and thrombin, *Biochemistry* 32, 2787, 1993.
18. Nakazawa, M., and Manabe, K., Differential exposure of tryptophan residues in the red and far-red light absorbing forms of phytochrome, as revealed by chemical modification, *Plant Cell Physiol.* 34, 1097, 1993.
19. Bray, M.R., Carriere, A.D., and Clarke, A.J., Quantitation of tryptophan and tyrosine residues in proteins by fourth-derivative spectroscopy, *Anal. Biochem.* 221, 278, 1994.
20. Hirasawa, M. et al., The effect of *N*-bromosuccinimide on ferredoxin:$NADP^+$ oxidoreductase, *Arch. Biochem. Biophys.* 320, 280, 1995.
21. Bandivadekan, K.R., and Deshpande, V.V., Structure-function relationship of xylanase: fluorimetric analysis of the tryptophan environment, *Biochem. J.* 315, 583, 1996.
22. Karchurin, A.M. et al., Tryptophan residues in alpha-galactosidase from *Trichoderma reesei*, *Biochemistry (Mosc.)* 63, 1183, 1998.
23. Martinez, A., Bella, J.L., and Stockert, J.C., Fluorogenic reaction of blood cells induced by *N*-bromosuccinimide, *Micron* 33, 399, 2002.

24. Falkenstein, E. et al., Chemical modification and structural analysis of the progesterone membrane binding protein from porcine liver membranes, *Mol. Cell. Biochem.*
218, 71, 2001.
25. Cardoso, O.A., and Filho, E.Y., Purification and characterization of a novel cellulase-
free zylanase from *Acrophialophora nainiana*, *FEMS Micobiol. Lett.* 223, 309, 2003.
26. Green, N.M., Avidin. III. The nature of the biotin binding site, *Biochem. J.* 89, 599,
1963.
27. Crestfield, A.M., Moore, S., and Stein, W.H., The preparation and enzymatic hydrolysis of reduced and *S*-carboxymethylated proteins, *J. Biol. Chem.* 238, 622, 1963.
28. Ohnishi, M., Kawagishi, T., Abe, T., and Hiromi, K., Stopped-flow studies on the
chemical modification with *N*-bromosuccinimide of model compounds of tryptophan
residues, *J. Biochem.* 87, 273, 1980.
29. Sartin, J.L., Hugli, T.E., and Liao, T.-H., Reactivity of the tryptophan residues in
bovine pancreatic deoxyribonuclease with *N*-bromosuccinimide, *J. Biol. Chem.* 255,
8633, 1980.
30. Gaitonde, M.K., and Dovey, T., A rapid method for the quantitative determination of
tryptophan in the intact protein, *Biochem. J.* 117, 907, 1970.
31. Sodak, L. et al., Rapid determination of tryptophan in beans (*Phaseolus vulgaris*) by
the acid ninhydrin method, *J. Agric. Food Chem.* 23, 1147, 1975.
32. Molnár-Perl, I., and Pinter-Szakacs, M., Spectrophotometric determination of tryptophan in intact proteins by the acid ninhydrin method, *Anal. Biochem.* 177, 1, 1989.
33. Mbithi-Mwikya, S. et al., Amino acid profiles after sprouting, autoclaving, and lactic
acid fermentation of finger millet (*Eleusine coracan*) and kidney beans (*Phaseolus
vulgaris* L.), *J. Agric. Food Chem.* 48, 3081, 2000.
34. Nguchi, D.D., Kuo, Y.-H., and Lambein, F., Food safety and amino acid balances in
processed cassava "cassettes," *J. Agric. Food Chem.* 50, 3042, 2002.
35. Nagaraja, P., Yathirajim, H.S., and Vasentha, R.A., Highly sensitive reaction of tryptophan with *p*-phenylenediamine, *Anal. Biochem.* 312, 157, 2003.
36. Daniel, V.W., III, and Trowbridge, C.G., The effect of *N*-bromosuccinimide upon
trypsinogen activation and trypsin catalysis, *Arch. Biochem. Biophys.* 134, 506, 1969.
37. Freisheim, J.H., and Huennekens, F.M., Effect of *N*-bromosuccinimide on dihydrofolate reductase, *Biochemistry* 8, 2271, 1969.
38. Poulos, T.L., and Price, P.A., The identification of a tryptophan residue essential to
the catalytic activity of bovine pancreatic deoxyribonuclease, *J. Biol. Chem.* 246,
4041, 1971.
39. Kosman, D.J., Ettinger, M.J., Bereman, R.D., and Giordano, R.S., Role of tryptophan
in the spectral and catalytic properties of the copper enzyme, galactose oxidase,
Biochemistry 16, 1597, 1977.
40. Inokuchi, N., Takahashi, T., Yoshimoto, A., and Irie, M., *N*-Bromosuccinimide oxidation of a glucoamylase from *Aspergillus saitoi*, *J. Biochem.* 91, 1661, 1982.
41. O'Gorman, R.B., and Matthews, K.S., *N*-Bromosuccinimide modification of lac
repressor protein, *J. Biol. Chem.* 252, 3565, 1977.
42. Kumar, G.K., Beegen, H., and Wood, H.G., Involvement of tryptophans at the catalytic
and subunit-binding domains of transcarboxylase, *Biochemistry* 27, 5972, 1988.
43. Keskar, S.S., Srinivasan, M.C., and Deshpande, V.V., Chemical modification of a
xylanase from a thermotolerant *Streptomyces*. Evidence for essential tryptophan and
cysteine residues at the active site, *Biochem. J.* 261, 49, 1989.
44. Bray, M.R., and Clarke, A.J., Identification of an essential tyrosyl residue in the
binding site of *Schizophyllum commune* xylanase A, *Biochemstry* 34, 2006, 1995.

45. Kim, Y.T., Muramatsu, T., and Takahashi, K., Leader peptides from *Escherichia coli*: overexpression, characterization, and inactivation by modification of tryptophan residues 300 and 310 with *N*-bromosuccinimide, *J. Biochem.* 117, 535, 1995.

46. Kim, Y.T., Muramatsu, T., and Takahashi, K., Identification of trp300 as an important residue for *Escherichia coli* leader peptidase activity, *Eur. J. Biochem.* 234, 358, 1995.

47. Ray, S., Mukerji, S., and Bhaduni, A., Two tryptophans at the active site of UDP-glucose-4-epimerase from *Kluyveomyces fragilis*, *J. Biol. Chem.* 270, 11383, 1995.

48. Xue, H. et al., Chemical modification of *Bacillus subtilis* tryptophanyl-tRNA synthetase, *Biochem. Cell. Biol.* 75, 709, 1997.

49. Doublié, S. et al., Tryptophanyl-tRNA synthetase crystal structure reveals an unexpected homology to tyrosyl-tRNA synthetase, *Structure* 3, 17, 1995.

50. Daghigh, F. et al., Chemical modification of rat liver arginase by *N*- bromosuccinimide. Reaction with His141. *Arch. Biochem. Biophys.* 327, 107, 1996.

51. Clottes, E., and Vial, C., Discrimination between the four tryptophan residues of MM-creatine kinase as the basis of the effect of *N*-bromosuccinimide on activity and spectral properties, *Arch. Biochem. Biophys.* 329, 97, 1996.

52. Josse, D. et al., Tryptophan residue(s) as major components of the human serum paraoxonase active site. *Chem. Biol. Interact.* 119, 1999.

53. Soulages, J.L., and Arrese, E.L., Interaction of the α-helices of apophorin III with the phospholipids acyl chains in discoidal lipoprotein particles: a fluorescence quenching study, *Biochemistry* 40, 14279, 2001.

54. Tanksale, A.M. et al., Evidence for tryptophan in proximity to histidine and cysteine as essential to the active site of an alkaline protease, *Biochem. Biophys. Res. Commn.* 270, 910, 2000.

55. Takahashi, T. et al., Identification of essential amino acid residues of an α-amylase inhibitor from *Phaseolus vulgaris* white kidney beans, *J. Biochem.* 126, 838, 1999.

56. Ramachandran, L.K., and Witkop, B., *N*-Bromosuccinimide cleavage of peptides, *Methods Enzymol.* 11, 283, 1967.

57. Feldhoff, R.C., and Peters, T., Jr., Determination of the number and relative position of tryptophan residues in various albumins, *Biochem. J.* 159, 529, 1976.

58. Shechter, Y., Patchornik, A., and Burstein, Y., Selective chemical cleavage of tryptophanyl peptide bonds by oxidative chlorination with *N*-chlorosuccinimide, *Biochemistry* 15, 5071, 1976.

59. Sakurai, J., and Nagahama, M., Role of one tryptophan residue in the lethal activity of *Clostridium perfringens* epsilon toxin, *Biochem. Biophys. Res. Commn.* 128, 760, 1985.

60. Brandt, W.F., and von Holt, C., The determination of the primary structure of histone F3 from chicken erythrocytes by automatic Edman degradation. 1. Cleavage and alignment of the fragments, *Eur. J. Biochem.* 46, 407, 1974.

61. Kmiecik, D. et al., Primary structure of the two variants of a sperm-specific histone H1 from the annelid *Platynereis dumerilii*, *Eur. J. Biochem.* 150, 359, 1985.

62. Kellerman, J. et al., Completion of the primary structure of human high-molecular-mass kininogen. The amino acid sequence of the entire heavy chain and evidence for its evaluation by gene triplication, *Eur. J. Biochem.* 154, 471, 1986.

63. Hoyer-Fender, S., and Crossbach, V., Histone H1 heterogeniety in the midge, *Chironomus thummi*. Structural comparison of the H1 variants in an organism where their intrachromasomal localization is possible, *Eur. J. Biochem.* 176, 139, 1988.

64. Coletti-Previero, M.-A., Previero, A., and Zuckerkandl, E., Separation of the proteolytic and esteratic activities of trypsin by reversible structural modification, *J. Mol. Biol.* 39, 493, 1969.

65. Previero, A., Coletti-Previero, M.-A., and Cavadore, J.C., A reversible chemical modification of the tryptophan residue, *Biochim. Biophys. Acta* 147, 453, 1967.

66. Holmgren, A., Reversible chemical modification of the tryptophan residues of thioredoxin from *Eschericia coli* B., *Eur. J. Biochem.* 26, 528, 1972.

67. Cooper, H.M., Jemmerson, R., Hunt, D.F., Griffin, P.R., Yates, J.R., III, Shabanowitz, J., Zhu, N.-Z., and Paterson, Y., Site-directed chemical modification of horse cytochrome c results in changes in antigenicity due to local and long-range conformation perturbations, *J. Biol. Chem.* 262, 11591, 1987.

68. Koshland, D.E., Jr., Karkhanis, Y.D., and Latham, H.G., An environmentally-sensitive reagent with selectivity for the tryptophan residue in proteins, *J. Am. Chem. Soc.* 86, 1448, 1964.

69. Horton, H.R., and Koshland, D.E., Jr., A highly reactive colored reagent with selectivity for the tryptophanyl residue in proteins, 2-hydroxy-5-nitrobenzyl bromide, *J. Am. Chem. Soc.* 87, 1126, 1965.

70. Barman, T.E., and Koshland, D.E., Jr., A colorimetric procedure for the quantitative determination of tryptophan residues in proteins, *J. Biol. Chem.* 242, 5771, 1967.

71. Moore, S., Amino acid analysis: aqueous dimethyl sulfoxide as solvent for the ninhydrin reaction, *J. Biol. Chem.* 243, 6281, 1968.

72. Fruchter, R.G., and Crestfield, A.M., Preparation and properties of two active forms of ribonuclease dimer, *J. Biol. Chem.* 240, 3868, 1965.

73. Clemmer, J.D., Carr, J., Knaff, D.B., and Holwerda, R.A., Modification of laccase tryptophan residues with 2-hydroxy-5-nitrobenzyl bromide, *FEBS Lett.* 91, 346, 1978.

74. Loudon, G.M., and Koshland, D.E., Jr., The chemistry of a reporter group: 2-hydroxy-5-nitrobenzyl bromide, *J. Biol. Chem.* 245, 2247, 1970.

75. Strohalm, M., Kodíek, M., and Pechar, M., Tryptophan modification by 2-hydroxy-5-nitrobenzyl bromide studied by MALDI-TOF mass spectreometry, *Biochem. Biophys. Res. Commn.* 312, 811, 2003.

76. Horton, H.R., and Tucker, W. P., Dimethyl (2-hydroxy-5-nitrobenzyl) sulfonium salts. Water-soluble environmentally sensitive protein reagents, *J. Biol. Chem.* 245, 3397, 1970.

77. Peyser, Y.M., Muhlrad, A., and Werber, M.M., Tryptophan-130 is the most reactive tryptophan residue in rabbit skeletal myosin subfragment-1, *FEBS Lett.* 259, 346, 1990.

78. Horton, H.R., and Koshland, D.E., Jr., Reactions with reactive alkyl halides, *Methods Enzymol.* 11, 556, 1967.

79. Horton, H.R., and Young, G., 2-Acetoxy-5-nitrobenzyl chloride. A reagent designed to introduce a reporter group near the active site of chymotrypsin, *Biochim. Biophys. Acta* 194, 272, 1969.

80. Uhteg, L.C., and Lundblad, R.L., The modification of tryptophan in bovine thrombin, *Biochim. Biophys. Acta* 491, 551, 1977.

81. Mole, J.E., and Horton, H.R., A kinetic analysis of the enhanced catalytic efficiency of papain modified by 2-hydroxy-5-nitrobenzylation, *Biochemistry* 12, 5285, 1973.

82. Chang, S.-M. T., and Horton, H.R., Structure of papain modified by reaction with 2-chloromethyl-4-nitrophenyl *N*-carbobenzoxylglycinate, *Biochemistry* 18, 1559, 1979.

83. Bhattacharya, S.K., and Dubey, A.K., The N-terminus of m5C-DNA methyltransferase *Msp*1 is involved in its topoisomerase activity, *Eur. J. Biochem.* 269, 2491, 2002.

84. Kim, H.S., Austin, J., and Hage, D.S., Identification of drug-binding sites on human serum using affinity capillary electrophoresis and chemically modified proteins as buffer additives, *Electrophoresis* 23, 956, 2002.

85. Fontana, A., and Scoffone, E., Sulfenyl halides as modifying reagents for polypeptides and proteins, *Methods Enzymol.* 25, 482, 1972.
86. Shechter, Y., Burstein, Y., and Patchornik, A., Sulfenylation of tryptophan-62 in hen egg-white lysozyme, *Biochemistry* 11, 653, 1972.
87. Bewley, T.A., Kawauchi, H., and Li, C.H., Comparative studies of the single tryptophan residue in human chorionic somatomammotropin and human pituitary growth hormone, *Biochemistry* 11, 4179, 1972.
88. Wilchek, M., and Miron, T., The conversion of tryptophan to 2-thioltryptophan in peptides and proteins, *Biochem. Biophys. Res. Commn.* 47, 1015, 1972.
89. Chersi, A., and Zito, R., Isolation of tryptophan-containing peptides by adsorption chromatography, *Anal. Biochem.* 73, 471, 1976.
90. Rubinstein, M., Schechter, Y., and Patchornik, A., Covalent chromatography: the isolation of tryptophanyl containing peptides by novel polymeric reagents, *Biochem. Biophys. Res. Commn.* 70, 1257, 1976.
91. Mollier, P., Chwetzoff, S., Bouet, F., Harvey, A.L., and Ménez, A., Tryptophan 110, a residue involved in the toxic activity but not in the enzymatic activity of notexin, *Eur. J. Biochem.* 185, 263, 1989.
92. Kuyama, H. et al., An approach to quantitative proteome analysis by labeling tryptophan residues, *Rapid Commn. Mass Spectrom.* 17, 1642, 2003.
93. Hansen, K.C. et al., Mass spectrometric analysis of protein mixtures at low levels using cleavable ^{13}C-isotope-coded affinity tag and multidimensional chromatography, *Mol. Cell. Proteom.* 2, 299, 2003.

11 The Chemical Cleavage of Peptide Bonds

Elucidation of the covalent structure of a protein requires the development of specific, reproducible methods for cleavage of the protein into fragments of a size amenable to structural analysis. Proteolytic enzymes such as trypsin, chymotrypsin, and pepsin have proven quite useful in the cleavage of specific peptide bonds in proteins. In addition to the use of specific proteases, the nature of certain amino acid residues has permitted the development of nonenzymatic chemical methods for the cleavage of certain peptide bonds.[1,2]

Cleavage of methionine-containing peptide bonds with CNBr[1–4] is likely the most widely used method for specific chemical cleavage of peptide bonds. In the reaction, a peptide bond is cleaved (Figure 11.1), in which methionine contributes the carboxyl moiety. Methionine is converted into homoserine lactone and homoserine during this process, with the loss of methyl thiocyanate. This means that, for example, if there is a ^{35}S-labeled methionine in a protein, the radiolabel will be lost during cyanogen bromide (CNBr) cleavage. Several examples of this have been noted. The reaction is reasonably quantitative, although, as indicated later, variable amounts of CNBr (CNBr) might be required. Second, the methionine content of most proteins is low enough[5] to obtain a reasonably small number of fragments, providing a distinct advantage in primary structure analysis. The major advances in mass spectrometry have resulted in the development of a top–down strategy for protein structure analysis that uses CNBr fragmentation.[6]

The chemistry of this reaction is straightforward, involving the nucleophilic attack of the thioether sulfur on the carbon in CNBr, followed by cyclization to form the iminolactone, which is hydrolyzed by water, resulting in cleavage of the peptide bond. At acidic pH, this reaction does not generally, in and by itself, affect any other amino acid with the exception of cysteine, which is converted to cysteic acid. Noted that one would rarely be working with a protein or peptide containing free sulfhydryl groups. The yield of cleavage is measured either by the loss of methionine or by the sum of homoserine and homoserine lactone after acid hydrolysis. This value is probably best determined by allowing complete conversion to homoserine with base at room temperature. Cleavage of peptide chains at methionine with CNBr proceeds best with a fully denatured protein in mild acid. Early work with this reaction used 0.1 M HCl as the solvent or 0.1 M HCl in 6 M guanidine hydrochloride. Most recent studies have used 70% formic acid, trifluoroacetic acid (TFA),[7] or an equal mixture of formic acid and TFA.[8] Use of formic acid has, on occasion, resulted in blocking amino-terminal residues via reaction with formaldehyde (present as a contaminant in formic acid).[9] Acetic acid can also be used as a solvent for this reaction. In general, the reaction proceeds effectively with a 20- to 100-fold molar excess of CNBr (added

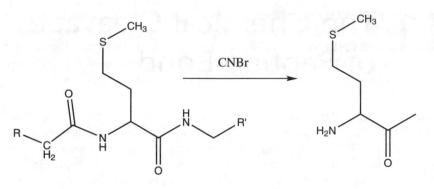

Methionine

Methionine

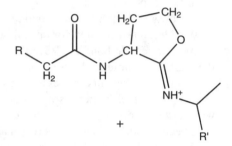

Methyl Thiocyanate

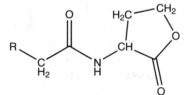

Peptide Homoserine Lactone

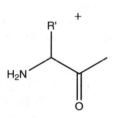

FIGURE 11.1 Cyanogen bromide cleavage of methionine peptide bonds with the formation of homoserine lactone and methyl isothiocyanate.

either as a solid to the protein or peptide dissolved in the solvent of choice). Solutions of acetic acid can also be used as a solvent for the CNBr reaction. The molar ratio of CNBr to methionine residues needs to be established for each peptide and protein under study. In the work on the structure of the pancreatic deoxyribonuclease, it was necessary to use a 3000-fold molar excess to cleave a particular methionine-serine peptide bond.[10] Met-123 in human serum albumin is converted to homoserine lactone by treatment with CNBr without concomitant peptide bond cleavage.[11] The conversion of methionine to methionine sulfoxide under conditions used for the CNBr cleavage has been reported.[12] With a 10-fold molar excess of CNBr for 22 h at ambient temperature, 1% conversion to methionine sulfoxide was observed in 70% formic acid; 8% conversion in 0.1 M HCl; 64% conversion in 0.1 M citrate, pH 3.5; and 97% conversion in 0.1 M phosphate, pH 6.5. CNBr does not cleave methionine-sulfoxide-containing peptide bonds.[13]

Simpson and Nice[14] developed a procedure for the *in situ* CNBr cleavage of proteins absorbed to the glass-fiber membranes developed for the gas-phase protein–peptide sequenator. The procedure was originally developed to determine internal sequence information from amino-terminal blocked proteins. In the procedure described, satisfactory sequence information was obtained from 2.2 nmol (24 µg) cytochrome c. CNBr (20 µl containing a 20-fold molar excess of reagent in 70% formic acid) was applied to the membrane. The membrane was then placed in a vacuum desiccator over CNBr/HCOOH for 16 h at ambient temperature. The membrane was air dried and taken to the sequenator.

Xu and Shively[15] described improvements on the electroblotting of proteins. They reported higher degrees of success with polyvinyldifluoride (PVDF) membranes. Transfer yields were markedly improved on pretreating the membranes with Polybrene®.

Scott and co-workers[16] described an additional approach to the CNBr cleavage of proteins on PDVF membranes, followed by elution of the reaction products from the membranes with 2% sodium dodecyl (SDS) sulfate per 1% Triton X-100 in 50 mM Tris, pH 9.2. (Using bovine serum albumin as a model protein, they obtained a 90% recovery of eluted peptides [10 µg protein applied to PVDF membrane].[17]) The heavy and light chains of the antibody were separated by electrophoresis and transferred to a PVDF membrane. After staining with Ponceau Red, the bands were excised and placed in a 500-µl Eppendorf tube. CNBr (150 µl; 150 µM in 70% formic acid) was added, and the reaction was allowed to proceed for 16 h at ambient temperature. The solvent was removed *in vacuo* and the peptides eluted (90 min at ambient temperature) with 75 µl of the Tris-detergent solvent as described previously. The peptides were separated by SDS-polyacrylamide gel electrophoresis and the bands subjected to gas-phase sequence analysis. Useful sequence information was obtained from 25 to 50 µg of parent immunoglobulin.

Sokolov and co-workers[18] described a modified method for direct CNBr cleavage directly within the polyacrylamide gel. After identification of the proteins by either staining or autoradiography, the gel is sliced and dried. The dried gel is taken into 30 µl CNBr (200 mg/ml in 70% HCOOH). The reaction is allowed to proceed for 16 h at ambient temperature or 3 h at 37°C. The gel is then dried, and the reaction products are separated by a second electrophoretic step. Drying the gel before the cleavage step is critical to avoid protein loss during this procedure. Jahnen and co-workers[19] have developed an alternative approach. They isolated the fragments from the CNBr cleavage of proteins on polyacrylamide gel (the gel slices were dried by lyophilization before the CNBr cleavage step) by either a second electrophoretic step or by HPLC after elution. The electrophoretic step is recommended over the HPLC step. More recent work by Robillard and co-workers[20] has provided methods for the in-gel cleavage of integral membrane proteins for subsequent analysis by MALDI-TOF mass spectrometry. The CNBr cleavage reaction was performed in 70% TFA (one small CNBr crystal dissolved in 200 to 300 μl 70% TFA was added to the gel slice) for 14 h in the dark at room temperature. The digested gel piece was sonicated for 5 min and then extracted twice by sonication with 30 μl 60% acetonitrile, 1% TFA and concentrated *en vacuo* before analysis.

There is continued work on the optimization of the CNBr reaction. Kaiser and Metzka[21] developed conditions for improving the efficiency of the CNBr cleavage at Met-Ser and Met-Thr peptide bonds. Decreasing or increasing the acid concentration improved cleavage yields in model peptides. For example, cleavage of PFAMSL was 60% in 0.1 *N* HCl, 0% in 0.1 *N* HCl in 100% acetonitrile, and 36% in 30% formic acid. Campanini and colleagues[22] showed that Tris buffer reacts with homoserine lactone to provide a product with a mass difference of 103 Da compared to the homoserine peptide. As noted previously, CNBr cleavage can yield a mixture of homoserine and homoserine lactone products. This can yield a poorly resolved doublet on mass spectrometric analysis. The Tris reaction provides easily resolved doublets. Dratz and co-workers[23] developed a useful method for the fragmentation and mass spectrometric analysis of integral membrane proteins. They studied the CNBr fragmentation of rhodopsin for subsequent MALDI-MS analysis. Rhodopsin in lauryldimethylamine oxidase was precipitated with trichloroacetic acid washed with ethanol/sonication (3×), and dissolved in 100% TFA containing a 500-fold molar excess (to methionine). The digestion reaction was allowed to proceed overnight in the dark with intermittent sonication.

CNBr has proved useful for the cleavage of selected fusion proteins.[24–31] The caveat is that the protein/peptide of interest must be stable to acidic conditions. This has made it more applicable to small peptides and proteins. CNBr fragments have been used in the preparation of a semisynthetic serine protease.[32]

A number of other methods for the chemical cleavage of specific peptide bonds have been used infrequently as a result of the considerable success obtained with the CNBr reaction. There is, however, continued interest in these various methods.

Partial acid hydrolysis is the oldest of the various chemical approaches to the cleavage of specific peptide bonds.[33–37] The general principle of partial acid hydrolysis is based on the use of dilute acid at a pH just adequate to maintain the β-carboxyl group of aspartic acid in the protonated form. Under these conditions, peptide bonds in which the carboxyl moiety is contributed by aspartic acid are cleaved 100-fold more rapidly than other peptide bonds. Piszkiewicz and workers[38] reported the cleavage of an Asp-Pro bond during amino acid sequencing.

The use of 0.03 *N* HCl *in vacuo* at 105°C for 20 h has been found to be satisfactory for the partial acid hydrolysis of proteins. Hulmes and Pan[39] demonstrated that gas-phase TFA preferentially cleaves peptide chains at the amino termini of threonine and serine (22°C for 1–15 days or 45°C for 2–3 days). In addition, cleavage at the carboxyl termini of aspartic acid was observed in solutions of TFA. The reactions were performed in sealed polypropylene tubes in which the TFA was introduced either as a gas or a liquid. In the gas-phase reactions, the sample was suspended above the bottom of the tube. Peptide samples used in this study included the synthetic N-acetyl vasoactive intestinal peptide, N-acetyl parathymosin, and N-peptide derived from a brain proteolipid. Other conditions for partial acid hydrolysis include the use of 0.25 M acetic acid at 110°C for 8 h *en vacuo* for a monoclonal antibody[40] and 0.25 M oxalic acid at 100°C *en vacuo* for 5 h for some small cyclic peptides.[41] In the latter study,[40] cleavage was observed at Asn-Gly, Gly-Glu, Glu-Ser, and Ser-Thr for circulin B and at Asn-Ser, Gly-Gln, and Glu-Ser for cyclopyschotride.

Oliyai and Barchardt[42] studied the degradation of an aspartic acid residue in a model peptide, VYPDGA. At acid pH (0.3 to 3.0; HCl or formic acid, 37°C), the predominant reaction was the hydrolysis of the Asp-Gly peptide bond. At pH 4 to 5 (formate or acetate, 37°C), the peptide isomerized to form the iso-Asp-containing hexapeptide.

The degradation of antiflammin 2 under acidic conditions has been studied.[43] Products of degradation were isolated by HPLC and characterized by mass spectrometry. Peptide bond cleavage occurred primarily at the carboxy termini of aspartic acid peptide bonds. The reaction was more rapid at pH 3.0 (37°C, sodium citrate) than at pH 2.37 (37°C, sodium phosphate) or pH 4.10 (37°C, sodium citrate).

Peptide bond cleavage at aspartic acid or the formation of isoaspartate poses a problem with the stability of protein therapeutics.[44] In studies on the epidermal growth factor,[45] it was found that 30% of the protein was cleaved at Asp3-Ser4 and the Edman degradation did not proceed beyond His-10. It was demonstrated the Asp-11 had isomerized to isoaspartic acid. Replacement of this residue with glutamic acid (D11E) improved protein stability. The cleavage of aspartic acid residues can also occur in matrix-assisted laser desorption mass spectrometry.[46,47]

A number of methods have been proposed for chemical cleavage at cysteine residues. One approach is based on the conversion of cysteine to dehydroalanine[48–50] and subsequent hydrolysis with either acid or base to release pyruvic acid and involves the conversion of cysteine to the dialkyl sulfonic salt with methyl bromide or methyl iodide at pH 6.0 and subsequent β-elimination in dilute bicarbonate with mild heating.[50] The use of 2,4-dinitrofluorobenzene for the modification of cysteine to form the S-dinitrophenyl derivatives at pH 5.6 has been reported.[51] The β-elimination of these derivatives was accomplished with sodium methoxide in methanol. Cleavage of the dehydroalanine-containing peptide bond was accomplished by heating (100°C) in dilute acid (0.01 M HCl) for 1 h. This reaction mixture was then lyophilized, treated with a volume of 0.1 M NaOH equivalent to the original volume of acid and one fifth volume of 30% hydrogen peroxide, and then heated at 37°C for 30 min. The reaction mixture was then neutralized with acetic acid, and excess peroxide was removed with catalase. Alternatively, cleavage can be accomplished with bromide or performic acid.

The cleavage of peptide bonds containing cystine (disulfide groups) has been examined in some detail.[51] In this reaction, cyanide reacts with cysteine to yield a sulfhydryl and a thiocyano group. The thiocyano-containing derivative cyclize at alkaline pH to form an acyliminothiazolidine ring, which then undergoes hydrolysis to cleave the peptide bond. Formation of the iminothiazolidine can be followed by absorbance at 235 nm. George Stark and co-workers in 1973[52] developed an application of this approach, which has been extensively referenced since then. S-cyanocysteine

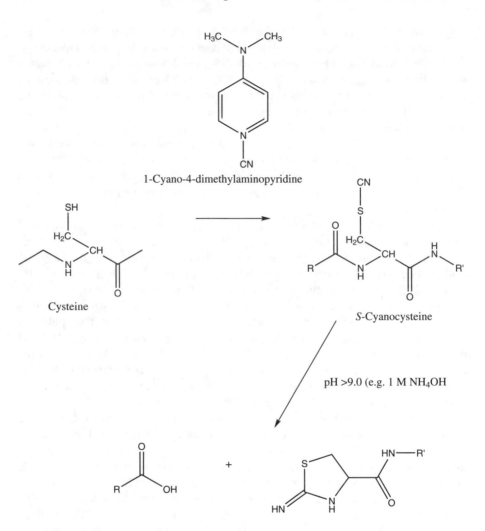

FIGURE 11.2 Cleavage of peptide bonds by cyanate.

is obtained by reaction of cysteine or cystine with 2-nitro-5-thiocyano-benzoic acid (Figure 11.2). Cleavage of the S-cyanocysteine is achieved by incubation in 0.1 M sodium borate, 6 M guanidine, pH 9.0, at 37°C, with the formation of 2-iminothia-zolidine-4-carboxyl peptides. Virtually 100% cleavage has been achieved for several proteins. This results in the formation of a free carboxyl group and a blocked amino-terminal peptide (2-iminothiazolidine-4-carboxyl) derivative. This reaction has been recently used to identify reactive sulfhydryl groups in phosphoglycerate kinase.[53] In this study, the two highly reactive cysteine residues (there are seven cysteine residues in the native protein) were modified with 5,5′-dithiobis-(2-nitrobenzoate). An excess of potassium cyanide (KCN) was added, resulting in the formation of S-cyanocysteine. The excess cyanide was removed by gel filtration. Incubation of the S-cyano protein in 6.0 M guanidine hydrochloride, pH 8.0, at 50°C resulted in peptide bond cleavage. Schaffer and Stark[54] proposed a catalyst prepared from nickel chloride and sodium borohydride for the conversion of 2-iminothiazolidine-4-carboxylate to alanine. These investigators also noted that cleavage could occur at phenylalanyl-seryl and phenylalanyl-threonyl peptide bonds. Lu and Gracy[55] employed 2-nitro-5-thiocyanobenzoic acid to convert the cysteinyl residues in human placental glucosephosphate isomerase to S-cyanocysteine, followed by cleavage at the modified cysteine residues. Conversion to the S-cyanocysteinyl derivative was accomplished with a five- to tenfold molar excess of 2-nitro-5-thiocyanobenzoic acid in 0.2 M Tris-acetate, 6 M guanidinium chloride, pH 9.0 (protein previously incubated with adequate dithiothreitol—fourfold molar excess over sulfhydryls in the protein), for 5 h at 37°C. The modified protein was dialyzed extensively against 10% acetic acid and lyophilized. Cleavage of S-cyanylated protein was achieved by incubation in 0.2 M Tris-acetate, 6 M guanidium chloride, pH 9.0, at 37°C for 2 h. The average extent of cleavage obtained was ca. 80%. Watson and colleagues[56] (Figure 11.2) recently used cyanylation combined with mass spectrometric analysis to determine the disulfide structure of sillucin. They used a combination of partial reduction and CN-induced cleavage. The peptide was partially reduced with phosphine and the resulting cysteine residues immediately cyanylated with 1-cyano-4-(dimethylamino)pyridinium tetrafluoroborate. The cyanylated peptides were isolated by HPLC and cleaved with aqueous ammonia.

Specific cleavage at tryptophanyl residues in peptides and proteins has been frequently used to obtain specific fragments. Cleavage of tryptophanyl peptide bonds with N-bromosuccinimide can occur as a side reaction of the N-bromosuccinimide oxidation of tryptophanyl residues, but it generally requires a substantial molar excess of reagent with mild acid.[57] This reaction is generally accomplished in 70% acetic acid. Although cleavage is generally restricted to tryptophanyl residues, cleavage can also occur at tyrosyl and histidine residues. Cleavage at tryptophanyl residues under these conditions generally occurs with an efficiency of 50 to 80% with peptides, but is substantially lower with proteins (10 to 50%).

BNPS-skatole (2-(2-nitrophenylsulfenyl)-3-methyl-3-bromoindolenine) has been used for the cleavage of peptide bonds involving tryptophan.[58,59] The reaction conditions are similar to those used for N-bromosuccinimide, and the reaction mechanism is similar in terms of the production of an active bromide. It is reported to be somewhat more selective than N-bromosuccinimide, but nonspecific cleavages do occur as does the conversion of methionine to methionine sulfoxide. The yield of peptide bond cleavage is similar to that reported with N-bromosuccinimide. Skatole has been used recently for the cleavage of the secretin receptor.[60]

The specific cleavage of tryptophanyl peptide bonds with N-chlorosuccinimide (NCS)[61] has been reported. The peptide bond cleavages obtained with NCS are much more specific than those achieved with N-bromosuccinimide or BPNS-skatole. The cleavage of tryptophanyl peptide bonds requires a twofold excess of NCS in 50% acetic acid under ambient conditions. Cleavage of other peptide bonds has not been detected under these conditions, but methionine is converted to methionine sulfoxide and cysteine to cystine. Model peptides were cleaved in ca. 40% yield, whereas for several proteins yields from 19 to 50% were reported. Mechanistically, the reaction proceeds as described for *N*-bromosuccinimide. The NCS should be recrystallized from ethyl acetate before use. Lischwe and Sung[62] examined the NCS cleavage of proteins in some detail. Cytochrome c was used as the model protein. The protein was dissolved in water (1 nMol/ml), and the NCS was dissolved in a buffer composed of 1.0 ml glacial acetic acid, 1 g urea, and 1 ml water. Four volumes of the protein solution were mixed with ten volumes of the NCS in this buffer (the resulting solution in 4.68 *M* with respect to urea in 27.5% acetic acid). Approximately 50% of the tryptophanyl peptide bonds were cleaved after 30 min of reaction, using a 10-fold molar excess of reagent. The oxidation of methionine (to methionine sulfoxide) and cysteine (to cysteic acid) occurs as a side reaction. More recently, the specific cleavage of tryptophanyl peptide bonds by NCS under these reaction conditions has been used to study epitope distribution in α-1-antiprotease inhibitor.[63]

NCS continues to see active use for the site-specific cleavage of proteins. Douady and co-workers[64] used NCS in acetic acid to cleave peptide bonds in the major polypeptide component of the light-harvesting complex from a brown alga (*Laminaria saccharina*). Droste and colleagues[65] used NCS for the fragmentation of adenyl cyclase type I in a study on the identification of an ATP-binding site. Kooi and co-workers[66] identified neutralizing peptides from a *Pseudomonas aeruginosa* elastase by using NCS cleavage to obtain fragments for defining the sites of interaction with monoclonal antibodies. Digestion with NCS was performed as previously described[62] and as discussed previously. Pliszka and co-workers[67] used NCS to identify EDC cross-links in subfragment 1 or skeletal muscle myosin. Cleavage with NCS was performed as described by Lischwe and Ochs[68] in polyacrylamide gel slices. The gel slices were first washed with water for 20 min, with one change of solvent. The gel slices were then washed with urea/H_2O/acetic acid (1 g/1 ml/1 ml) for 20 min, with one change of solvent. Cleavage was accomplished with NCS (15 m*M*) in urea/H_2O/acetic acid for 30 min.[62]

Burstein and Patchornik[69] advanced the use of 2,4,6-tribromo-4-methyl-cyclo-hexadione (TBC) for the cleavage of tryptophanyl peptide bonds in proteins. The cleavage is fairly specific for tryptophanyl residues, but modification of other amino acid residues was noted (tyrosine, methionine, cysteine, etc.). Optimal conditions for the cleavage reaction were a threefold excess of reagent at pH 3.0 at ambient conditions for 15 min. Generally, 60 to 80% acetic acid is used as the solvent and the reaction is allowed to proceed in the dark. There is ca. 50% cleavage of tryptophanyl-containing peptide bonds with synthetic peptides such as *N*-benzyloxy-carbonyl-tryptophanyl-glycine, whereas 5 to 60% yields are reported with proteins such as lysozyme.

The cleavage of protein at asparaginyl-glycyl peptide bonds with hydroxylamine[70] has proved useful in selective circumstances. The reaction is generally performed in the presence of 6 *M* guanidium chloride, pH 9.0, with 2 *M* NH_2OH. The pH of the solution is maintained either with a pH-stat or with 0.2 *M* potassium carbonate. Generally, as with other means of peptide bond cleavage, optimal results are obtained with the reduced and alkylated protein. The reaction yields a new amino-terminal amino acid and aspartyl hydroxyamate.

Cleavage at peptide bonds in which the carboxyl group is contributed by tryptophan occurs on reaction with *o*-iodosobenzoic acid. Mahoney and co-workers[71,72] have studied the reaction in detail. The reaction can be reasonably specific for tryptophan, although some modification of methionine to form methionine sulfoxide is observed. The reaction is performed in 60 to 80% acetic acid in the presence of a denaturing agent such as guanidine. The occasional modification of tyrosyl residues seen with some preparations of o-iodosobenzoic acid has been shown to be a property of o-iodoxybenzoic acid contamination of certain o-iodosobenzoic acid preparations.[72] Pretreatment of the o-iodosobenzoic acid preparations with p-cresol obviates cleavage at tyrosyl peptide bonds. There has been limited use of o-iodosobenzoic acid in the past decade.[73,74]

Schepartz and Cuenoud[75] reported the specific cleavage of calmodulin with a reagent based on the structure of trifluoroperazine. In these experiments, the trifluoroperazine-ethylenediaminetetraacetic acid reagent (TFE-EDTA), ferric ions, and dithiothreitol were required for effective peptide bond cleavage in 10 mM Tris, pH 7.2. In a directly related approach, Hoyer and co-workers[76] attached EDTA to biotin. The resulting derivative was used to place either a cupric ion or a ferric ion close to the biotin binding site on streptavidin. Relatively specific cleavage was obtained in solvent 50 mM borate, pH 7.5, containing 20 μM protein, 20 μM cupric chloride, and 20 μM biotin-EDTA.

Rubio and colleagues[77] in 1992 described the peptide bond cleavage at nucleotide-binding sites in proteins. Cleavage of carbamoyl phosphate synthetase I as determined by the appearance of fragments on SDS-PAGE was observed with 50 μM ATP in 50 mM HEPES, 0.1 M NaCl, 0.25 mM FeCl$_3$, 31 mM ascorbate, pH 7.2, at 37°C. Activity is lost under these conditions. Activity is also lost when the ATP is replaced with acetylglutamate, but fragmentation of the protein is not observed. Subsequent work from this laboratory[78] demonstrated that the cleavage is catalyzed by an ATP–Fe complex and that the carbamoyl phosphate synthetase I is cleaved at seven unique peptide bonds in a domain containing an ATP-binding site. Csermely and co-workers[79] extended these observations in studies with the heat shock protein 90 (Hsp 90). They demonstrated a unique cleavage site with either ATP or ADP, using the solvent conditions described previously[77] with 1 mM nucleotide. Cleavage occurs in the absence of ATP or ADP, but the fragmentation pattern is different from that observed in the presence of the nucleotides. The peptide bond cleavage observed in the presence of ATP or ADP is not seen with protein preincubated with SDS. Geldanonycin, an antagonist of ATP binding, blocks the amino-terminal cleavages observed in the presence of ATP or ADP but does not affect the cleavages observed in the absence of nucleotide. Subsequent work from this group[80] extended the understanding of the specificity of the reaction. A second ATP-binding site was found in the carboxy-terminal domain of Hsp 90. The carboxy-terminal site is more promiscuous than the amino-terminal binding site. Cleavage can be seen at this site with CTP, GTP, and UTP, as well as some nucleotide analogues. CTP was able to support a limited amount of cleavage in the amino-terminal domain.

Several groups have reported the oxidative cleavage of peptide bonds by ascorbate/H$_2$O$_2$ in the presence of iron. In 1990, Rana and Meares[81] modified the single sulfydryl group in bovine serum albumin with a bromoacetamido derivative of EDTA. Freshly prepared FeSo$_4$ in 0.1 M phosphate, pH 7.0 was added to the modified albumin and allowed to stand at room temperature for 25 min. Excess iron was removed with exogenous EDTA and the iron chelate isolated by spin gel filtration. HEPES buffer containing 20 mM sodium ascorbate and 5 mM EDTA was added and after 10 sec, the reaction was stopped with SDS containing 0.1 M thiourea. Bovine serum albumin was cleaved between Ala-150 and Pro-151 and between Ser-190 and Ser-191. Subsequent work extended these observations to human carbonic anhydrase, in which the bromoacetamido chelate was attached to Cys-212 and cleavage occurred between Leu-189 and Asp-190.[82,83] In a modification of the procedure, the chelator [1-(p-bromoacetimido-benzyl)-EDTA] was loaded with iron before the alkylation reaction. The iron-chelate protein was exchanged into 0.1 M phosphate, pH 7.0. Cleavage was accomplished by the sequential addition of 10 mM ascorbate in 50 mM HEPES, pH 7.0, followed by 5 mM H_2O_2 in 50 mM HEPES, pH 7.0. The reaction was terminated after 10 sec by adding 25% (w/v) sodium dodecyl sulfate/2-mercaptoethanol/glycerol/2 M Tris-HCl, pH 6.8/0.2% (w/v) bromophenol blue.[83] The synthesis and characterization of the iron chelator have been reported.[84] This technique has been used for protein footprinting with subunit II or cytochrome bd.[85] and $Escherichia coli$ RNA polymerase.[86] The iron chelate has also been attached to DNA via phosphorothioate groups to identify DNA-binding sites in proteins.[87]

Heyduk and Heyduk[88] developed a solution-based approach for peptide bond cleavage with an iron–EDTA complex, using bacterial cAMP receptor protein. The reaction was performed at room temperature in 10 mM MOPS, pH 7.2, containing 10 mM $MgCl_2$ and 250 mM NaCl. Freshly prepared solutions of $FeSO_4$, ascorbate, EDTA, and H_2O_2 are added to a final concentration of 1, 20, 2, and 1 mM, respectively. Lower concentrations can be used. The reaction was allowed to proceed for 10 min and the reaction terminated by adding SDS/glycerol/Tris/2-mercaptoethanol/bromophenol blue (sample buffer for electrophoresis). Samples can be frozen in the sample buffer before electrophoresis. In subsequent work, Heyduk and colleagues[89] extended the use of this reaction to study RNA polymerase α-subunit and further defined the reaction as mediated via hydroxyl radicals. This reaction has been used to a limited extent[90] since its development. This process does generate hydroxyl radicals in solutions and the modification of amino acid residues can be expected in addition to peptide bond cleavage.[91] Peptide bond cleavage with other metal ion (Cu^{2+}, Co^{3+}, Pt^{2+}) complexes have been repored.[92]

REFERENCES

1. Spande, T.F. et al., Selective cleavage and modification of peptides and proteins, *Adv. Protein Chem.* 24, 97, 1970.
2. Smith, B.J., Chemical cleavage of proteins, *Methods Mol. Biol.* 32, 197, 1994.
3. Gross, E., The cyanogen bromide reaction, *Methods Enzymol.* 11, 238, 1967.
4. Spande, T.F., and Witkop, B., CNBr cleavage of peptides and proteins. In *CRC Handbook of Biochemistry—Selected Data for Molecular Biology*, 2nd ed. (Sober, H.A., and Harte, R.A., Eds.). CRC Press, Boca Raton, FL, 1970, p. C-137.
5. Tristram, G.R., and Smith, R.H., Amino acid composition of certain proteins. In *The Proteins*, 2nd ed. (Neurath, H., Ed.). Academic Press, New York, 1963, p. 45.

6. Kelleher, N.L. et al., Top down versus bottom up protein characterization by tandem high-resolution mass spectrometry, *J. Am. Chem. Soc.* 121, 806, 1999.

7. Morrison, J.R., Fidge, N.N., and Grego, H., Studies on the formation, separation, and characterization of CNBr fragments of human A1 apolipoprotein, *Anal. Biochem.* 186, 145, 1990.

8. Shively, J.E., Reverse-phase HPLC isolation and microsequence analysis. In *Methods of Protein Microcharacterization* (Shively, J.E., Ed.). Humana Press, Clifton, NJ, 1986, p. 65.

9. Shively, J.E., Hawke, D., and Jones, B.N., Microsequence analysis of peptides and proteins. III. Artifacts and the effects of impurities on analysis, *Anal. Biochem.* 120, 312, 1982.

10 Liao, T.-H., Salnikow, J., Moore, S., and Stein, W.H., Bovine pancreatic deoxyribonuclease A. Isolation of CNBr peptides; complete covalent structure of the polypeptide chain, *J. Biol. Chem.* 248, 1489, 1973.

11. Doyen, N., and LaPresle, C., Partial non-cleavage by CNBr of a methionine-cystine bond from human serum albumin and bovine α-lactalbumin, *Biochem. J.* 177, 251, 1979.

12. Joppich-Kuhn, R., Corkill, J.A., and Giese, R.W., Oxidation of methionine to methionine sulfoxide as a side reaction of CNBr cleavage, *Anal. Biochem.* 119, 73, 1982.

13. Hollemeyer, K., Heinzle, E., and Tholey, A., Identification of oxidized methionine residues in peptides containing two methionine residues by derivatization and matrix-assisted laser desorption.ionization mass spectrometry, *Proteomics* 2, 1524, 2002.

14. Simpson, R.J., and Nice, E.C., In situ CNBr cleavage of N terminally blocked proteins in a gas-phase sequencer, *Biochem. Int.* 8, 787, 1984.

15. Xu, Q.-Y., and Shively, J.E., Microsequence analysis of peptides and proteins. VIII. Improved electroblotting of proteins onto membranes and derivatized glass-fiber sheets, *Anal. Biochem.* 170, 19, 1988.

16. Scott, M.G., Crimmins, D.L., McCourt, D.W., Tarrand, J.J., Eyerman, M.C., and Nahm, M.H., A simple *in situ* CNBr cleavage method to obtain internal amino acid sequence of proteins electroblotted to polyvinyldifluoride membranes, *Biochem. Biophys. Res. Commn.* 155, 1353, 1988.

17. Szewczyk, B., and Summers, D.F., Preparative elution of proteins blotted to immobilon membranes, *Anal. Biochem.* 168, 48, 1988.

18. Sokolov, B.P., Sher, B.M., and Kalinin, V.N., Modified method for peptide mapping of collagen chains using CNBr-cleavage of protein within polyacrylamide gels, *Anal. Biochem.* 176, 365, 1989.

19. Jahnen, W., Ward, L.D., Reid, G.E., Moritz, R.L., and Simpson, R.J., Internal amino acid sequencing of proteins by *in situ* CNBr cleavage in polyacrylamide gels, *Biochem. Biophys. Res. Commn.* 166, 139, 1990.

20. van Montfort, B.A. et al., Improved in-gel approaches to generate peptide maps of integral membrane proteins with matrix-assisted laser desorption/ionization time-of-flight mass spectrometry, *J. Mass Spectrom.* 37, 322, 2002.

21. Kaiser, R., and Metzka, L., Enhancement of cyanogen bromide cleavage yields for methionyl-serine and methionyl-threonine peptide bonds, *Anal. Biochem.* 266, 1, 1999.

22. Compagnini, A. et al., Improved accuracy in the matrix-assisted laser desorption/ionization-mass spectrometry determinations of the molecular mass of cyanogen bromide fragments of proteins by post-cleavage reaction with tris(hydroxymethyl)aminomethane, *Proteomics* 1, 967, 2001.

23. Kraft, P., Mills, J., and Dratz, E., Mass spectrometric analysis of cyanogen bromide fragments of integral membrane proteins at the picomole level: application to rhodopsin, *Anal. Biochem.* 292, 76, 2001.

24. Olson, H. et al., Production of a biologically active variant form of recombinant human secretin, *Peptides* 9, 301, 1988.

25. Forsberg, G., Baastrup, B., Brobjer, M., Lake, M., Jörnvall, H., and Hartmanis, M., Comparison of two chemical cleavage methods for preparation of a truncated form of recombinant human insulin-like growth factor I from a secreted fusion protein, *BioFactors* 2, 105, 1989.

26. Husken, D., Beckers, T., and Engels, J.W., Overexpression in *Escherichia coli* of a methionine-free designed interleukin-2 receptor (Tac protein) based on a chemically cleavable fusion proteins, *Eur. J. Biochem.* 193, 387, 1990.

27. Myers, J.A. et al., Expression and purification of active recombinant platelet factor 4 from a cleavable fusion protein, *Protein Expr. Purif.* 2, 136, 1991.

28. Dobeli, H. et al., Recombinant fusion proteins for the industrial production of disulfide bridge containing peptides: purification, oxidation without concatmer formation, and selective cleavage. *Protein Expr. Purif.* 12, 404, 1998.

29. Rais-Beghdadi, C. et al., Purification of recombinant proteins by chemical removal of the affinity tag, *Appl. Biochem. Biotechnol.* 74, 95, 1998.

30. Fairlie, W.D. et al., A fusion protein system for the recombinant production of short disulfide-containing peptides, *Protein Expr. Purif.* 26, 171, 2002.

31. Rodriguez, J.C., Wong, L., and Jennings, P.A., The solvent in CNBr cleavage reactions determines the fragmentation efficiency of ketosteroid isomerase fusion proteins used in the production of recombinant peptides, *Protein Expr. Purif.* 28, 224, 2003.

32. Pál, G. et al., The first semi-synthetic serine protease made by native chemical ligation, *Protein Expr. Purif.* 29, 185, 2003.

33. Sanger, F., The terminal peptides of insulin, *Biochem. J.* 45, 563, 1949.

34. Sanger, F., and Tuppy, H., The amino-acid sequence in the phenylalanyl chain of insulin. I. The identification of lower peptides from partial hydrolysates, *Biochem. J.* 49, 463, 1951.

35. Sanger, F., and Thompson, E.O.P., The amino acid sequence in the glycyl chain of insulin. I. The identification of lower peptides from partial hydrolysates, *Biochem. J.* 53, 353, 1953.

36. Schultz, J., Cleavage at aspartic acid, *Methods Enzymol.* 11, 255, 1967.

37. Tsung, C.M., and Fraenkel-Conrat, H., Preferential release of apartic acid by dilute acid treatment of tryptic peptides, *Biochemistry* 4, 793, 1965.

38. Piszkiewicz, D., Landon, M., and Smith, E.L., Anomalous cleavage of aspartyl-proline bond during amino acid sequence determination, *Biochem. Biophys. Res. Commn.* 40, 1173, 1970.

39. Hulmes, J.D., and Pan, Y.-C.E., Selective cleavage of polypeptides with trifluoroacetic acid: applications for microsequencing, *Anal. Biochem.* 197, 368, 1991.

40. Bundle, D.R. et al., Molecular recognition of a *Salmonella* trisaccharide epitope by monoclonal antibody Se155-4, *Biochemistry* 33, 5172, 1994.

41. Tam, J.P., and Lu, Y.-L., A biomimetic strategy I the synthesis and fragmentation of cyclic proteins, *Protein Sci.* 7, 1583, 1998.

42. Oliyai, C., and Borchardt, R.T., Chemical pathways of peptide degradation. IV. Pathway, kinetics, and mechanism of degradation of an aspartyl residue in a model hexapeptide, *Pharm. Res.* 10, 95, 1993.

43. Ye, J.M. et al., Degradation of antiflammin 2 under acidic conditions, *J. Pharm. Sci.* 85, 695, 1996.

44. Manning, M.C., Patel, K., and Borchardt, R.T., Stability of protein pharmaceuticals, *Pharm. Res.* 6, 903, 1989.

45. George-Nascimento, C. et al., Replacement of a labile aspartyl residue increases the stability of human epidermal growth factor, *Biochemistry* 29, 9584, 1990.

46. Yu, W., Vath, J.E., Huberty, M.C., and Martin, S.A., Identification of the facile gas-phase cleavage of the Asp-Pro and Asp-Xxx peptide bonds in matrix-assisted laser desorption time-of-flight mass spectrometry, *Anal. Chem.* 65, 3015, 1993.

47. Gu, C.G. et al., Selective gas-phase cleavage at the peptide bond terminal to aspartic acid in fixed-charge derivatives of Asp-containing peptides, *Anal. Chem.* 72, 5804, 2000.

48. Witkop, B., and Ramachandran, L.K., Progress in non-enzymatic selective modification and cleavage of proteins, *Metabolism* 13, 1016, 1964.

49. Patchornik, A., and Sokolovsky, M., Nonenzymatic cleavages of peptide chains at the cysteine and serine residues through their conversion into dehydroalanine. I. Hydrolytic and oxidative cleavage of dehydroalanine residues, *J. Am. Chem. Soc.* 86, 1206, 1964.

50. Sokolovsky, M., Sadeh, T., and Patchornik, A., Nonenzymatic cleavages of peptide chains at the cysteine and serine residues though their conversion to dehydroalanine (DHAL). II. The specific chemical cleavage of cysteinyl peptides, *J. Am. Chem. Soc.* 86, 1212, 1964.

51. Catsimpoolas, N., and Wood, J.L., Specific cleavage of cystine peptides by cyanide, *J. Biol. Chem.*, 241, 1790, 1966.

52. Jacobson, G.R., Schaffer, M.H., Stark, G.R., and Vanaman, T.C., Specific chemical cleavage in high yield at the amino peptide bonds of cysteine and cystine residues, *J. Biol. Chem.* 248, 6583, 1973.

53. Minard, P., Desmadril, M., Ballery, N., Perahia, D., Mouawad, L., Hall, L., and Yon, J.M., Study of the fast-reacting cysteines in phosphoglycerate kinase using chemical modification and site-directed mutagenesis, *Eur. J. Biochem.* 185, 419, 1989.

54. Schaffer, M.H., and Stark, G.R., Ring cleavage of 2-iminothiazolidine-4-carboxylates by catalytic reduction. A potential method for unblocking peptides formed by specific chemical cleavage at half-cysteine residues, *Biochem. Biophys. Res. Commn.* 71, 1040, 1076.

55. Lu, H.S., and Gracy, R.W., Specific cleavage of glucosephosphate isomerase at cysteinyl residues using 2-nitro-5-thiocyanobenzoic acid: analyses of peptides eluted from polyacrylamide gels and localization of active site histidyl and lysyl residues, *Arch. Biochem. Biophys.* 212, 347, 1981.

56. Qi, J. et al., Determination of the disulfide structure of sillucin, a highly knotted, cysteine-rich peptide, by cyanylation/cleavage mass mapping, *Biochemistry* 40, 4531, 2001.

57. Ramachandran, L.K., and Witkop, B., *N*-Bromosuccinimide cleavage of peptides, *Methods Enzymol.* 11, 283, 1967.

58. Fontana, A., Modification of tryptophan with BNPS-skatole (2-(2-nitrophenylsulfe-nyl)-3-methyl-3-bromoindolenine), *Methods Enzymol.* 25, 419, 1972.

59. Halliday, J.A., Bell, K., and Shaw, D.C., The complete amino acid sequence of feline-beta-globulin II and a partial revision of beta-lactoglobulin II sequence, *Biochim. Biophys. Acta* 1077, 25, 1981.

60. Zang, M. et al., Spatial approximation between a photolabile residue in position 13 of secretin and the amino terminus of the secretin receptor, *Mol. Pharmacol.* 63, 993, 2003.

61. Shechter, Y., Patchornik, A., and Burstein, Y., Selective chemical cleavage of tryptophanyl peptide bonds by oxidative chlorination with *N*-chlorosuccinimide, *Biochemistry* 15, 5071, 1976.

62. Lischwe, M.A., and Sung, M.T., Use of *N*-chlorosuccinimide/urea for selective cleavage of tryptophanyl peptide bonds in proteins, *J. Biol. Chem.* 252, 4976, 1977.

63. Zhu, X.-J., and Chan, S.K., The use of monoclonal antibodies to distinguish several chemically modified forms of human alpha-1-proteinase inhibitor, *Biochem. J.* 246, 19, 1987.

64. Douady, D., Rousseau, B., and Caron, L., Fucoxanthin-chlorophyll a/c light-harvesting complexes of *Laminaria saccharina*: partial amino acid sequences and arrangement in the thylakoid membranes, *Biochemistry* 33, 3165, 1994.

65. Droste, M., Mollner, S., and Pfeuffer, T., Localisation of an ATP-binding site on adenyl cyclase type I after chemical and enzymatic fragmentation, *FEBS Lett.* 391, 208, 1996.

66. Kooi, C., Hodges, R.S., and Sakol, P.A., Identification of neutralizing epitopes on *Pseudomonas aeruginosa* elastase and effects of cross-reaction with other thermolysin-like proteases, *Infect. Immun.* 65, 472, 1997.

67. Pliszka, B., Karczewska, E., and Wawro, B., Nucleotide-induced movements in the myosin head near the converter region, *Biochim. Biophys. Acta* 1481, 55, 2000.

68. Lischwe, M.A., and Ochs, D., A new method for partial peptide mapping using *N*-chlorosuccinimide/urea and peptide silver staining in sodium dodecyl sulfate-polyacrylamide gels, *Anal. Biochem.* 127, 453, 1982.

69. Burstein, Y., and Patchornik, A., Selective chemical cleavage of tryptophanyl peptide bonds in peptides and proteins, *Biochemistry* 11, 4641, 1972.

70. Bornstein, P., and Balian, G., Cleavage at Asn-Gly bonds with hydroxylamine, *Methods Enzymol.* 47, 132, 1977.

71. Mahoney, W.C., and Hermodson, M.A., High yield cleavage of tryptophanyl peptide bonds by *o*-iodosobenzoic acid, *Biochemistry* 18, 3810, 1979.

72. Mahoney, W.C., Smith, P.K., and Hermodson, M.A., Fragmentation of proteins with *o*-iodosobenzoic acid: chemical mechanism and identification of *o*-iodoxybenzoic acid as a reactive contaminant that modifies tyrosyl residues, *Biochemistry* 20, 443, 1981.

73. Aragon-Ortiz, F., Mentele, R., and Auerswald, E.A., Amino acid sequence of a lectin-like protein from *Lachesis muta stenophyrs* venom, *Toxicon* 34, 763, 1996.

74. Yamagami, T., and Ishiguro, M., Complete amino acid sequences of chitnase-1 and -2 from bulbs of genes *Tulipa*, *Biosci. Biotechnol. Biochem.* 62, 1253, 1998.

75. Schepartz, A., and Cuenoud, B., Site-specific cleavage of the protein calmodulin using a trifluoroperazine-based affinity reagent, *J. Am. Chem. Soc.* 112, 3247, 1990.

76. Hoyer, D., Cho, H., and Schultz, P.G., A new strategy for selective protein cleavage, *J. Am. Chem. Soc.* 112, 3249, 1990.

77. Alonso, E. et al., Oxidative inactivation of carbamoyl phosphate synthetase (ammonia). Mechanisms and sites of oxidation, degradation of the oxidized enzyme, and inactivation by glycerol, EDTA, and thiol protecting agents, *J. Biol. Chem.* 267, 4524, 1992.

78. Alonso, E., and Rubio, V., Affinity cleavage of carbamoyl-phosphate synthetase localizes regions of the enzyme interacting with the molecule of ATP that phosphorylates carbamate, *Eur. J. Biochem.* 229, 377, 1995.

79. Söti, C., Rácz, A., and Csermely, P., A nucleotide-dependent molecular switch controls ATP binding at the C-terminal domain of Hsp-90, *J. Biol. Chem.* 277, 7066, 2002.

80. Söti, C. et al., Comparative analysis of the ATP-binding site of Hsp 90 by nucleotide affinity cleavage of the C-terminal binding site, *Eur. J. Biochem.* 270, 2421, 2003.

81. Rana, T.M., and Meares, C.F., Specific cleavage of a protein by an attached iron chelate, *J. Am. Chem. Soc.* 112, 2457, 1990.

82. Rana, T.M., and Meares, C.F., Iron chelate mediates proteolysis: protein structure dependence, *J. Am. Chem. Soc.* 113, 1859, 1991.
83. Rana, T.M., and Meares, C.F., Transfer of oxygen from an artificial protease to peptide carbon during proteolysis, *Proc. Natl. Acad. Sci.USA* 88, 10579, 1991.
84. Greiner, D.P. et al., Synthesis of the protein cutting reagent iron (s)-1-(*p*-bromoace-timidobenzyl)ethylenediaminetetraacetate and conjugation to cysteine side chains, *Bioconjugate Chem.* 8, 44, 1997.
85. Ghaim, J.B. et al., Proximity mapping the surface of a membrane protein using an artificial protease. Demonstration that the quinine-binding domain of subunit I is near the N-terminal of region of subunit II of cytochrome *bd*, *Biochemistry* 34, 11311, 1995.
86. Grenier, D.P. et al., Binding of the δ protein to the core subunits of *Escherichia coli* RNA polymerase, studied by iron-EDTA protein footprinting, *Proc. Natl. Acad. Sci.* 93, 71, 1996.
87. Schmidt, B.D., and Meares, C.F., Proteolytic DNA for mapping protein-DNA inter-actions, *Biochemistry* 41, 4186, 2002.
88. Heyduk, E., and Heyduk, T., Mapping protein domains involved in macromolecular interactions: a novel protein footprinting approach, *Biochemistry* 33, 9643, 1994.
89. Heyduk, T. et al., Determinants of RNA polymerase α subunit for interacting with ' and δ subunits: hydroxyl radical footprinting, *Proc. Natl. Acad. Sci.* 93, 10162, 1996.
90. Schwanbeck, R. et al., Point mutations within AT-hook domains of the HMG1YL1 affect binding to gene promoter but not to four-way junction DNA, *Biochemistry* 39, 14419, 2000.
91. Malekinia, S.D., and Downard, K.M., Unfolding of apomyoglobin helices by syn-chrotron radiolysis and mass spectrometry, *Eur. J. Biochem.* 268, 5578, 2001.
92. Grant, K.B., and Pattabhi, S., Use of a fluorescence microplate reader for the detection and characterization of metal-assisted peptide hydrolysis, *Anal. Biochem.* 289, 196, 2001.

12 The Chemical Cross-Linking of Peptide Chains

This chapter is placed last for what I consider to be a good reason: site-specific chemical cross-linking is an application of the various reagents that have been discussed in the preceding chapters and issues regarding reactivity and specificity have been considered. It is simply impossible to cover the many applications of protein cross-linking in a single chapter, and the reader is directed to several excellent reviews.[1-9] Biological cross-linking[10,11] will not be considered except in situations wherein dityrosine formation is an approach to zero-length cross-linking.[12,13]

Site-specific chemical cross-linking (intramolecular or intermolecular) can be a valuable tool in the study of protein structure. It is of value only when used in a well-planned experiment, taking into the consideration the issues raised in Chapter 1. Although not essential, it is quite useful to assess the effect of the monofunctional reagent on the proteins in question. This is of particular importance if the measurement of protein function or activity is of concern in the proposed study. Monneron and d'Alayer[14] compared the effect of a monofunctional reagent (methylacetimidate) and a bifunctional reagent (dimethylsuberimidate) on the activity of membrane-bound adenylate cyclase. The chemistry of the reaction of the two reagents with lysine residues should be identical. The modification reactions were performed in 50 mM triethanolamine, 10%(w/v) sucrose, 5 mM MgCl$_2$, pH 8.1. The activity was measured before and after adding the effectors sodium fluoride or guanylylimido-diphosphate. When these effectors were added before adding the imidate reagent, the addition of dimethylsuberimidate simulated activity with either effector, whereas methylacetimidate had no effect. Both imidates modestly stimulated activity in the absence of added effector. Previous work[15] from this group on this enzyme system demonstrated that both formaldehyde and glutaraldehyde stimulated adenyl cyclase activity in cerebellar synaptosomes. Reaction with glutaraldehyde abolished the ability of either effector to stimulate enzyme activity. Formaldehyde had no effect on stimulation with either sodium fluoride or guanylylimidodiphosphate.

Cross-linkage can be analyzed by a variety of techniques. The application of mass spectrometry is discussed later. Most studies on intermolecular cross-linking have used SDS-PAGE. Structural analysis of intramolecular cross-linking usually involves the study of peptide maps from experimental and control proteins. Care must be taken in the size analysis of intramolecularly linked species in any denaturing medium (i.e., sodium dodecyl sulfate or guanidine hydrochloride), because unless the cross-linking reagent is cleaved by reduction, the intramolecularly cross-linked protein will not denature properly and will probably give falsely low-molecular-weight results.[16,17]

Intramolecular cross-linking can be enhanced by the following reaction conditions:

1. Low protein concentration (<0.1 mg/ml)
2. High net charge on protein
3. High ratio of protein reactive sites to reagent concentration

Zero-length cross-linking[18] is a procedure that joins peptide chains via existing functional groups such that a spacer group is not used. Examples include the covalent linkage of proteins to nucleic acids via photochemical reactions,[4] the formation of 3,3'-dityrosine mediated by tetranitromethane, and isopeptide bond formation. Kodedek and co-workers.[19,20] developed a novel approach to zero-length cross-linking via dityrosine formation. In this approach, the photolysis of ruthenium(II) tris-bipyridyl dictation in the presence of ammonium persulfate was used to generate Ru(III), a one-electron oxidant. Ru(III) is thought to remove a hydrogen atom from the phenyl ring of tyrosine, generating a tyrosine radical that then either reacts with a proximate nucleophile such as lysine or cysteine or reacts with a proximate tyrosine residue to generate a 3,3'-dityrosine cross-link. The cross-linking reaction is performed in 15 mM sodium phosphate, 100 mM NaCl, 0.125 mM Ru(bpy)$_3$Cl$_2$ containing the protein of interest at a concentration between 0.01 and 20 μM. Ammonium persulfate is added to a concentration of 2.5 mM immediately before irradiation. Irradiation (>380 nm) is performed for 0.5 sec and the sample quenched by adding the electrophoresis loading buffer (0.2 M Tris, 8% SDS, 2.88 M 2-mercaptoethanol, 40% glycerol, 0.4% xylene cyanol, 0.4% bromophenol blue and heated to 95°C for 5 min before electrophoretic analysis. This technique has been used in the study of a variety of proteins, including hormone-sensitive lipase,[21] macrophage inflammatory protein 1α,[22] and tRNA-specific adenosine deaminase (tadA).[23]

Isopeptide bond formation mediated by carbodiimide[24,25] (Figure 12.1) is an extensively used approach. This technique has been applied to the cross-linking of heavy meromyosin and F-actin[26,27] and components of the *Azotobacter vinelandii* nitrogenase complex.[28] More recently, EDC-mediated zero-length cross-linking has been used to study the interaction between skeletal myosin light chain 1 and F actin,[29] *Escherichia coli* T-protein and methylenetetrahydropteroyl-tetraglutamate,[30] the calmodulin-melittin complex,[31] and a complex between biotin protein ligase and biotin carboxyl carrier protein.[32] In the study of the calmodulin-melittin complex,[31] zero-length cross-linking was combined with limited proteolysis to develop a model for the complex. The cross-linking experiments were performed in 50 mM MES, 30 μM CaCl$_2$, pH 7.0, at 25°C. Equimolar (15 nMol) quantities of calmodulin and melittin were incubated in this solvent for 10 min before adding EDC (10-fold molar excess to calmodulin carboxyl groups). The reaction was allowed to proceed for 2 h and then terminated by adding trifluoroacetic acid. The cross-linked protein was purified by HPLC (C4 column) and effluent fractions assayed by electrospray mass spectrometry. Further analysis by mass spectrometry followed proteolysis of the cross-linked complexes following CNBr cleavage. These results were compared with results obtained from the analysis of fragments obtained from limited proteolysis,

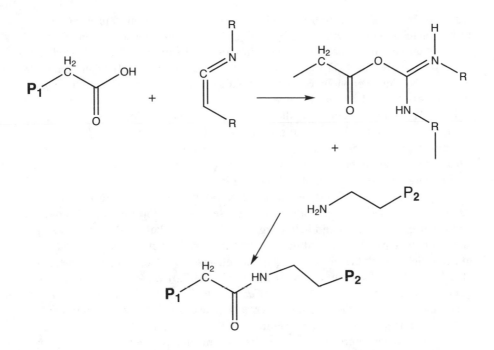

Isopeptide Bond for Zero-Length Cross-Linking

FIGURE 12.1 Isopeptide bond formation via carbodiimide as an example of zero-length cross-linking.

using trypsin, chymotrypsin, elastase, endoproteinase Glu-C, endoproteinase Asn-N, and subtilisin.

A more recent modification of this technique[33] involves a two-step procedure in which one protein is first incubated with a water-soluble carbodiimide and N-hydroxy-succinimide, resulting in the formation of a N-oxysuccinimide ester. The second step involves condensation with lysine residues in the second component of the reaction. This procedure has been popular since its development by Grabarek and Gergely in 1990.[33] This approach activates one of the protein components before the second protein component is added. This system has the advantage that only one of the components is exposed to the cross-linking reagents. The first protein is activated in a stepwise manner with the reaction of EDC with the carboxyl groups to yield the acylisourea, which in turn reacts with the N-hydroxysuccinimide to form an activated ester that is relatively stable and can be separated from the other components (acyl urea, unreacted N-hydroxysuccinimide) before adding the second protein component. This approach allows considerable flexibility in the design of the actual cross-linking reaction. Lindberg and colleagues[34] reported the following procedure for this reaction for the cross-linking of profilin and β/γ actin. Profilin was taken into 10 mM HEPES, 0.1 M CaCl$_2$, pH 7.5, at a concentration of 2 mg/ml. EDC and sulfo-N-hydroxysuccinmide were added to a final concentration of 6 mM and 15 mM, respectively, and the reaction allowed to stand 20 min at room temperature. This activation reaction was terminated by adding DTT to a final concentration of 20 mM to an equal volume of actin (1 mg/ml) in ice-cold 10 mM HEPES, 0.1 mM ATP, 0.1 mM CaCl$_2$, 0.5 mM DTT, pH 7.5, and the mixture immediately taken to a G-25 Sephadex® column previous equilibrated with the same buffer. This step removes the unreacted N-hydroxysuccinimide and the acylurea product from the hydrolysis of EDC. The protein fraction was allowed to stand overnight in ice. The profilin–actin product was separated from excess profilin by gel filtration on Sephacryl®-300. This combined reagent approach as developed by Graberek and Gergely[33] has seen considerable application in the development of artificial protein-based matrices for tissue engineering[35–38] and endothelial cell seeding.[39,40]

Winters and Day[41] developed a novel zero-length cross-linking procedure to identify intermolecular salt bridges between self-associating proteins. A covalent bond between a basic amino acid (Arg, His, Lys) side chain and an acidic amino acid (Glu or Asp) side chain was obtained in a reaction mediated by cyanogen. This was used to determine the salt bridges involved in the self-association of lysozyme. Lysozyme (1 mM in 0.015 M sodium citrate, 0.2 M NaCl, pH 7.0) was placed in a glass vial sealed with a septum. Cyanogen gas (*Caution:* toxic, perform reaction in a hood) was bubbled through the solution. Intermolecular cross-links were determined by mass spectrometry, followed separation by gel electrophoresis.

As with other approaches to the site-specific modification of proteins, mass spectrometry has been of increasing importance in the study of both intra- and intermolecularly cross-linked proteins. Bennett, Kussman, and co-workers[42] combined thiol-cleavable cross-linking reagents such as 3,3′-dithio-bis(succinimidypropionate) with mass spectrometry to determine intermolecular protein contacts in two associating protein systems (ParR, homodimeric DNA protein; CD28-I$_g$G and CD80-F$_{ab}$). Cross-linking was accomplished in phosphate-buffer saline, pH 7.5, at room temperature. In the case of CD28-I$_g$G and CD80-F$_{ab}$, the cross-linking reaction was terminated by adding 1.0 M Tris, pH 7.5. The cross-linked proteins were isolated and digested with trypsin (overnight, 37°C) and the digests divided into two equal portions. One portion was taken directly to mass spectrometric analysis and the second was reduced with DTT to cleave the cross-link before mass spectrometric analysis. Comparison of the maps obtained by mass spectrometry allowed the identification of regions of intermolecular contact. Several groups have developed isotopically labeled cross-linking reagents to improve the sensitivity of detection of low-abundance cross-linked peptides.[43,44] The use of mass spectrometry has allowed the development of solution chemistry approaches to the definition of three-dimensional (tertiary) protein structure.[9, 45–47]

Reagents that cross-link proteins by consecutive Michael reactions have been described.[48] These studies made the point that because the cross-linkage reaction is driven by consecutive Michael additions, eventually the most thermodynamically stable cross-link will be established, which can be subsequently stabilized by reduction. This method is also referred to as equilibrium transfer alkylation and the cross-linking agents as equilibrium transfer alkylation cross-link (ETAC) reagents. Although the overall reaction takes a considerable period of time (30 to 40 h), the equilibrium nature of the process most likely enhances the specificity of modification. These investigators explored the reaction of pancreatic ribonuclease with 2-(*p*-nitrophenyl) allyl-4-nitro-3-caboxyphenyl sulfide (twofold molar excess with respect to ribonuclease in 0.1 *M* sodium phosphate, pH 10.5 at 37°C, 36 h, cross-link stabilized by sodium dithionite, fivefold molar excess). No reaction occurred at pH 8.0. Analysis of the reaction mixture showed 61% monomer, 21% dimer, 10% trimer, and a trace of tetramer. The monomer fraction was characterized; the predominant cross-links occurred between Lys-7 and Lys-37 and between Lys-31 and Lys-41. A more complex series of reagents have been described by this group for use as a molecular yardstick.[49] Further insight into the chemistry of this process (α,α'-annelation) can be obtained from subsequent publications by this group.[50,51] Despite what appears to be some distinct advantages with ETAC reagents, they have had little use since their development. Wilbur and colleagues[52] used a trifunctional ETAC reagent to cross-link Fab' fragments with a diagnostic or therapeutic agent. A monoclonal IgG$_1$ molecule was digested with pepsin to yield the disulfide-linked F(ab)$_2$ moiety and an Fc fragment. The F(ab)$_2$ moiety was then reduced and coupled in series via an ETAC reagent with the diagnostic agent. The chemistry provides for the specific attachment of a large diagnostic or therapeutic agent to a monoclonal antibody F(ab')$_2$ moiety.

The remainder of our consideration involves intermolecular cross-linking, which includes the cross-linking of identical promoters to form homopolymers (e.g., cross-linkage of identical subunits in an oligomeric protein).[53,54] Cross-linkage to form heteropolymers includes studies on protein–protein interactions (this could result in homopolymers in self-associating systems),[55,56] studies on multienzyme complexes,[57–59] and protein–ligand interactions with cell membrane receptors.[60–66] One of the uses of photoactivated affinity cross-linking reagents is the characterization of the interactions between biologically active peptides and proteins and cell surface receptors. The work of Ji and co-workers[61,67] provides an excellent example (Figure 12.2) of this approach. They developed photoactivatable bifunctional cross-linking agents based on a stable, photo-activatable aromatic azide and either an imido ester (methyl 4-azidobenzoate) or an *N*-hydroxysuccinimide ester (of azidosalicylic acid). This basic chemistry has been used for developing reagents such as 5-[*N*-(*p*-azido-benzoyl)-3-aminoallyl]-dUMP for the photoaffinity labeling of DNA[68] and photoaffinity label derivatives of polyamines.[69–71]

Glutaraldehyde should cross-link proteins with the formation of α,-Schiff bases, which should be a readily reversible process in the absence of reduction of the Schiff base. This is not the case, as Richards and Knowles[72] summarize. These investigators noted that the reaction of proteins with glutaraldehyde was essentially irreversible even without reduction, and in the absence of reduction, there was a loss of lysine on amino acid analysis following acid hydrolysis. These investigators proposed the formation of a complex reagent resulting from aldol condensation of glutaraldehyde, which would then react with the protein. It seems clear that the chemistry of the reaction of glutaraldehyde is complex. The use of glutaraldehyde has received considerable attention in the study of the properties of protein crystals in solution.[73–77] The rationale of these studies has been to show that the properties of a protein in

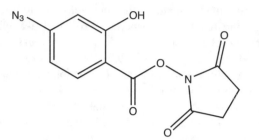

N-Hydroxysuccinimide ester of 4-azido-salicyclic acid

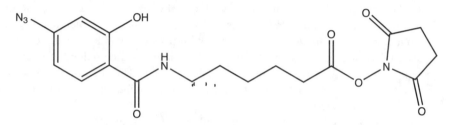

N-Hydroxysuccinimide ester of N-(4-azidosalicyl)-6-aminocaproic acid

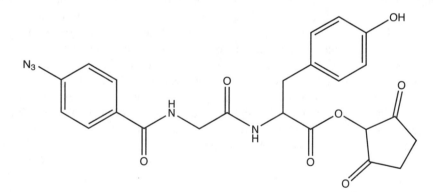

N-Hydroxysuccinimide ester of 4-azidobenzoylglycyltyrosine

FIGURE 12.2 Some examples of photoactivated aromatic azide/N-hydroxysuccinimide ester as heterobifunctional cross-linking agents.

crystalline form are similar to those for the protein in solution.[77] Although glutaraldehyde has a fair degree of specificity for the ε-amino group of lysine, reaction has also occurred with other nucleophilic functional groups in proteins, such as the sulfhydryl group of cysteine, the imidazole ring of histidine, and the phenolic hydroxyl group of tyrosine.[78] There is little current use of glutaraldehyde for protein structure analysis; there is some interest in the use of glutaraldehyde to improve enzyme stability.[79,80] The majority of work on proteins is in the area of hydrogel development[81,82] and related biologies.[83–86]

Dutton and co-workers[87] introduced homobifunctional imidoesters. These reagents have the advantage that the reaction with the protein results in charge preservation of the lysine residue modified. This class of reagents is highly specific for primary amines. Buffer effects on the reaction have not been extensively investigated, except to specify that the use of potential competing nucleophiles (e.g., Tris, imidazole) should be avoided; ammonium acetate is frequently used to quench the reactions. Most studies have used 0.02 to 0.1 M triethanolamine in the range of pH 8.0 to 9.0. It has been suggested that the amidation reaction is enhanced by the presence of triethanolamine in studies on the reaction of methyl-4-mercaptobutyrimidate.[63] Sinha and Brew[88] developed a useful procedure employing the prior trace labeling of the protein with acetic anhydride. Because the reactions with acetic anhydride and imidoester are mutually exclusive, fragmentation and subsequent determination of specific radioactivity at specific lysine residues allow the identification of the sites of reaction with bifunctional imidoesters.

A family of bis-imidates can be used for cross-linking proteins. The members differ in the carbon chain length between the imido ester functions. A brief survey of the literature in this area suggests that dimethylsuberimidate (Figure 12.3) is the most widely used member of this family. As with monofunctional reagents, these reagents react with the unprotonated ε-amino group of lysine; at low pH (≤8), there is considerable specificity in reaction, whereas at higher pH lysine reactivity increases with a concomitant increase in cross-link density. Ohen and Wagner[89] cross-linked 14-3-3 proteins to histones in 0.1 M triethanolamine, 4 mi EDTA, pH 8.5, for 30 min at 37°C. Gotte and co-workers[90] cross-linked aggregates of ribonuclease A at pH 8.0 for 15 min at 21°C, the reaction being terminated with an excess of ammonium acetate. These conditions are taken from an elegant study by Stanford Moore and co-workers on the cross-linking of ribonuclease A with several bifunctional imidates.[91] This work demonstrated that at pH 8.0, there is one crosslink in ribonuclease A with dimethyl suberimidate with one other amino group with a monosubsitution. It is interesting that the yield of dimer could not be extended beyond 20%. Introduction of a cross-linking reagent that could be subsequently cleaved, 3,3′-dithiobis-(succinimidylpropionate),[92] (Figure 12.4) has proved to be of considerable use in the study of proteins.[93–98]

Although homobifunctional reagents have been quite useful, the majority of work in this area uses heterobifunctional reagents. Of particular interest has been the use of photoactivatable derivatives. An example of this type of derivative is methyl-3-(p-azidophenyl)dithiopropioimidate.[60] In these experiments, epidermal growth factor was reacted in the dark at pH 8.5 (0.1 M triethanolamine, 0.2 M NaCl, pH 8.5) for reaction of the imido ester formation with lysine. This reaction was terminated by adding ammonium acetate. Photoactivation in the presence of mouse 3T3 cells resulted in the specific labeling of a cell surface protein. This reagent can be cleaved by reduction, permitting isolation of such a cell surface protein free of ligand.

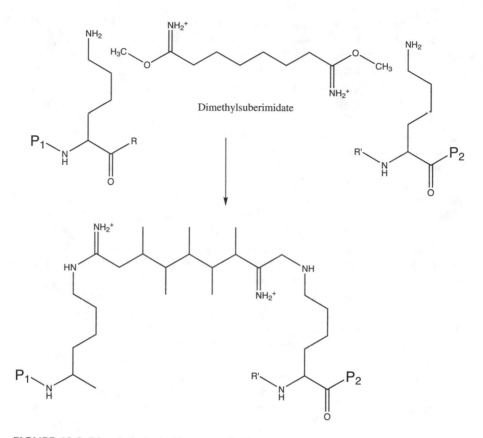

Dimethylsuberimidate

FIGURE 12.3 Dimethylsuberimidate cross-linking of lysine residues in proteins.

A series of homo- and heterobifunctional reagents that form acid-labile cross-links in proteins has been reported.[99] The ability to be cleaved by mild acid (pH 5.0) is based on the presence of *ortho* ester, acetal, and ketal functional groups. The synthesis of 4-(6-formyl-3-azidophenoxy)butyrimidate has been reported.[100] The imido function can react with the amino groups on a protein (0.1 M sodium borate, pH 8.6). Cross-linking can occur either with nitrene formation from the azido functional group on irradiation or via reductive alkylation using the aldehyde formation. In either instance, radiolabel can be introduced with sodium borotritiide (NaB[^{3}T]$_4$) with conversion of the free aldehyde to an alcohol or by reducing the Schiff base formed during reductive alkylation. Sulfosuccinimidyl-2-(p-azidosalicylamido)-1,3-dithiopropionate has been used to identify the receptor for phytohemagglutinin in mononuclear cells.[101] A related approach involves the use of p-nitrophenyl-3-diazo-pyruvate.[102] This reagent reacts with amines to form the corresponding pyruvamide derivatives. On photolysis at 300 nm, a ketene amide is formed, which is highly reactive with nucleophiles and produces malonic acid derivatives.

Chong and Hodges[103,104] reported studies of a complex heterobifunctional affinity reagent, N-(4-azidobenzoylglycyl)-5-(2-thiopyridyl)-cysteine (AGTC) (Figure 12.5). This reagent is incorporated into a protein via disulfide exchange with sulfhydryl

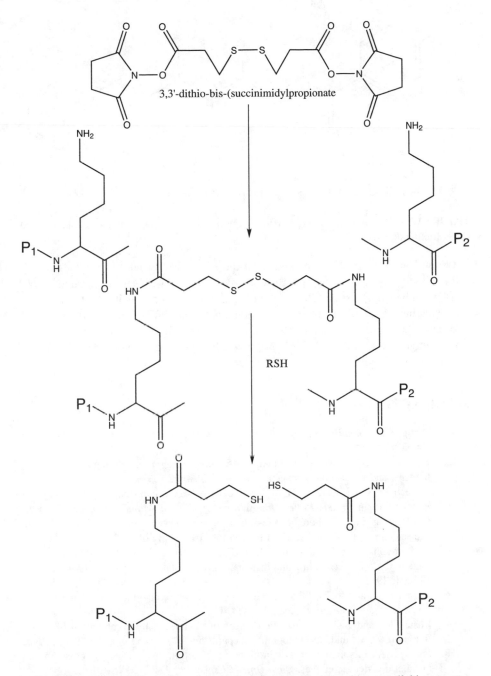

FIGURE 12.4 3,3′-Dithio-*bis*-(succinimidylpropionate) as a cleavable cross-linking agent.

groups. This is accomplished without effect on the aziodobenzyl function. Incorporation into the protein can be monitored by the release of pyridine-2-thionine (λ_{max} = 343 nm). Photolysis (>300 nm) can be followed by a decrease in absorbance at

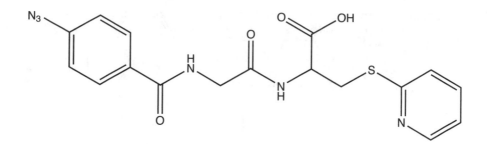

N-(4-azidobenzoylglycyl)-S-(2-thiopyridyl)cysteine

FIGURE 12.5 N-(4-Azidobenzoylglycyl)-S-(2-thiopyridyl)cysteine: a complex photoactivated cross-linking agent.

270 nm. The probe can be removed from the sulfhydryl group after photolysis by mild reducing agents. Ultee and Bosch[105] synthesized a related compound, N,N'-bis(4-azidobenzoyl)cystine. This reagent has the advantage of using highly radioactive cysteine as a radiolabel. A hydrazide analogue, S-(2-thiopyridyl)-L-cysteine hydrazide has been developed[106] to couple with the aldehyde function formed in carbohydrate after mild oxidation (periodate). Ebright and co-workers[107] developed S-[2-(4-azidosalicylamido)ethylthio]-2-thiopyridine.

REFERENCES

1. Wong, S.S., *Chemistry of Protein Conjugation and Cross-Linking*, CRC Press, Boca Raton, FL, 1991.
2. Brinkley, M., A brief survey of methods for preparing protein conjugates with dyes, haptens, and cross-linking reagents, *Bioconj. Chem.* 3, 2, 1992.
3. Nadeau, O.W., and Carlson, G.M., Chemical cross-linking in studying protein-protein interactions. In *Protein-Protein Interactions, A Cloning Manual* (Golemis, E., Ed.) Cold Spring Harbor Laboratory Press, Cold Spring Harbor, New York, 2002, Chap. 6.
4. Meisenheimer, K.M., and Koch, T.H., Photocross-linking of nucleic acids to associated proteins, *Crit. Rev. Biochem. Mol. Biol.* 32, 101, 1997.
5. Fágáin, C.O., Understanding and increasing protein stability, *Biochim. Biophys. Acta* 1252, 1, 1995.
6. Desantis, G., and Jones, J.B., Chemical modification of enzymes for enhanced functionality, *Curr. Opin. Biotechnol.* 10, 324, 1999.
7. Löster, K., and Josi, D., Analysis of protein aggregates by combination of cross-linking reactions and chromatographic separations, *J. Chromatogr. B.* 699, 43, 1997.
8. Fancy, D.A., Elucidation of protein-protein interactions using chemical cross-linking or label transfer techniques, *Curr. Opin. Chem. Biol.* 4, 28, 2000.
9. Sinz, A., Chemical cross-linking and mass spectrometry for mapping three-dimensional structures of proteins and protein complexes, *J. Mass Spectrom.* 38, 1225, 2003.
10. Yang, S., Litchfield, J.E., and Baynes, J.W., AGE-breakers cleave model compounds but do not break Maillard crosslinks in skin and tail collagen from diabetic rats, *Arch. Biochem. Biophys.* 412, 42, 2003.

11. Heinecke, J.W., Tyrosyl radical production by myeloperoxidase: a phagocytic pathway for lipid peroxidation and tyrosine cross-linking of proteins, *Toxicology* 177, 11, 2002.

12. Fancy, D.A., and Kodedek, T., Chemistry for the analysis of protein-protein interactions: rapid and efficient cross-linking triggered by long wavelength light, *Proc. Natl. Acad. Sci. USA* 96, 6020, 1999.

13. Ottersbach, K., and Graham, G.J., Aggregation-independent modulation of proteoglycan binding by neutralization of C-terminal acidic residues in the chemokine macrophage inflammatory protein 1α, *Biochem. J.* 354, 447, 2001.

14. Monneron, A., and d'Alayer, J., Effects of imido-esters on membrane-bound adenylate cyclase, *FEBS Lett.* 122, 241, 1980.

15. Monneron, A., and d'Alayer, J., Effects of crosslinking agents on adenylate cyclase regulation, *FEBS Lett.* 109, 75, 1980.

16. Ruoho, A., Bartlett, P.A., Dutton, A., and Singer, S.J., A disulfide-bridge bifunctional imidoester as a reversible cross-linking reagent, *Biochem. Biophys. Res. Commn.* 63, 417, 1975.

17. Steele, J.C.H., Jr., and Nielson, T.B., Evidence of cross-linked polypeptides in SDS gel electrophoresis, *Anal. Biochem.* 84, 218, 1978.

18. Kunkel, G.R., Mehrabian, M., and Martinson, H.G., Contact-site cross-linking agents, *Mol. Cell. Biochem.* 34, 3, 1981.

19. Fancy, D.A., and Kodedek, T., Chemistry for the analysis of protein-protein interactions: rapid and efficient cross-linking triggered by long wavelength light, *Proc. Natl. Acad. Sci. USA* 96, 6020, 1999.

20. Amini, F. et al., Using oxidative crosslinking and proximity labeling to quantitatively characterize protein-protein and protein-peptide complexes, *Chem. Biol.* 10, 1115, 2003.

21. Shen, W.-J. et al., Hormone-sensitive lipase functions as an oligomer, *Biochemistry* 39, 2392, 2000.

22. Ottersbach, K., and Graham, G.J., Aggregation-independent modulation of proteoglycan binding by neutralization of C-terminal acidic residues in the chemokine inflammatory protein 1α, *Biochem. J.* 354, 447, 2001.

23. Wolf, J. Gerber, A.P., and Keller, W., tadA, an essential tRNA-specific adenosine deaminase from *Escherichia coli*, *EMBO J.* 21, 3841, 2002.

24. Sheehan, J.C., and Hlavka, J.J., The use of water-soluble and basic carbodiimides in peptide synthesis, *J. Org. Chem.* 21, 439, 1956.

25. Sheehan, J.C., and Hlavka, J.J., The cross-linking of gelatin using a water-soluble carbodiimide, *J. Am. Chem. Soc.* 79, 4528, 1957.

26. Onishi, H., Maita, T., Matsuda, G., and Fujiwara, K., Evidence for the association between two myosin heads in rigor acto-smooth muscle heavy meromyosin, *Biochemistry* 28, 1898, 1989.

27. Onishi, H., Maita, T., Matsuda, G., and Fujiwara, K., Carbodiimide-catalyzed cross-linking sites in the heads of gizzard heavy meromyosin attached to F-actin, *Biochemistry* 28, 1905, 1989.

28. Willing, A., and Howard, J.B., Cross-linking site in *Azotobacter vinelandii* complex, *J. Biol. Chem.* 265, 6596, 1990.

29. Andreev, O.A. et al., Interaction of the N-terminus of chicken skeletal essential light chain 1 with F-actin, *Biochemistry* 38, 2480, 1999.

30. Okamura-Ikeda, K., Fujiwara, K., and Motokawa, Y., Identification of the folate binding sites on *Escherichia coli* T-protein of the glycine cleavage system, *J. Biol. Chem.* 274, 17471, 1999.

31. Scaloni, A. et al., Topology of the calmodulin-melittin complex, *J. Mol. Biol.* 277, 945, 1998.

32. Clarke, D.J. et al., Biotinylation in the hyperthermophile *Aquivex aeolicus*. Isolation of a cross-linked BPL:BCCP complex, *Eur. J. Biochem.* 270, 1277, 2003.

33. Grabarek, Z., and Gergely, J., Zero-length cross-linking procedure with the use of active esters, *Anal. Biochem.* 185, 131, 1990.

34. Nyman, T. et al., A cross-linked profilin-actin heterodimer interferes with elongation of the fast-growing end of F-actin, *J. Biol. Chem.* 277, 15828, 2002.

35. Kuijpers, A.J. et al., Cross-linking and characterization of gelatin matrices for biomedical applications, *J. Biomater. Sci. Polym. Ed.* 11, 225, 2000.

36. Chung, T.W. et al., Preparation of alginate.galactosylated chitosan scaffold for hepatocyte attachment, *Biomaterials* 23, 2827, 2002.

37. Wang, X.H. et al., Crosslinked collagen/chitosan matrix for artificial livers, *Biomaterials* 24, 3213, 2003.

38. Cui, Y.L. et al., Biomimetic surface modification of poly(L-lactic acid) with gelatin and its effect on articular chondrocytes *in vitro*, *J. Biomed. Mat. Res.* 66A, 770, 2003.

39. Wissink, M.J. et al., Immobilization of heparin to EDC/NHS-crosslinked collagen. Characterization and *in vitro* evaluation. *Biomaterials* 22, 151, 2001.

40. van Wachen, P.B. et al., *In vivo* biocompatibility of carbodiimide-crosslinked collagen matrices: effects of crosslink density, heparin immobilization, bFGF loading, *J. Biomat. Mat. Res.* 55, 368, 2001.

41. Winters, M.S., and Day, R.A., Identification of amino acid residues participating in intermolecular salt bridges between self-associating proteins, *Anal. Biochem.* 309, 48, 2002.

42. Bennett, K.L. et al., Chemical cross-linking with thiol-cleavable reagents, combined with differential mass spectrometric peptide mapping: a novel approach to assess intermolecular protein contacts, *Protein Sci.* 9, 1503, 2000.

43. Pearson, K.M., Pannell, L.K., and Fales, H.M., Intramolecular cross-linking experiments on cytochrome C and ribonuclease A using an isotope multiplet method, *Rapid Commn. Mass Spectrom.* 16, 149, 2002.

44. Collins, C.J. et al., Isotopically labeled crosslinking reagents: resolution of mass degeneracy in the identification of crosslinked peptides, *Bioorg. Med. Chem. Lett.* 13, 4023, 2003.

45. Dihazi, G.H., and Sinz, A., Mapping low-resolution three-dimensional protein structures using chemical cross-linking and Fourier transform ion cyclotron resonance mass spectrometry, *Rapid Commn. Mass Spectrom.* 17, 2005, 2003.

46. Back, J.W. et al., Chemical cross-linking and mass spectrometry for protein structure modeling, *J. Mol. Biol.* 331, 303, 2003.

47. Fujii, N., A novel protein crosslinking reagent for the determination of moderate resolution protein structures by mass spectrometry (MS3-D), *Bioorg. Med. Chem. Lett.* 14, 427, 2004.

48. Mitra, S., and Lawton, R.G., Reagents for the cross-linking of proteins by equilibrium transfer alkylation, *J. Am. Chem. Soc.* 101, 3097, 1979.

49. Brocchini, S.J., Eberle, M., and Lawton, R.G., Molecular yardsticks. Synthesis of extended equilibrium transfer alkylating cross-link reagents and their use in the formation of macrocycles, *J. Am. Chem. Soc.* 110, 5211, 1988.

50. Chen, Y.S., Kampf, J.W., and Lawton, R.G., Aromatic stacking in folded architecture through hydrogen bonding, *Tetrahedron Lett.* 38, 5781, 1997.

51. Brocchini, S.J., and Lawton, R.G., Titantium chelation in regioselective Michael additions to conjugated dienones and trienones, *Tetrahedron Lett.* 38, 6319, 1997.

52. Wilbur, D.S. et al., Monoclonal antibody Fab' fragment cross-linking using equilibrium transfer alkylation reagents. A strategy for site-specific conjugation of diagnostic and therapeutic agents with F(ab')$_2$ fragments, *Bioconj. Chem.* 5, 220, 1994.

53. Davies, G.E., and Stark, G.R., Use of dimethyl suberimidate, a cross-linking reagent, in studying the subunit structure of oligomeric proteins, *Proc. Natl. Acad. Sci. USA* 66, 651, 1970.

54. Carpenter, F.H., and Harrington, K.T., Intermolecular cross-linking of monomeric proteins and cross-linking of oligomeric proteins as a probe of quaternary structure. Application to leucine aminopeptidase (bovine lens), *J. Biol. Chem.* 247, 5580, 1972.

55. Tarvers, R.C., Noyes, C.M., Roberts, H.R., and Lundblad, R.L., Influence of metal ions on prothrombin self-association. Demonstration of dimer formation by intermolecular cross-linking with dithiobis(succinimidyl propionate), *J. Biol. Chem.* 257, 10708, 1982.

56. Lewis, R.V. et al., Photoactivated heterobifunctional cross-linking reagents which demonstrate the aggregation state of phospholipase A2, *Biochemistry* 16, 5650, 1977.

57. DeAbreu, R.A. et al., Cross-linking studies with the pyruvate dehydrogenase complexes from *Azotobacter vinelandii* and *Escherichia coli*, *Eur. J. Biochem.* 97, 379, 1979.

58. Baskin, L.S., and Yang, C.S., Cross-linking studies of cytochrome P-450 and reduced nicotinamide adenine dinucleotide phosphate-cytochrome P-450 reductase, *Biochemistry* 19, 2260, 1980.

59. Austin, C., Boehm, M., and Tooze, S.A., Site-specific cross-linking reveals a differential direct interaction of class 1, 2, and 3 ADP-ribosylation factors with adaptor protein complexes 1 and 3, *Biochemistry* 41, 4669, 2002.

60. Das, M., Miyakawa, T., Fox, C.F., Pruss, R.M., Aharonov, A., and Herschman, H.R., Specific radiolabeling of a cell surface receptor for epidermal growth factor, *Proc. Natl. Acad. Sci. USA* 74, 2790, 1977.

61. Ji, T.H., A novel approach to the identification of surface receptors. The use of photosensitive heterobifunctional cross-linking reagent, *J. Biol. Chem.* 252, 1566, 1977.

62. Pilch, P.R., and Czech, M.P., Interaction of cross-linking agents with the insulin effector system of isolated fat cells. Covalent linkage of 125I-insulin to a plasma membrane receptor protein of 140,000 daltons, *J. Biol. Chem.* 254, 3375, 1979.

63. Birnbaumer, M.F., Schrader, W.T., and O'Malley, B.W., Chemical cross-linking of chick oviduct progesterone-receptor subunits by using a reversible bifunctional cross-linking agent, *Biochem. J.* 181, 201, 1979.

64. Kasuga, M., Van Obberghen, E., Nissley, S.P., and Rechler, M.M., Demonstration of two subtypes of insulin-like growth factor receptors by affinity cross-linking, *J. Biol. Chem.* 256, 5305, 1981.

65. Rebois, R.V., Omedeo-Sale, F., Brady, R.O., and Fishman, P.H., Covalent cross-linking of human chorionic gonadotropin to its receptor in rat testes, *Proc. Natl. Acad. Sci. USA* 78, 2086, 1981.

66. Woltjer, R.L., Weclas-Henderson, L., Papyannopoulos, J.A., and Staros, J.V., High-yield covalent attachment of epidermal growth factor to its receptor by kinetically controlled, stepwise affinity cross-linking, *Biochemistry* 31, 7341, 1992.

67. Ji, T.H., and Ji, I., Macromolecular photoaffinity labeling with radioactive photoactiveable heterobifunctional reagents, *Anal. Biochem.* 121, 286, 1982.

68. Persinger, J., and Bartholomew, B., Mapping the contacts of yeast TFIIIB and RNA polymerase III at various distances from the major groove of DNA by DNA photoaffinity labeling, *J. Biol. Chem.* 271, 33039, 1996.

69. Felschow, D.M. et al., Selective labelling of cell-surface polyamine-binding proteins on leukemic and solid-tumor cell types using a new polyamine photoprobe, *Biochem. J.* 328, 889, 1997.

70. Aramrantos, I., and Kalpaxis, D.L., Photoaffinity polyamines: interactions with AcPhe-tRNA free in solution or bound at the P-site of *Escherichia coli* ribosomes, *Nucleic Acids Res.* 28, 3733, 2000.

71. Aramarantos, I. et al., Effects of two photoreactive spermine analogues on peptide bond formation and their application for labeling proteins in *Escherichia coli* functional ribosomal complexes, *Biochemistry* 40, 76441, 2001.

72. Richards, F.M., and Knowles, J.R., Glutaraldehyde as a protein cross-linking reagent, *J. Mol. Biol.* 37, 231, 1968.

73. Quiocho, F.A., and Richards, F.M., Intermolecular cross-linking of a protein in the crystalline state: carboxypeptidase A, *Proc. Natl. Acad. Sci. USA* 52, 833, 1964.

74. Wong, C., Lee, T. J., Lee, T.Y., Lu, T.H., and Hung, C.S., Intermolecular cross-linking of a protein crystal—acid protease from *Endothia parasitica*—in 2.7 M ammonium sulfate solution, *Biochem. Biophys. Res. Commn.* 80, 886, 1978.

75. Spillburg, C.A., Bethune, J.L., and Vallee, B.L., Kinetic properties of crystalline enzymes. Carboxypeptidase A, *Biochemistry* 16, 1142, 1977.

76. Wong, C., Lee, T.J., Lee, T.Y., Lu, T.H., and Hung, C.S., The structure of acid protease from *Endothia parasitica* in cross-linked form at 3.5 Å resolution, *Biochem. Biophys. Res. Commn.* 80, 891, 1978.

77. Tüchsen, E., Hvidt, A., and Ottesen, M., Enzymes immobilized as crystals. Hydrogen isotope exchange of crystalline lysozyme, *Biochimie* 62, 563, 1980.

78. Habeeb, A.F.S.A., and Hiramoto, R., Reaction of proteins with glutaraldehyde, *Arch. Biochem. Biophys.* 126, 16, 1968.

79. Govardhan, C.P., Crosslinking of enzymes for improved stability and performance, *Curr. Opin. Biotechnol.* 10, 331, 1999.

80. DeSantis, G., and Jones, J.B., Chemical modification of enzymes for enhanced functionality, *Curr. Opin. Biotechnol.* 10, 324, 1999.

81. Stevens, K.R. et al., *In vivo* biocompatibility of gelatin-based hydrogels and interpenetrating networks, *J. Biomater. Sci. Polym. Ed.* 13, 1353, 2002.

82. Yamamoto, M., Takahashi, Y., and Tabata, Y., Controlled release of biodegradable hydrogels enhances the ectopic bone formation of bone morphogenetic system, *Biomaterials* 24, 4375, 2003.

83. Freiss, W., Collagen-biomaterial for drug delivery, *Eur. J. Pharm. Biopharm.* 45, 113, 1998.

84. Nimni, M.E., Glutaraldehyde fixation revisited, *Long Term. Eff. Med. Implants* 11, 151, 2001.

85. Taqieddin, E., and Amiji, M., Enzyme immobilization in novel alginate-chitosan core-shell microcapsules, *Biomaterials* 25, 1937, 2004.

86. Jaturanpinyo, M. et al., Preparation of bionanoreactor based on core-shell structured polyion complex micelles entrapping trypsin in the core cross-linked with glutaraldehyde, *Bioconjug. Chem.* 15, 344, 2004.

87. Dutton, A., Adams, M., and Singer, S.J., Bifunctional imidoesters as cross-linking reagents, *Biochem. Biophys. Res. Commn.* 23, 730, 1966.

88. Sinha, S.K., and Brew, K., A label selection procedure for determining the location of protein-protein interaction sites by cross-linking with bisimidoesters. Application to lactose synthase, *J. Biol. Chem.* 256, 4193, 1981.

89. Chen, F., and Wagner, P.D., 14-3-3 proteins bind to histones and affect both histone phosphorylation and dephosphorylation, *FEBS Lett.* 347, 128, 1994.

90. Gotte, G. et al., Cross-linked trimers of bovine ribonuclease A: activity on double-stranded RNA and antitumor action, *FEBS Lett.* 415, 38, 1997.
91. Wang, D., Wilson, G., and Moore, S., Preparation of cross-linked dimers of pancreatic ribonuclease, *Biochemistry* 15, 600, 1976.
92. Lomant, A.J., and Fairbanks, G., Chemical probes of extended biological structures: synthesis and properties of the cleavable protein cross-linking reagent [35S] dithio-bis(succinimidyl propionate), *J. Mol. Biol.* 104, 243, 1976.
93. Tarvers, R.C., Noyes, C.M., Roberts, H.R., and Lundblad, R.L., Influence of metal ions on prothrombin self-association. Demonstration of dimer formation by intermolecular cross-linking with dithiobis(succinimidylpropionate), *J. Biol. Chem.* 257, 10708, 1982.
94. Hantula, J., and Bamford, D.H., Chemical cross-linking of bacteriophage 6 nucleocapsid proteins, *Virology* 165, 482, 1988.
95. Cornell, R., Chemical cross-linking reveals a dimeric structure for CTP: phosphocholine cytidylyltransferase, *J. Biol. Chem.* 264, 9077, 1989.
96. Wiland, E., Siemieniako, B., and Trzeciak, W.H., Binding of low mobility group protein from rat liver chromatin with histones studied by chemical cross-linking, *Biochem. Biophys. Res. Commn.* 166, 11, 1990.
97. D'Souza, S.E., Ginsberg, M.H., Lam, S.C.-T., and Plow, E.F., Chemical cross-linking of the arginyl-glycyl-aspartic acid peptides to an adhesion receptor on platelets, *J. Biol. Chem.* 263, 3943, 1988.
98. Rexin, M., Busch, W., and Gehring, U., Chemical cross-linking of heteromeric glucocorticoid receptors, *Biochemistry* 27, 5593, 1988.
99. Srinivasachar, K., and Neville, D.M., Jr., New protein cross linking reagents that are cleaved by mild acid, *Biochemistry* 28, 2501, 1989.
101. Maassen, J.A., Cross-linking of ribosomal proteins by 4-(6-formyl-3-azidophenoxy)butyrimidate. A heterobifunctional cleavable cross-linker, *Biochemistry* 18, 1288, 1979.
101. Shephard, E.G., de Beer, F.C., von Holt, C., and Hapgood, J.P., The use of S-sulfosuccinimidyl-2-(p-azidosalicylamido)-1,3'-dithiopropionate as a cross-linking reagent to identify cell surface receptors, *Anal. Biochem.* 168, 306, 1988.
102. Goodfellow, V.S., Settineri, M., and Lawton, R.G., p-Nitrophenyl 3-diazopyruvate and diazopyruvamides, a new family of photoactivatable cross-linking bioprobes, *Biochemistry* 28, 6346, 1989.
103. Chong, P.C.S., and Hodges, R.S., A new heterobifunctional cross-linking reagent for the study of biological interactions between proteins. I. Design, synthesis and characterization, *J. Biol. Chem.* 256, 5064, 1981.
104. Chong, P.C.S., and Hodges, R.S., A new heterobifunctional cross-linking reagent for the study of biological interactions between proteins. II. Application to the troponin C-troponin I interaction, *J. Biol. Chem.* 256, 5071, 1981.
105. Ultee, M.E., and Basch, R.S., N,N'-bis(4-azidobenzoyl)cystine: a cleavable photoaffinity reagent, *Anal. Biochem.* 149, 331, 1985.
106. Zara, J.J. et al., A carbohydrate-directed heterobifunctional cross-linking reagent for the synthesis of immunoconjugates, *Anal. Biochem.* 194, 156, 1991.
107. Ebright, Y.W. et al., S-[2-(4-azidosalicylamindo)ethylthio]-2-thiopyridine: radiodinatable, cleavable, photoactivatible cross-linking agent, *Bioconj. Chem.* 7, 380, 1996.

Appendix I

Some Reagents Used in the Modification of Proteins

Comp	Amino Acid Modified	M.W.
Acetaldehyde	Lysine, α-amino groups	44.05
Acetic acid	Solvent	60.05
Acetic anhydride	Lysine, tyrosine	102.1
Acetone (2-propanone; dimethylketone)	Solvent	58.1
Acetonitrile	Solvent	41.1
N-acetyl-imidazole (1-acetylimidazole)	Tyrosine, lysine	110.1
Arsanlic acid (4-amino-phenyl arsonic acid)	Tyrosine	217.0
Benzene	Organic solvent	78.1
Bolton–Hunter reagent [N-succinimidyl-3-(4-hydroxyphenyl) propionate; 3-p-(hydroxyphenyl)-propionic acid, N-hydroxysuccinide ester]	Iodination reagent	263.6
BNPS-skatole [2-(2-nitrophenylsulfenyl)-3-methyl-3-bromoindolenine]	Tryptophan, tyrosine	363.2
Bromoacetamide	Cysteine, histidine, lysine	138
Bromoacetic acid	Cysteine, histidine	139
N-Bromosuccimide	Tryptophan, cysteine methionine	178
2,3-Butanedione	Arginine, can also participate in photoinactivation	86.1
Chloramine T (N-chloro-4-methyl benzenesulfonamide, sodium salt)	Oxidizing agent, catalyzes iodination of proteins	227.7
Chloroacetamide	Cysteine	93.5
Chloroacetic acid	Cysteine	94.5
N-Chlorosuccinimide	Tryptophan, peptide bond cleavage	133.5
Citraconic anhydride (methylmaleic anhydride)	Lysine, α-amino groups	112.1
Cyanogen bromide	Methionine, peptide bond cleavage	105.9
1,2-Cyclohexanedione	Arginine	112.1
1-Cyclohexyl-3-(2- morpholinyl-4-ethyl) carbodiimide, p-toluene-sulfonate (CMC metho-p-toluenesulfonate, morpho CDI)	Carboxyl groups	423.8
Cystamine (2-mercaptoethylamine; 2-aminoethanethiol)	Coupled to carboxyl groups to yield a sulfhydryl function; also used as an antioxidant	152.3
Cysteine	Reducing agent	121.1

Dabsyl chloride [4-(dimethylamino)-azobenzene-4'-sulfonyl chloride]	Lysine, α-amino groups	323.8
Dansyl chloride [5-dimethylamino-1-naphthalene-sulfonyl chloride]	Lysine, α-amino groups	269.7
N,N'-Dicyclohexyl carbodiimide	Carboxyl groups	206.3
Diethyl pyrocarbonate (diethyl oxydiformate; ethoxyformic anhydride)	Histidine, tyrosine, lysine	162.1
Dimethyl formamide	Solvent	73.1
Dimethyl sulfoxide	Organic solvent	78.1
5,5'-Dithiobis-(2-nitrobenzoic acid) Ellman's reagent; DTNB)	Sulfhydryl groups	396.4
p-Dioxane (1,4-dioxane)	Solvent	88.1
2,2'-Dithiopyridine also includes 2,2'-dithio-bis-(5-nitropyridine)	Sulfhydryl groups	293.2
1,4-Dithiothreitol (Cleland's reagent)	Reducing agents	154
Ethanolamine (2-aminoethanol; monoethanolamine)	Buffer	61.1
1-Ethyl-3-(3-dimethyl aminoprophyl) carbodiimide hydrochloride	Carboxyl groups	191.8
Ethyl acetamide	Lysine, α-amino groups	87.2
Ethyl acetate	Organic solvent	88.1
Ethyl alcohol (ethanol)	Organic solvent	46.1
N-Ethylmaleimide	Sulfhydryl groups	125.1
1-Fluoro-2,4-dinitrobenzene (2,4-dinitrofluorobenzene; Sanger's reagent)	Lysine, α-amino groups	186.1
Glutaraldehyde (1,5-pentanedial)	Lysine	100.1
Glyceraldehyde (2,3-dihydroxy propanal)	Lysine, α-amino groups	90.1
Guanidine hydrochloride	Denaturing agent	95.5
Guanidine thiocyanate	Denaturing agent	181.2
Hydrogen chloride (muriatic acid)	Mineral acid	36.51
Hydrogen peroxide	Oxidizing agent, cysteine, methionine	34.0
Hydroxylamine	Peptide bond cleavage, also hydrolysis of O-acetyl tyrosine	33.1
2-Hydroxy-5-nitrobenzyl bromide (Koshland's reagent)	Tryptophan, cysteine	232.1
N-Hydroxysuccinimide	Lysine, α-amino groups	115.1
Imidazole (1,3-diaza-2,4-cyclopentanediene)	Buffer	68.1
Iodine (I_2)	Tyrosine	253.8
Iodoacetamide	Cysteine, histidine	185
Iodoacetic acid	Cysteine, histidine	186
Maleic anhydride	Lysine	98.1
2-Mercaptoethanol (β-mercaptoethanol)	Reducing agent	78.1
Methanol (methyl alcohol)	Organic solvent	32
O-Methylisourea sulfate	Lysine	172.2
Methylene blue (3,7-bis(dimethyl-amino) phenothiazine-5-onium chloride	Photooxidizing reagent	319.9
Ninhydrin (1, 2, 3-triketohydrindene monohydrate)	Arginine, also used as reagent for amino acid detection	178.1

2-Nitrophenyl-sulfenyl chloride (2-nitrobenzene-sulfenyl chloride; *o*-nitrophenyl-sulfenyl chloride)	Tryptophan	189.6
Phenylglyoxal	Arginine, also potential for photooxidation	113
Pyridoxal-5′-phosphate	Lysine, α-amino groups	247.2
Sodium borohydride	Reducing agent for Schiff bases, also reduces disulfide bonds	37.8
Sodium cyanate	Lysine, α-amino groups	65.0
Sodium cyanide	Cysteine, peptide bond cleavage	49.0
Sodium cyanoborohydride	Reducing agent for Schiff bases	62.8
Sodium periodate	Oxidizing agent	213.9
Sodium sulfite	Cystine, cleavage of disulfide bonds	126.0
Sodium tetrathionate	Cystine, cysteine	270.3
Succinic anhydride	Lysine, α-amino groups, protein cross-linking	100.1
Tetranitromethane	Tyrosine, cysteine, tryptophan	196.0
Tri-*n*-butyl phosphine	Reducing agent for disulfide bonds	266.3
Trichloroacetic acid	Protein precipitant	163.4
Triethanolamine [tris-(2-hydroxyethyl)amine]	Buffer	149.2
Trifluoroacetic acid	HPLC solvent, counter-ion	114.0
Tris-(hydroxymethyl) aminomethane		121.1
2,4,6-Trinitro benzenesulfonic acid (TNBS) (picryl sulfonic acid)	Lysine, α-amino groups	293.2

Appendix II
Assay of Protein Concentration in Solution

The determination of the concentration of chemically modified proteins is critical for the characterization of such molecules. The modification process can create difficulties with the use of certain techniques such as absorbance at 280 nm. For example, the modification of tryptophan or tyrosine can markedly change the absorption characteristics, as these residues are the primary contributors to absorbance at 280 nm. Phenylalanine and cystine make only minor contributions to such absorbance.) The reader is recommended to recent reviews of protein assay methods.[1-4] Quantitative amino analysis, discussed at the end of the appendix, is recommended as the method of choice while recognizing that the Kjehldahl method[5] is the gold standard.

The purpose of this short section is to describe some commonly used techniques for the determination of protein concentration. Care must be taken with the use of these techniques as several of the more frequently used techniques depend on protein quality as well as quantity.

MICROPLATE BIURET ASSAY

This procedure is based on the macroprocedure originally developed by Gornall and co-workers[6] and modified by Jenzano and co-workers.[7]

REAGENT PREPARATION

Dissolve 90.0 g NaOH in 300 ml deionized H_2O. (The reaction is exothermic, so cooling with an ice bath is recommended.) While this solution is cooling after dissolving NaOH, dissolve 1.5 g $CuSO_4 \cdot 5H_2O$ and 6.0 g sodium potassium tartrate in 500 ml deionized H_2O. After the NaOH solution has cooled to ambient (room) temperature, combine the two solutions and take to a final volume of 1.0 1 with deionized H_2O.

STANDARD

In general, a standard solution of bovine serum albumin (3.0 mg/ml established by spectroscopy) in 0.15 M NaC1 is used; on occasion, we have used the Pierce Albumin

Standard (Cat. No. 23210; 2.0 mg/ml in 0.9% NaCl with sodium azide as stabilizer). The choice of solvent is critical and should reflect the solvent used for the sample and reaction blank. There are significant solvent effects in this reaction. Tris-based buffers and other amino and guanidino compounds react with the biuret reagent. If another protein standard is used, the concentration should be validated by a primary method such as the Kjehldahl or quantitative amino acid analysis.

BLANK

In general, we use 0.15 *M* NaC1 as the blank, but the previous comments regarding solvent selection must be considered. Unless there is an exceptional situation, we use only a reaction blank and not an optical blank.

PROCEDURE

Add 160 µl biuret reagent to a 40-µl sample, standard, or blank. Allow to stand at ambient temperature for 20 min and then read on microplate reader at 540 nm.

COMMENTS

Although this technique is generally quite reliable and is not significantly influenced by protein quality, it is insensitive when compared with other techniques (c.f. Ref. 2). See selected references on the use of this method with various proteins.[8–19]

BCA (BICINCHONINIC ACID) ASSAY FOR PROTEIN

This procedure can be used for dye-binding assays based on the original procedure of Smith and co-workers[20] as illustrated by Jenzano and co-workers.[7] This procedure is a modification of the reaction of Lowry and co-workers,[21] but it is significantly easier and somewhat more sensitive (see Ref. 20).

REAGENT PREPARATION

We use the commercial reagent as supplied by the Pierce Chemical Company (Rockford, IL). Specifically, the BCA protein assay reagent is obtained (Cat. No. 23225). For the assay, one part reagent B is mixed with 50 parts of reagent A. This reagent (working reagent) is stable for at least 4 h at room temperature. The mention of a specific vendor is for information and is not to be construed as a recommendation.

STANDARD

A commercially available bovine serum albumin standard is used. The use of Pierce Albumin Standard (Cat. No. 23210; 2.0 mg/ml in 0.9% NaCl with sodium azide as a stabilizer) is recommended. Appropriate dilutions of this standard are used as indicated later. It might be necessary to alter these standards, but this is easily done with the BioRad Microplate Reader or other similar instrumentation. As indicated in the discussion of the biuret reaction, solvent selection is critical and potential interferences must be considered. The choice of a standard protein is critical and the concentration of said standard should be validated by a primary method such as the Kjehldahl or UV absorption, if appropriate.

BLANK

In general, we use 0.15 M NaCl as the blank, but solvent selection should reflect the solvent used for the sample.

PROCEDURE

Add 100 μl reagent to a 5-μl sample, standard, or blank, allow to stand for 30 min at 37°C (incubator), and then read on microplate reader at 570 nm (562 nm recommended by the manufacturer, but 570 nm effective).

COMMENTS

This reaction is quite sensitive, but it reflects qualitative differences in proteins. As a reflection of the dependence on protein quality,[1,2] it is critical to select a standard that is qualitatively similar, if not identical, with the samples. This is obviously difficult when the assay is used with heterogeneous mixtures such as saliva or serum. See selected references to the use of this method.[22-40]

DYE-BINDING ASSAY FOR PROTEIN USING COOMASSIE BRILLIANT BLUE G-250

The dye-binding assay for proteins using Coomassie Brilliant Blue G-250 is likely the most sensitive and most extensively used protein assay at this time. It is also extremely easy to perform. The technique, as noted later, is extremely dependent on the quality of the protein. The procedure given below is based on the procedure of Bradford[41] as illustrated by Jenzano and co-workers.[7]

REAGENT PREPARATION

For most procedures, we use the commercial reagent as supplied by the Pierce Chemical Company (Rockford, IL). Variations are in accordance with respective modification of procedure.

STANDARD

In general, a standard solution of bovine serum albumin (3.0 mg/ml established by spectroscopy) in 0.15 M NaCl is used. Solvent choice must reflect the solvent used for the sample and blank. The choice of a standard is critical because of the dependence of the spectral shift on the quality of the protein (see Refs. 1 to 3).

BLANK

As with the other assays, we use 0.15 *M* NaCl as the blank, but solvent choice must reflect the solvent used with the sample. If this is not possible, it is necessary to include sample solvent as a sample and correct final absorbance readings for any contribution by solvent.

PROCEDURE

Add 195 µl reagent to a 5-µl sample, standard, or blank (see attached sheet for system and format sheets), allow to stand for 15 min at 37°C (incubator), and then read on microplate reader at 595 nm.

COMMENTS

As noted above, this assay technique is likely the most sensitive and facile of the currently available procedures. Rigorous application of the dye-binding assay to the quantitative determination of a broad spectrum of proteins is difficult because of the marked influence of protein quality on the reaction. This is reflected by various studies attempting to modify the assay system to eliminate dependence on the quality of the protein.[22–49] Literature citations are provided showing a broad application of this technique.[49–69]

New techniques for the analysis of protein concentration continue to be developed.[70] It is also useful consider the reaction of other chemicals such as pharmaceuticals with the assay system.[71]

AMINO ACID ANALYSIS

The current technology for total amino acid analysis has the required sensitivity for use in the analysis of protein concentration[4,72–75] and has been suggested as an approach to the determination of total protein concentration.[74,75] With a well-characterized biopharmaceutical such as a growth factor or cytokine, the concentration of the protein can be determined by measuring the amounts of specific stable and abundant amino acids such as alanine and lysine with reference to an added internal standard such as norleucine.[76] The reader is recommended to the excellent review by Friedman[4] on the reaction of ninhydrin with amino acids and proteins. For practical purposes, I recommend quantitative amino acid analysis as the primary method of choice for the determination of protein concentration. Other methods can be used as long as such methods are validated for accuracy by a primary method such as Kjehldahl or quantitative amino acid analysis.

REFERENCES

1. Lundblad, R.L., and Price, N.C., Protein concentration determination. The Achilles' heel of cGMP? *BioProcess Int.* January 1, 2004.
2. Sapan, C.V. et al., Colorimetric protein assay techniques, *Biotechnol. Appl. Biochem.* 29, 99, 1999.

3. Brush, M., Array of assays, standard techniques and new methods provide alternatives for the quantitation of proteins in solution, *The Scientist* 14, 23, 2000.

4. Friedman, M., Applications of the ninhydrin reaction for analysis of amino acids, peptides, and proteins to agricultural and biomedical sciences, *J. Agric. Food Chem.* 52, 385, 2004.

5. Kjeldahl, J.Z., Neue Methode zur Bestimmung des Stickstoffs in Organischen Körpern, *Zeitschrift für Analytische Chemie* 22, 366, 1883.

6. Gornall, A.G., Bardawill, C.J., and David, M.M., Determination of serum proteins by means of the biuret reaction, *J. Biol. Chem.* 177, 751, 1949.

7. Jenzano, J.W., Hogan, S.L., Noyes, C.M., Featherstone, G.L., and Lundblad, R.L., Comparison of five techniques for the determination of protein content in mixed human saliva, *Anal. Biochem.* 159, 370, 1986.

8. Arneberg, P., Quantitative determination of albumin, lysozyme, and amylase. A comparison of seven analytical methods, *Scand. J. Dent. Res.* 78, 435, 1970.

9. Wolf, R.O., and Taylor, L.L., A comparative study of saliva protein analysis, *Arch. Oral Biol.* 9, 135, 1964.

10. Skurk, A., Mlynski, G., and Fendel, K., Methoden der Quantitativen Eiweiszbestimmung im Menschlichen Parotisspeichel, *Acta Oto-Laryngol.* 71, 71, 1971.

11. Itzhaki, R.F., and Gill, D.M., A micro-biuret method for estimating proteins, *Anal. Biochem.* 9, 401, 1964.

12. Eckfeldt, J.H., Kershaw, M.J., and Dahl, I.I., Direct analysis for urinary protein with biuret reagent, with use of urine ultrafiltrate blanking: comparison with a manual biuret method involving trichloroacetic acid precipitation, *Clin. Chem.* 30, 443, 1984.

13. Schlabach, T.D., Postcolumn detection of serum proteins with the biuret and Lowry reactions, *Anal. Biochem.* 139, 309, 1984.

14. de Keijzer, M.H. et al., Infusion of plasma expanders may lead to unexpected results in urinary protein assays, *Scand. J. Clin. Lab. Invest.* 59, 133, 1999.

15. Batic, M., and Raspor, P., Effect of cultivation mode on a bioprocess for chromium yeast biomass enrichment, *Pflugers Arch.* 439 (Suppl. 3), R73, 2000.

16. Mendler, M.H., In patients with cirrhosis, serum albumin determination should be carried out by immunonephelometry rather than by protein electrophoresis, *Eur. J. Gastroenterol. Hepatol.* 11, 1405, 1999.

17. Guobing, X. et al., Application of an improved biuret method to the determination of total protein in urine and cerebrospinal fluid without concentration step by use of the Hitachi 7170 Auto-Analyzer, *J. Clin. Lab. Anal.* 15, 161, 2001.

18. Eppel, G.A. et al., Variability of standard clinical protein assays in the analysis of a model urine solution of fragmented albumin, *Clin. Biochem.* 33, 487, 2000.

19. Greive, K.A. et al., Protein fragments in urine have been considerably underestimated by various protein assays. *Clin. Chem.*, 47, 1717, 2001.

20. Smith, P.K., Krohn, R.I., Hermanson, G.T., Mallia, A.K., Gartner, F.H., Provenzano, M.D., Fujimoto, E.K., Goeke, N.M., Olson, B.J., and Klenk, D.C., Measurement of protein using bicinchoninic acid, *Anal. Biochem.* 150, 76, 1985.

21. Lowry, O.H., Rosebrough, N.J., Farr, A.L., and Randall, R.J., Protein measurement with the Folin phenol reagent, *J. Biol. Chem.* 193, 265, 1951.

22. Kirazov, L.P., Venkov, L.G., and Kirazov, E.P., Comparison of the Lowry and the Bradford assays as applied for protein estimation of membrane-containing fractions, *Anal. Biochem.* 208, 44, 1993.

23. Zhang, J.-X., and Halling, P.J., pH and buffering in the bicinchoninic acid (4.4′-dicarboxy- 2,2′-biquinoline) protein assay, *Anal. Biochem.* 188, 9, 1990.

24. Stich, T.M., Determination of protein covalently bound to agarose supports using bicinchoninic acid, *Anal. Biochem.* 191, 343, 1990.

25. Baker, W.L., Potential interference of hydrogen peroxide in the 2,2′-bicinchoninic acid protein assay, *Anal. Biochem.* 192, 212, 1991.

26. Hill, H.D., and Straka, J.G., Protein determination using bicinchoninic acid in the presence of sulfydryl reagents, *Anal. Biochem.* 170, 203, 1988.

27. Wicchelman, K.J., Braun, R.D., and Fitzpatrick, J.D., Investigation of the bicinchoninic acid protein assay: identification of the groups responsible for color formation, *Anal. Biochem.* 175, 231, 1988.

28. Goldschmidt, R.C., and Kimelberg, H.K., Protein analysis of mammalian cells in monolayer culture using the bicinchoninic assay, *Anal. Biochem.* 177, 41, 1989.

29. Brown, R.E., Jarvis, K.L., and Hyland, K.J., Protein measurement using bicinchoninic acid: elimination of interfering substances, *Anal. Biochem.* 180, 136, 1989.

30. Tuszynski, G.P., and Murphy, A., Spectrophotometric cell-adhesion assay with bicinchoninic acid, *Anal. Biochem.* 184, 189, 1990.

31. Lacy, M.J. et al., Adaptation of commercial immunoassays to analysis of complex biological substances, *J. Immunoassay Immunochem.* 23, 1, 2002.

32. Soltys-Robitalle, C.E. et al., The relationship between contact lens surface charge and *in-vitro* protein deposition, *Biomaterials* 22, 3257, 2001.

33. Youan, B.B. et al., Protein release profiles and morphology of biodegradable microspheres containing an oily core, *J. Control. Release* 76, 313, 2001.

34. Liu, Y. et al., Biomimetic coprecipitation of calcium phosphate and bovine serum albumin on titanium alloy, *J. Biomed. Mater. Res.* 57, 327, 2001.

35. Shugars, D.C. et al., Salivary concentration of secretory leukocyte protease inhibitor, an antimicrobial protein, is decreased with advanced age, *Gerontology* 47, 246, 2001.

36. Keith, E.O. et al., Adhesion of tear proteins to contact lenses and vials, *Biotechnol. Appl. Biochem.* 34, 5, 2001.

37. Hayakawa, K. et al., Protein determination by high-performance gel-permeation chromatography: applications to human pancreatic juice, human bile and tissue homogenate, *J. Chromatogr. B* 754, 65, 2001.

38. Sokoloff, R.L. et al., A dual-monoclonal antibody sandwich assay for prostate-specific membrane antigen: levels in tissues, seminal fluid and urine. *Prostate* 43, 150, 2000.

39. Chattaraj, S.C., and Das, S.K., Interference of thimersol in the bicinchoninic acid protein microassay, *Boll. Chim. Farm.* 139, 30, 2000.

40. Diogo, M.M. et al., Purification of a cystic fibrosis plasmid vector for gene therapy using hydrophobic interaction chromatography, *Biotechnol. Bioeng.* 68, 576, 2000.

41. Bradford, M.M., A rapid and sensitive method for the determination of microgram quantities of protein utilizing the principle of protein-dye binding, *Anal. Biochem.* 72, 248, 1976.

42. Wilmsatt, D.K., and Lott, J.A., Improved measurement of urinary total protein (including light-chain proteins) with a Coomassie Brilliant Blue G-250-sodium dodecyl sulfate reagent, *Clin. Chem.* 33, 2100, 1987.

43. Tal, M., Silberstein, A., and Nusser, E., Why does Coomassie Brilliant Blue R interact differently with different proteins, *J. Biol. Chem.* 260(18), 9976, 1985.

44. Pierce, J., and Suelter, C.H., An evaluation of the Coomassie Brilliant Blue G-250 dye-binding method for quantitative protein determination, *Anal. Biochem.* 81, 478, 1977.

45. Read, S.M., and Northcote, D.H., Minization of variation in the response to different proteins of the coomassie blue G dye-binding assay for protein, *Anal. Biochem.* 116, 53, 1981.

46. Sedmak, J.J., and Grossberg, S.F., A rapid, sensitive, and versatile assay for protein using coomassie brilliant blue G250, *Anal. Biochem.* 79, 544, 1977.
47. Stoscheck, C.M., Increased uniformity in the response of the Coomassie blue G protein assay to different proteins, *Anal. Biochem.* 184, 111, 1990.
48. Zor, T., and Selinger, Z., Linearization of the Bradford protein assay increases its sensitivity, *Anal. Biochem.* 236, 302, 1996.
49. Ng, V., and Cho, P, The relationship between total tear protein concentrations determined by different methods and standards, *Graefes Arch. Clin. Exp. Ophthalmol.* 238, 571, 2000.
50. Pollard, H.B., Menard, R., Brant, H.A., Pazoles, C.J., Creutz, C.E., and Ramu, A., Applications of Bradford's protein assay to adrenal gland subcellular fractions, *Anal. Biochem.* 86, 761, 1978.
51. Khan, M.Y., and Newman, S.A., An assay for heparin by decrease in color yield (DECOY) of a protein-dye-binding reaction, *Anal. Biochem.* 187, 124, 1990.
52. Gotham, S.M., Fryer, P.J., and Paterson, W.R., The measurement of insoluble proteins using a modified Bradford assay, *Anal. Biochem.* 173, 353, 1988.
53. Spector, T., Refinement of the Coomassie blue method of protein quantitation. A simple and linear spectophotometric assay for <0.5 to 50 μg of protein, *Anal. Biochem.* 86, 142, 1978.
54. Hanggi, D., and Carr, P., Analytical evaluation of the purity of commercial preparations of cibacron blue F3GA and related dyes, *Anal. Biochem.* 149, 91, 1985.
55. Compton, S.J., and Jones, C.G., Mechanism of dye response and interference in the Bradford protein assay, *Anal. Biochem.* 151, 369, 1985.
56. Lea, M.A., Grasso, S.V., Hu, J., and Seilder, N., Factors affecting the assay of histone H1 and polylysine by binding of coomassie blue G, *Anal. Biochem.* 141, 390, 1984.
57. Wise, B.L., The quantitation and fractionantion of proteins in cerebrospinal fluid, *Am. J. Med. Technol.* 48, 821, 1982.
58. Pollard, H.B., Menard, R., Brant, H.A., Pazoles, C.J., Creutz, C.E., and Ramu, A., Applications of Bradford's protein assay to adrenal gland subcellular fractions, *Anal. Biochem.* 86, 761, 1978.
59. Almog, R., and Berns, D., A sensitive assay for proteins and biliproteins, *Anal. Biochem.* 114, 336, 1981.
60. Kain, S.R., and Henry, H.L., Quantitation of proteins bound to polyvinylidene difluoride membranes by elution of Coomassie brilliant blue R-250, *Anal. Biochem.* 189, 169, 1990.
61. Minamide, L.S., and Bamburg, J.R., A filter paper dye-binding assay for quantitative determination of protein without interference from reducing agents or detergents, *Anal. Biochem.* 190, 66, 1990.
62. Said-Fernández, S., González-Garza, M.T., Mata-Cárdenas, B.D., and Navarro-Marmolejo, L., A multipurpose solid-phase method for protein determination with Coomassie brilliant blue G-250, *Anal. Biochem.* 191, 119, 1990.
65. Kirazov, L.P., Venkov, L.G., and Kirazov, E.P., Comparison of the Lowry and the Bradford assays as applied for protein estimation of membrane-containing fractions, *Anal. Biochem.* 208, 44, 1993.
66. Demir, N. et al., Carbonic anhydrase from bovine bone, *Prep. Biochem. Biotechnol.* 31, 33, 2001.
67. Zhu, G., and Schwendeman, S.P., Stabilization of proteins encapsulated in cylindrical poly (lactide-co-glycide) implants: mechanism of stabilization by basic additives, *Pharm. Res.* 17, 351, 2000.

68. Golovan, S. et al., Characteriztion and overproduction of the *Escherichia coli* App A encoded bifunctional enzyme that exhibits both phytase and acid phosphatase activities, *Can. J. Microbiol.* 46, 59, 2000.
69. Frankel, A.E. et al., High-level expression and purification of the recombinant diphtheria fusion toxin DTGM for Phase I clinical trials, *Protein Express. Purif.* 16, 190, 1999.
70. Grant, P.G. et al., α Particle energy loss measurement of microgram depositions of biomolecules, *Anal. Chem.* 75, 4519, 2003.
71. Williams, K.M. et al., An evaluation of protein assays for quantitative determination of drugs, *J. Biochem. Biophys. Methods* 57, 45, 2003.
72. Weiss, M. et al., Effect of the hydrolysis of method on the determination of the amino acid composition of proteins, *J. Chromatogr. A* 795, 263, 1998.
73. Fountoulakis, M., and Lahm, H.-W., Hydrolysis and amino acid composition analysis for proteins, *J. Chromatogr. A* 826, 109, 1998.
74. Schegg, K.M. et al., Quantitation and identification of proteins by amino acid analysis: ABRF-96AAA collaborative trial, *Tech. Protein Chem.* 7, 207, 1997.
75. Anders, J.C. et al., Using amino acid analysis to determine absorptivity constants. A validation case study using bovine albumin, *Biopharm. Int.* February, 30, 2003.
76. Price, N.C. The determination of protein concentration. In *Enzymology Labfax* (Engel, P.C., Ed.). Bios Scientific Publishers, Oxford, 1996, pp. 34–41.

Index

A

T